THE
THEORY OF METALS

BY

A. H. WILSON
M.A., F.R.S.

SECOND EDITION

CAMBRIDGE
AT THE UNIVERSITY PRESS
1965

CAMBRIDGE UNIVERSITY PRESS
Cambridge, New York, Melbourne, Madrid, Cape Town,
Singapore, São Paulo, Delhi, Tokyo, Mexico City

Cambridge University Press
The Edinburgh Building, Cambridge CB2 8RU, UK

Published in the United States of America by Cambridge University Press, New York

www.cambridge.org
Information on this title: www.cambridge.org/9780521279000

© Cambridge University Press 1936, 1953

First edition 1936
Second edition 1953
Reprinted 1954, 1958, 1965
First paperback edition 2011

A catalogue record for this publication is available from the British Library

ISBN 978-0-521-06815-4 Hardback
ISBN 978-0-521-27900-0 Paperback

CONTENTS

PREFACE

When the first edition of this book was being prepared in 1936 the basic physical principles of the modern theory of metals had been established and many satisfactory qualitative conclusions had been drawn. There were, however, still many gaps in the strict mathematical theory, and much work remained to be done before it could be seen whether theory and experiment were in quantitative agreement. Further, in many cases the experimental data were either completely lacking or insufficiently precise. Since that time the mathematical formulation of the theory has been largely completed, and it is now possible to compare the quantitative predictions of the theory with much more extensive experimental data. The result is to a certain extent disappointing in that, as the theory has gained in mathematical precision, it has become obvious that there are serious quantitative discrepancies, and that the present theory, in its details, only applies strictly to the monovalent metals, though, even for these, some properties are still not quantitatively explained. On the other hand, it is clear that the physical principles are soundly based, and that the difficulties are essentially mathematical ones due to the complexity of the many-body problems which must be solved. In some cases, moreover, it is possible to foresee how to improve the theory, though the mathematical calculations required (probably mainly of a numerical character) will be formidable. It is just one hundred years since Wiedemann and Franz laid the first foundations of the theory of metals by discovering the law which bears their name, connecting the electrical and thermal conductivities of metals. To-day only the explanation of the phenomenon of superconductivity seems to lie beyond our grasp, and even this may be due to the inadequacies of the mathematical techniques employed rather than to the omission of some vital physical principle.

In view of the many papers which have appeared in the last fifteen years, it is impossible to give a comprehensive account of all parts of the subject while keeping the book to a reasonable length, and a considerable amount of selection has been necessary. Moreover, the existence of a large number of review articles has reduced the necessity for giving detailed references to the older work, which in itself would be a formidable task. I have therefore endeavoured to introduce all the important physical principles and to give applications of them which are of wide interest. I have omitted all reference to surface effects (with the sole exception of a brief description of the conductivity of thin films), since the physical ideas involved are very different from those which form the main theme of the book. This has meant omitting the theory of semi-conducting rectifying contacts and of other phenomena such as the photo-electric effect, which are of great technical importance, but adequate accounts of these are available elsewhere. For the same reason the chapter on superconductivity in the first edition has not been retained, and I have also deleted the chapter on optical properties, partly because these depend so much on the surface of a metal and partly because no significant advance in the theory has been made since 1936. These omissions have been made reluctantly, but they have made it possible to devote more space to other parts of the subject, thereby improving, I hope, the exposition of what is essentially a complex branch of theoretical physics, if more than a superficial acquaintance with the subject is required. The most difficult parts have been introduced in such a way that the continuity of ideas is not vitally affected if they are omitted.

My best thanks are due to Dr E. H. Sondheimer for considerable assistance in preparing the manuscript and reading the proofs.

A. H. W.

Chapter I

HISTORICAL INTRODUCTION

DRUDE'S THEORY

1·1. An early attempt to explain the passage of electricity through metals was made by Weber in 1875. He imagined that a molecule of a metal consisted of a number of electrically charged particles in relative motion to one another and that the molecule was not completely stable, so that from time to time one or more particles would leave the molecule and move through the metal until they were captured by another molecule. The particles would then take part in the motion of the capturing molecule until they were once more ejected to continue their journey through the metal. The discovery of the electron by J. J. Thomson in 1897 and the partial light it threw on the constitution of matter resulted in better-founded attempts to discover the mechanism of metallic conduction. The first contribution was made by Riecke in 1898, but his work was superseded by Drude in 1900 who was able to give a theoretical derivation of the empirical law discovered by Wiedemann and Franz in 1853, namely, that the ratio of the electrical and thermal conductivities at a given temperature is the same for all metals.

Drude's theory assumed that in a metal there are free electrons forming a kind of 'electron gas'. The electrons have an average velocity c_0 and pursue random motions through the metal due to collisions with the metal ions, which owing to their large mass can be considered to be immovable. If an electric field $\mathscr{E}$ (measured in electrostatic units) is imposed, the motions of the electrons will no longer be entirely random and an electric current will be set up in the direction of the electric field. Let $-\epsilon$ and m be the charge (in electrostatic units) and the mass of an electron, and let 2τ be the average time that an electron spends between two collisions with the metal ions. Then the acceleration of an electron due to the electric field is $-\epsilon\mathscr{E}/m$ and, if u_0 is the velocity component in the direction of the field of an electron immediately after a collision, the average velocity component of the electron between this collision and the next will be $u_0 - \tau\epsilon\mathscr{E}/m$. The electric current is parallel to the electric field, and the current density is given by

$$j = \Sigma\{-\epsilon(u_0 - \tau\epsilon\mathscr{E}/m)\}, \tag{1·1·1}$$

where the summation is over all the free electrons in a unit volume. We assume that the electrons are scattered by the metal ions with random velocities so that $\Sigma u_0 = 0$, and (1·1·1) then becomes

$$j = ne^2 \tau \mathscr{E}/m,$$

where n is the number of free electrons per unit volume. The electrical conductivity σ is defined to be $j/\mathscr{E}$ and is therefore given by

$$\sigma = ne^2 \tau/m. \tag{1·1·2}$$

To determine the thermal conductivity of a metal due to its free electrons, consider a metal in which there is a temperature gradient $\partial T/\partial x$. The average velocity c_0 and the energy E of an electron are now functions of the coordinate x. Consider the electrons reaching unit area in the plane $x = x_0$ and colliding with a metal ion there. An electron whose direction of motion makes an angle θ with the x-axis will have made its previous collision with an ion lying in the plane $x = x_0 - 2\tau c_0 \cos\theta$ and will have an energy $E(x_0 - 2\tau c_0 \cos\theta)$. Also the number of free electrons per unit volume whose directions of motion lie in a solid angle $d\omega$ is $n\,d\omega/4\pi$, so that the number whose directions of motion make angles with the x-axis lying between θ and $\theta + d\theta$ is $\frac{1}{2}n\sin\theta\,d\theta$. The number of free electrons satisfying this condition and crossing the unit area in the plane x_0 in time dt is equal to the number contained in a cylinder with unit base in the plane x_0 and with slant height $c_0\,dt$, i.e. with perpendicular height $c_0\cos\theta\,dt$; hence the volume of the cylinder is $c_0\cos\theta\,dt$. The energy flux across unit area in the plane x_0 in time dt is therefore

$$dt \int_0^\pi E(x_0 - 2\tau c_0 \cos\theta)\,\tfrac{1}{2}n c_0 \cos\theta \sin\theta\,d\theta$$

$$= dt \int_0^\pi \left\{ E(x_0) - 2\tau c_0 \cos\theta \left(\frac{\partial E}{\partial x}\right)_{x=x_0} \right\} \tfrac{1}{2}n c_0 \cos\theta \sin\theta\,d\theta$$

$$= -\tfrac{2}{3} dt\, n\tau c_0^2 (\partial E/\partial x)_{x=x_0}. \tag{1·1·3}$$

In a metal of uniform composition E can only depend upon x through the temperature T, and $\partial E/\partial x = (dE/dT)\,\partial T/\partial x$. The thermal current density is therefore

$$w = -\tfrac{2}{3} n\tau c_0^2 \frac{dE}{dT}\frac{\partial T}{\partial x},$$

and the thermal conductivity κ, defined by $w/(-\partial T/\partial x)$, is

$$\kappa = \tfrac{2}{3} n\tau c_0^2 dE/dT = \tfrac{2}{3}\tau c_0^2 C_v, \tag{1·1·4}$$

where $C_v = n\,dE/dT$ is the specific heat of the free electrons (per unit volume) at constant volume.

1·11. The equations (1·1·2) and (1·1·4) form the basis of Drude's theory, and, though they are based on a crude model and an unsophisticated method of calculation, both of which have long been superseded, they nevertheless merit careful consideration. According to these formulae, the Wiedemann-Franz ratio is

$$\frac{\kappa}{\sigma} = \frac{2mc_0^2}{3\epsilon^2}\frac{dE}{dT},$$ (1·11·1)

and the Wiedemann-Franz law follows if the energy (and therefore the mean velocity) of the free electrons is the same for all metals. In the classical dynamical theory of gases this is an immediate consequence of the principle of the equipartition of energy, according to which $E \equiv \frac{1}{2}mc_0^2 = \frac{3}{2}kT$, k being Boltzmann's constant. With this assumption, the right-hand side of (1·11·1) becomes $3(k/\epsilon)^2 T$, verifying the rule formulated in 1872 by Lorenz that the Wiedemann-Franz ratio is proportional to the absolute temperature. According to Drude's theory the value of the Lorenz number L, defined to be $\kappa/(\sigma T)$, is

$$L \equiv \frac{\kappa}{\sigma T} = 3\left(\frac{k}{\epsilon}\right)^2.$$ (1·11·2)

The numerical value of L given by (1·11·2) is $2·48 \times 10^{-13}$ e.s.u., which is in excellent agreement with the value $2·72 \times 10^{-13}$ e.s.u. obtained by averaging the observed results for a number of metals at $18°$ C. (see also § 8·31).

1·12. The successful prediction of the Lorenz number was to a certain extent offset by the difficulty in reconciling the formulae (1·1·2) and (1·1·4) separately with the observed facts. At ordinary temperatures σ is proportional to $1/T$, so that $n\tau$ should also vary as $1/T$. Now the 'free path' l of the electrons is equal to $2\tau c_0$ and ought to be of the order of the distance between the metal atoms, and therefore sensibly independent of the temperature. This would mean that n should vary as $1/T^{\frac{1}{2}}$, and it is very difficult to see how the number of free electrons could decrease with increasing temperature. For this and other reasons, efforts were made, particularly by Lorentz in 1905 and by Bohr in 1911, to improve the details of Drude's theory in the hope of obtaining a more reasonable expression for σ.

LORENTZ'S THEORY

1·2. Drude's theory is based on an elementary form of calculation which, when applied to the dynamical theory of gases, is known to give incorrect numerical values. Lorentz therefore reinvestigated the problem using the full statistical theory devised by Maxwell and Boltzmann and also investigated the dynamics of the collision process more carefully. He made the following five assumptions:

(1) The atoms of a metal are rigid spheres. During collisions with the free electrons they behave like perfectly elastic particles.

(2) The collisions of the electrons with each other may be neglected.

(3) For a given metal, the number n of free electrons per unit volume is a determinate function of the temperature, even when the temperature varies from point to point. In a stationary state the number of electrons emitted by the atoms is equal to the number which enter them.

(4) The atoms occupy only a small volume of the metal.

(5) If the metal is not homogeneous, or if there is a temperature gradient, the variations are small over a distance of the order of the mean free path l of the electrons. Also if external fields act on the electrons, the velocities acquired by the electrons in moving a distance l must be small compared with the mean velocity of thermal agitation.

The first assumption is unnecessarily restrictive and need not be introduced until a late stage in the calculation.

Boltzmann's Equation

1·3. In a strict analysis of transport problems it is seldom possible to deal solely with average properties such as the mean velocity and the mean energy, and it is necessary to determine the distribution of the particles both as regards position and as regards their velocities. Furthermore, the distribution function depends explicitly upon the mechanism of interaction between the particles and the fields and cannot, as in equilibrium problems, be determined once for all. The fundamental equation determining the distribution function is an integro-differential equation known as the Boltzmann equation which can be derived as follows.

Let (u, v, w) be the velocity of an electron, and at time t let

$$f(u,v,w;\ x,y,z;\ t)\,d\tau\,d\lambda \qquad (1\cdot3\cdot1)$$

be the number of electrons in the element of volume $d\tau = dx\,dy\,dz$ which have their velocities lying in the range $d\lambda = du\,dv\,dw$. Further, let (X, Y, Z) be the acceleration of an electron between collisions. From the fifth assumption the acceleration may be considered to be constant in time. Then an electron which at time t is at the point (x, y, z) and has velocity (u, v, w) would at time $t + dt$, if there were no collisions, be at the point

$$(x + u\,dt,\ y + v\,dt,\ z + w\,dt)$$

and have a velocity

$$(u + X\,dt,\ v + Y\,dt,\ w + Z\,dt).$$

Also the electrons which were in the volume $d\tau$, with their velocities in $d\lambda$, now fill a volume $d\tau'$ and have their velocities in $d\lambda'$. But, since the velocities and accelerations of all these electrons are approximately the same, we must have, neglecting higher order terms,

$$d\tau = d\tau', \quad d\lambda = d\lambda',$$

though, of course, these elements are taken about different points of the phase space. Now let
$$[\partial f/\partial t]_{\text{coll.}}\, dt\, d\tau\, d\lambda$$

be the net number of electrons forced into $d\tau\, d\lambda$ by collisions in time dt. Then we have

$$f(u + X\,dt, v + Y\,dt, w + Z\,dt;\; x + u\,dt, y + v\,dt, z + w\,dt;\; t + dt)$$
$$= f(u, v, w;\; x, y, z;\; t) + [\partial f/\partial t]_{\text{coll.}}\, dt,$$

that is, $\quad \dfrac{\partial f}{\partial t} + X\dfrac{\partial f}{\partial u} + Y\dfrac{\partial f}{\partial v} + Z\dfrac{\partial f}{\partial w} + u\dfrac{\partial f}{\partial x} + v\dfrac{\partial f}{\partial y} + w\dfrac{\partial f}{\partial z} = \left[\dfrac{\partial f}{\partial t}\right]_{\text{coll.}}, \quad (1\cdot3\cdot2)$

which is Boltzmann's equation for the distribution function. For a steady state we must have $\partial f/\partial t = 0$.

1·31. *The collision operator and the time of relaxation.* Since binary collisions of electrons are neglected, the number of electrons per unit volume with velocities lying in the range $d\lambda$ which have their velocities changed by a collision to lie in the range $d\lambda'$ in time dt must be proportional to $f(\mathbf{v}, \mathbf{r}, t)$ and to the interaction between an electron and metal ion. It can therefore be written as
$$f(\mathbf{v}, \mathbf{r}, t)\, d\lambda\, \mathscr{W}(\mathbf{v}, \mathbf{v}', \mathbf{r})\, d\lambda'\, dt,$$

where $\mathscr{W}$ is independent of f and depends only on the electron-ion interaction. Similarly, the corresponding number of electrons ejected from the velocity range $d\lambda'$ and into the range $d\lambda$ by collisions in time dt is

$$f(\mathbf{v}', \mathbf{r}, t)\, d\lambda'\, \mathscr{W}(\mathbf{v}', \mathbf{v}, \mathbf{r})\, d\lambda\, dt.$$

The net difference between the above two quantities, integrated with respect to $d\lambda'$, determines $[\partial f/\partial t]_{\text{coll.}}$, so that

$$\left[\frac{\partial f}{\partial t}\right]_{\text{coll.}} = \iiint \{ f(\mathbf{v}', \mathbf{r}, t)\, \mathscr{W}(\mathbf{v}', \mathbf{v}, \mathbf{r}) - f(\mathbf{v}, \mathbf{r}, t)\, \mathscr{W}(\mathbf{v}, \mathbf{v}', \mathbf{r}) \} \, du'\, dv'\, dw'.$$
$$(1\cdot31\cdot1)$$

The equilibrium distribution function f_0 must be such that $[\partial f_0/\partial t]_{\text{coll.}} = 0$, and so
$$\frac{\mathscr{W}(\mathbf{v}, \mathbf{v}', \mathbf{r})}{\mathscr{W}(\mathbf{v}', \mathbf{v}, \mathbf{r})} = \frac{f_0(\mathbf{v}', \mathbf{r})}{f_0(\mathbf{v}, \mathbf{r})}.$$

Now according to classical mechanics f_0 is a function of the energy only, and is in fact the Maxwell function. Also the collisions with the metal ions are supposed to be elastic, so that the energy of the electrons is conserved and $f_0(\mathbf{v}, \mathbf{r}) = f_0(\mathbf{v}', \mathbf{r})$. Hence $\mathscr{W}(\mathbf{v}, \mathbf{v}', \mathbf{r}) = \mathscr{W}(\mathbf{v}', \mathbf{v}, \mathbf{r})$ and

$$\left[\frac{\partial f}{\partial t}\right]_{\text{coll.}} = \iiint \{ f(\mathbf{v}', \mathbf{r}, t) - f(\mathbf{v}, \mathbf{r}, t) \} \, \mathscr{W}(\mathbf{v}, \mathbf{v}', \mathbf{r}) \, du'\, dv'\, dw'. \quad (1\cdot31\cdot2)$$

This expression for $[\partial f/\partial t]_{\text{coll.}}$ cannot be simplified further, and in the presence of a completely general field it always involves an integral operator even for the simplest interactions. However, for certain types of external fields and for certain interactions it can be shown that $[\partial f/\partial t]_{\text{coll.}}$ is of the form

$$\left[\frac{\partial f}{\partial t}\right]_{\text{coll.}} = -\frac{f(\mathbf{v}, \mathbf{r}) - f_0}{\tau(\mathbf{v}, \mathbf{r})}, \qquad (1\cdot31\cdot3)$$

and τ is then defined to be the time of relaxation. Its physical significance is that, if the distribution function f is set up by a system of external forces which are suddenly removed, the rate of approach to equilibrium is given by

$$\frac{\partial}{\partial t}(f - f_0) = -\frac{f - f_0}{\tau},$$

i.e.
$$(f - f_0)_t = (f - f_0)_{t=0}\, e^{-t/\tau}.$$

The time of relaxation is clearly of the same order of magnitude as the time between collisions, but the numerical factor connecting them can only be determined when an exact definition of the latter time has been adopted.

All transport problems are immensely simplified when a time of relaxation exists, and for the present we shall assume its existence, reserving for later discussion the conditions under which this simplifying assumption is justified and the extra complications which occur when no universal time of relaxation can be defined.

SOLUTION OF THE BOLTZMANN EQUATION ASSUMING A TIME OF RELAXATION

1·4. We shall restrict the present investigation to the determination of the electrical and thermal conductivities, and it is then sufficient to consider the external forces to be due to an electric field $\mathscr{E}$ and a temperature gradient $\partial T/\partial x$ along the x axis. The Boltzmann equation in this case is

$$-\frac{\epsilon\mathscr{E}}{m}\frac{\partial f}{\partial u} + u\frac{\partial f}{\partial x} = -\frac{f - f_0}{\tau}. \qquad (1\cdot4\cdot1)$$

In practice the electric fields and temperature gradients available are so small that terms involving their squares and products can be neglected. The solution of $(1\cdot4\cdot1)$ is therefore obtained at once by substituting $f = f_0$ on the left-hand side, and is

$$f = f_0 - \tau\left(-\frac{\epsilon\mathscr{E}}{m}\frac{\partial f_0}{\partial u} + u\frac{\partial f_0}{\partial x}\right). \qquad (1\cdot4\cdot2)$$

The electric and thermal current densities are in the x direction and are given by

$$j = -\epsilon\iiint uf\,du\,dv\,dw = \iiint \tau\left(-\frac{\epsilon^2\mathscr{E}}{m}u\frac{\partial f_0}{\partial u} + \epsilon u^2\frac{\partial f_0}{\partial x}\right)du\,dv\,dw \qquad (1\cdot4\cdot3)$$

and
$$w = \iiint uEf\,du\,dv\,dw = \iiint \tau E\left(\frac{\epsilon\mathcal{E}}{m}u\frac{\partial f_0}{\partial u} - u^2\frac{\partial f_0}{\partial x}\right) du\,dv\,dw, \qquad (1\cdot4\cdot4)$$

where $E = \tfrac{1}{2}m(u^2 + v^2 + w^2)$ is the energy of a free electron.

To evaluate these expressions we must know both f_0 and τ. For f_0 we assume the Maxwell distribution

$$f_0 = n\left(\frac{m}{2\pi kT}\right)^{\frac{3}{2}} e^{-E/kT}, \qquad (1\cdot4\cdot5)$$

while we assume that τ is given by

$$\tau = \mu c^p, \qquad (1\cdot4\cdot6)$$

c being the velocity $(u^2 + v^2 + w^2)^{\frac{1}{2}}$ of an electron. If we substitute these expressions into $(1\cdot4\cdot3)$ and $(1\cdot4\cdot4)$ and use the fact that

$$\iiint u^2 F(c)\,du\,dv\,dw = \frac{1}{3}\iiint (u^2 + v^2 + w^2)\,F(c)\,du\,dv\,dw = \tfrac{4}{3}\pi\int c^4 F(c)\,dc,$$

we find after some manipulation that

$$j = \left\{\epsilon^2\mathcal{E} + \epsilon kT\frac{\partial}{\partial x}\left(\log\frac{n}{T^{\frac{3}{2}}}\right)\right\}\mathcal{K}_1 + \frac{\epsilon}{T}\frac{\partial T}{\partial x}\mathcal{K}_2, \qquad (1\cdot4\cdot7)$$

and
$$w = \left\{-\epsilon\mathcal{E} - kT\frac{\partial}{\partial x}\left(\log\frac{n}{T^{\frac{3}{2}}}\right)\right\}\mathcal{K}_2 - \frac{1}{T}\frac{\partial T}{\partial x}\mathcal{K}_3, \qquad (1\cdot4\cdot8)$$

where
$$\mathcal{K}_s = \frac{2^{\frac{1}{2}p+3}\mu n}{3\pi^{\frac{1}{2}}m^{\frac{1}{2}p+1}}(kT)^{s+\frac{1}{2}p-1}\int_0^\infty x^{2s+p+2}\,e^{-x^2}\,dx$$

$$= \frac{2^{\frac{1}{2}p+2}\mu n}{3\pi^{\frac{1}{2}}m^{\frac{1}{2}p+1}}(kT)^{s+\frac{1}{2}p-1}\Gamma(s + \tfrac{1}{2}p + \tfrac{3}{2}). \qquad (1\cdot4\cdot9)$$

1·41. In a metal of uniform composition n is constant and only T can depend upon x. To determine the electrical conductivity σ we put $\partial T/\partial x = 0$, which gives
$$\sigma = j/\mathcal{E} = \epsilon^2\mathcal{K}_1. \qquad (1\cdot41\cdot1)$$

The conditions under which the thermal conductivity is measured are such as to make the electric current $j = 0$. This necessitates the existence of an electric field which is built up by the electrons which accumulate at the ends of the specimen whose conductivity is being measured. Elimination of $\mathcal{E}$ between equations $(1\cdot4\cdot7)$ and $(1\cdot4\cdot8)$ with $j = 0$ gives

$$\kappa = \frac{w}{-\partial T/\partial x} = \frac{\mathcal{K}_1\mathcal{K}_3 - \mathcal{K}_2^2}{\mathcal{K}_1 T}. \qquad (1\cdot41\cdot2)$$

Also
$$L \equiv \frac{\kappa}{\sigma T} = \frac{\mathcal{K}_1\mathcal{K}_3 - \mathcal{K}_2^2}{\epsilon^2\mathcal{K}_1^2 T^2}. \qquad (1\cdot41\cdot3)$$

If we now substitute the particular values of $\mathscr{X}_s$ given by (1·4·9) we finally obtain

$$\sigma = \frac{4\mu n\epsilon^2}{3\pi^{\frac{1}{2}}m}\left(\frac{2kT}{m}\right)^{\frac{1}{2}p}\Gamma(\tfrac{1}{2}p+\tfrac{5}{2}),\qquad (1\cdot41\cdot4)$$

$$\kappa = \frac{4\mu nk^2T}{3\pi^{\frac{1}{2}}m}\left(\frac{2kT}{m}\right)^{\frac{1}{2}p}\Gamma(\tfrac{1}{2}p+\tfrac{7}{2}),\qquad (1\cdot41\cdot5)$$

and

$$L \equiv \frac{\kappa}{\sigma T} = \frac{p+5}{2}\left(\frac{k}{\epsilon}\right)^2 \qquad (1\cdot41\cdot6)$$

(using in the reduction the relation $\Gamma(z+1)=z\Gamma(z)$). If we define a mean free path l by the relation

$$\sigma = \frac{4}{3}\frac{n\epsilon^2 l}{(2\pi mkT)^{\frac{1}{2}}}, \qquad (1\cdot41\cdot7)$$

then

$$l = \Gamma(\tfrac{1}{2}p+\tfrac{5}{2})\mu(2kT/m)^{\frac{1}{2}p+\frac{1}{2}}, \qquad (1\cdot41\cdot8)$$

the definition being such that l is constant when $p=-1$. These formulae are naturally similar to those obtained by Drude except that the coefficients are calculated by an exact method, and in addition the index p occurs, which can for the moment be considered to be an adjustable parameter.

1·42. On examining the formulae (1·41·4) to (1·41·6) we find that it is impossible to fit all the facts by a single choice of p. In order to obtain the correct temperature variation of σ it is necessary to choose $p=-2$, and this makes the value of L much too small. A value of p equal to 1 would give L reasonably correctly, but then σ increases with T instead of decreasing. The wrong temperature variation of σ would seem to be a greater defect than an incorrect numerical value for L, but it is necessary to examine the collision mechanism to see if the required value of p is physically plausible.

THE EXISTENCE OF A TIME OF RELAXATION WHEN THE IONS BEHAVE AS ELASTIC SPHERES

1·5. We can now examine the consequences of the first and most specialized of Lorentz's assumptions, namely, that the metal ions behave as fixed elastic spheres of radius R.

Consider an electron which before and after impact with an ion has velocities (u, v, w) and (u', v', w'). The number of such electrons which hit an element of the surface of the ion subtending a solid angle $d\omega$ at the centre during time dt is the number of electrons in a cylinder whose slant height is $c\,dt$ and whose base is the element of surface. If the direction of motion of the electron makes an angle θ with the inward normal to the surface at the point of contact, the volume of the cylinder is $R^2 d\omega c\, dt \cos\theta$, so that the number of impacts per unit time by electrons whose velocities lie in the range $d\lambda$ is $R^2 d\omega c \cos\theta f(u, v, w)\, d\lambda$. Similarly, the number of impacts by electrons whose velocities after impact lie in $d\lambda$ is $R^2 d\omega c \cos\theta f(u', v', w')\, d\lambda$, and if there are N atoms per unit volume we have

$$\left[\frac{\partial f}{\partial t}\right]_{\text{coll.}} = NR^2 c \int \{f(u', v', w') - f(u, v, w)\} \cos\theta\, d\omega. \tag{1·5·1}$$

An examination of equation (1·4·2) suggests that the solution of the Boltzmann equation in the present case is of the form

$$f = f_0(c) + u\chi(c). \tag{1·5·2}$$

This is readily verified as follows.

Let (l', m', n') be the direction cosines of the outward normal $\mathbf{n}$ at the point of impact. Then, since the normal velocity of an electron is reversed on impact,

$$\mathbf{v} = \mathbf{v}' + 2(\mathbf{v} \cdot \mathbf{n})\, \mathbf{n},$$

i.e.

$$(u, v, w) = (u', v', w') + 2c\cos\theta\,(l', m', n').$$
$$\tag{1·5·3}$$

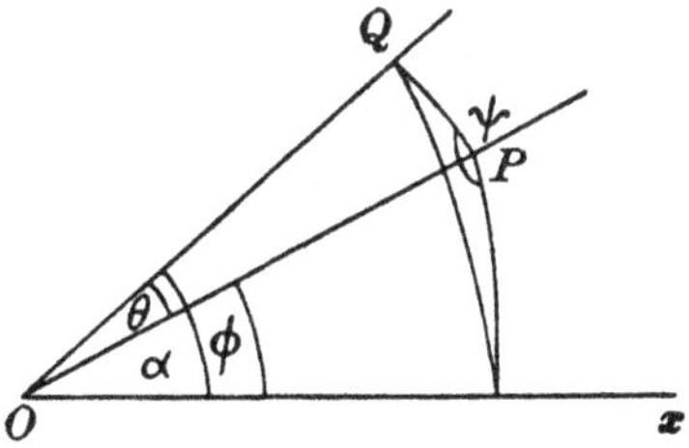

Fig. I 1.

On substituting (1·5·2) and (1·5·3) into (1·5·1) we obtain

$$[\partial f/\partial t]_{\text{coll.}} = -2NR^2 c^2 \chi(c) \int l' \cos^2\theta\, d\omega.$$

To evaluate this, consider fig. I 1. From a point O draw OP parallel to the direction of c and cutting the unit sphere, centre O, in P. Let OQ be parallel to the normal to the surface of the atom cut off by $d\omega$. Then with the notation in the figure

$$l' = \cos\alpha, \quad u = c\cos\phi.$$

Also

$$\cos\alpha = \cos\theta\cos\phi + \sin\theta\sin\phi\cos\psi.$$

We now integrate round OP as polar axis, instead of round Ox, the polar angles being θ and ψ. Therefore

$$\int l' \cos^2\theta\, d\omega = \int_0^{\frac{1}{2}\pi} \sin\theta\, d\theta \int_0^{2\pi} \cos^2\theta\,(\cos\theta\cos\phi + \sin\theta\sin\phi\cos\psi)\, d\psi$$

$$= \tfrac{1}{2}\pi\cos\phi = \tfrac{1}{2}\pi u/c.$$

We therefore have

$$[\partial f/\partial t]_{\text{coll.}} = -\pi NR^2 uc\chi(c) = -\pi NR^2 c(f - f_0), \tag{1·5·4}$$

which is of the form (1·31·3) with τ given by μc^p, where

$$\mu = 1/(\pi N R^2), \quad p = -1. \tag{1·5·5}$$

We have therefore verified that in this case a time of relaxation exists and we have evaluated μ and p. The explicit forms of equations (1·41·4) to (1·41·6) are now as follows:

$$\sigma = \frac{4n\epsilon^2}{3\pi N R^2}\left(\frac{1}{2\pi m k T}\right)^{\frac{1}{2}}, \tag{1·5·6}$$

$$\kappa = \frac{8nk^2T}{3\pi N R^2}\left(\frac{1}{2\pi m k T}\right)^{\frac{1}{2}}, \tag{1·5·7}$$

$$L = 2(k/\epsilon)^2. \tag{1·5·8}$$

The more general case of a central force between an electron and an ion varying as the inverse sth power of the distance can be treated similarly though the calculation is somewhat more complicated (see, for example, Jeans, 1925, p. 227). The result is that

$$p = \frac{4}{s-1} - 1, \tag{1·5·9}$$

the elastic sphere corresponding to the limiting case $s = \infty$. In general μ does not depend upon the temperature, but in the particular case of inverse square-law forces μ contains a factor $1/\log(1 + 36d^2k^2T^2 e_1^{-2}e_2^{-2})$, where the force between an electron and an ion is $e_1 e_2/r^2$ and $2d$ is the mean distance apart of the ions. This factor arises because of the long range of electrostatic forces (Chapman and Cowling, 1939, p. 179). The quantity d, which is the largest effective impact parameter for scattering by a particular ion, is not well defined, and the number 36 which occurs in the logarithm is therefore by no means precise.

CRITICISM OF LORENTZ'S THEORY

1·6. It is immediately obvious from the results of the preceding sections that Lorentz's theory is untenable, since attempts to adjust the parameters to fit one set of data produce serious anomalies in all the other expressions. In order to obtain the correct temperature variation of σ it is necessary to choose $p = -2$, which necessitates $s = -3$, a wholly impossible choice. It is therefore impossible to explain the temperature dependence of the conductivities. On the other hand, to reproduce the correct numerical value of L requires $p = 1$ and therefore $s = 3$, which is not unreasonable; but then σ would have to vary as $T^{\frac{1}{2}}$. The best that one can do is to take the smallest physically permissible value of p, namely, the value -1 corresponding to the ions behaving like elastic spheres. This gives a value for the Lorenz number which is too small by a factor of about $\frac{2}{3}$ and predicts that σ should vary as $T^{-\frac{1}{2}}$ instead of as T^{-1}.

There are other, still more serious objections to Lorentz's theory which did not become apparent for some years after his work appeared. The quantity $1/(\pi N R^2)$ can be considered to be the mean free path l of the electrons, and, if the physical picture put forward by Lorentz is correct, l must be of the order of the interatomic distance, that is, of the order of 5×10^{-8} cm. Then, in order to obtain the correct order of magnitude of σ at room temperatures it is necessary to assume that there is something like one free electron per atom. At first sight this might appear to be a point in favour of the theory, since if any electrons are to become free we should expect all the valency electrons to do so. However, in 1912 Debye showed that the specific heats of solids could be satisfactorily accounted for by considering only the thermal vibrations of the atoms, omitting any contribution from the electrons either free or bound. If there are N atoms per unit volume, then, since each atom has three degrees of freedom and possesses both kinetic and potential energy, the specific heat per unit volume should be $3Nk$ at ordinary temperatures. If, in addition, there are n free electrons per unit volume, their contribution to the specific heat should be $\frac{3}{2}nk$, and if n and N are of the same order of magnitude the specific heats of metals should be much greater than those of insulators, in contradiction with the law of Dulong and Petit, which states that the specific heat of all solids at constant volume tends to the limiting value $3Nk$ at high temperatures.

1·61. *Later developments.* Many suggestions have been made from time to time in order to reconcile the Drude-Lorentz theory with experiment, but as these were not based upon any plausible model they did not gain general acceptance. Inspection of equation (1·1·4) shows that the free electrons in a metal could give a small contribution to the specific heat and a large contribution to the thermal conductivity if the number of free electrons is sufficiently small and the free path sufficiently long. In the 1910's, however, there was no basis for postulating free paths of the requisite length or with the requisite temperature variation.

One of the most serious objections to these *ad hoc* hypotheses was provided by the theory of the Hall effect. According to Lorentz's theory the value of the Hall coefficient is $R = -3\pi/(8n\epsilon c)$ (see p. 234), so that measurements of the Hall coefficient should give n unambiguously. Now for silver $R = -9 \times 10^{-25}$ in Gaussian units; this gives $n = 10^{23}$ per c.c. which is rather more than the number of atoms per c.c. It is therefore clear that if the number of free electrons is taken to be small the Hall coefficient must reach an impossibly high value.

In 1914 Königsberger pointed out that the most obvious explanation of the origin of the free electrons in a metal is that they are produced by the thermal dissociation of the atoms, and that if this is so their number should be given by a formula of the type $n_0 e^{-\lambda/T}$. A small enough value of λ would lead to a sensibly constant number of free electrons at ordinary temperatures, but, however small λ might be, the conductivity would have to decrease exponentially to zero as the temperature is decreased, in contradiction with the observed facts. Königsberger put forward his suggestion as an explanation of the anomalous resistance curves of a number of substances such as titanium which were then (wrongly) classed as anomalous or semi-conductors. As we shall see later, Königsberger's hypothesis is correct for real semi-conductors, and for these substances alone the Drude-Lorentz theory is valid. Nevertheless, Königsberger's work received scant attention at the time.

The Drude-Lorentz theory was applied by various authors to all the magnetic and thermal effects associated with metallic conduction. There were a few apparent successes coupled with a large number of failures and the subject became a very confused one, as is well exemplified by the report of the Solvay conference which was held in 1923. No real advance was made until 1927 when Pauli used quantum statistical mechanics to explain away one of the more recently discovered anomalous properties of metals, namely, the existence of only a weak paramagnetism in the alkali metals.

In 1925 Pauli had enunciated the exclusion principle, which states that not more than two electrons can have the same values for all their three (translational) quantum numbers (and if there are two electrons, their spins must be in opposite directions). In 1926 Fermi and Dirac independently derived a new form of statistical mechanics based upon the exclusion principle, which reduced to the Maxwellian form in the limit of low particle densities. At that time Pauli was preoccupied with formulating the general theory of the spin of the electron, and an offshoot of this work was the application of the Fermi-Dirac statistics to the calculation of the paramagnetism of a free electron gas. It was, however, not until 1928 that anyone thought of re-examining the theory of the transport phenomena, and it was left to Sommerfeld to show that many of the most serious difficulties of the Drude-Lorentz theory could be reconciled.

The Fermi-Dirac Statistics†

1·7. Consider an assembly of structureless particles moving independently in a box. There must be some interactions to ensure thermodynamic equilibrium, but we assume the interactions to be vanishingly small. Then, if the particles obey the exclusion principle, the average occupation number of a quantum state with energy E is $f_0(E)$, where $f_0(E)$ is the Fermi function defined by

$$f_0(E) = \frac{1}{e^{(E-\zeta)/kT} + 1}. \tag{1·7·1}$$

Here ζ is a parameter determined by the total number of particles present, k is Boltzmann's constant and T is the absolute temperature. If the particles are electrons, there are two possible directions of the spin of an electron and, if these are not differentiated, the average occupation of a quantum state is $2f_0(E)$.

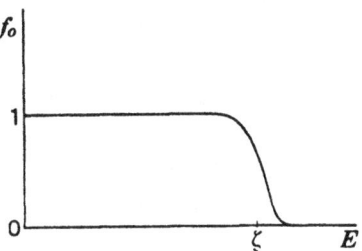

If ζ/kT is large and negative, $f_0(E)$ is the Boltzmann function, and comparison with the Maxwell distribution then shows that

$$e^{-\zeta/kT} = 2(2\pi m k T)^{\frac{3}{2}}/(nh^3)$$

Fig. I 2. The Fermi function.

(see equation (A 3·4)).

If ζ is positive and large compared with kT the assembly is said to be degenerate. For values of E less than ζ, $f_0(E)$ is effectively unity. As E approaches ζ from below, $f_0(E)$ decreases from unity, is equal to $\frac{1}{2}$ when $E = \zeta$ and falls exponentially to zero for larger values of E, being then effectively $e^{-(E-\zeta)/kT}$ (see fig. I 2).

Integrals involving $f_0(E)$ are of frequent occurrence, but are in general only required for the case $\zeta \gg kT$. They can then be evaluated by use of the asymptotic formula

$$-\int_0^\infty \phi(E) \frac{\partial f_0}{\partial E} dE = \phi(\zeta) + 2 \sum_{n=1}^\infty c_{2n}(kT)^{2n} \frac{d^{2n}\phi(\zeta)}{d\zeta^{2n}}, \tag{1·7·2}$$

where

$$c_{2n} = \sum_{s=1}^\infty \frac{(-1)^{s+1}}{s^{2n}} \tag{1·7·3}$$

$$\left(c_2 = \frac{\pi^2}{12}, \ c_4 = \frac{7\pi^4}{720}\right).$$

† Proofs of the statements made in this section will be found in the appendix.

SOMMERFELD'S MODEL OF A METAL

1·8. The picture which Sommerfeld used was a very crude one. He imagined a metal to consist of a number of electrons, without mutual interaction, moving in a region of constant negative potential energy. The number of electrons per unit volume n, and the inner potential energy $-W$ of the electrons, were treated as parameters, and no theory was given of them.

The motion of an electron in the metal is determined by the Schrödinger equation

$$\nabla^2\psi + \frac{8\pi^2 m}{h^2}(E+W)\psi = 0 \qquad (1\cdot8\cdot1)$$

and by certain boundary conditions. It is obvious that most properties of the metal, which are volume properties, such as the specific heat, are not affected by the exact shape of the metal boundary, and so we try to eliminate the boundary as much as possible. The obvious method of doing this is to consider an infinite metal, but this is not always convenient, since convergence difficulties arise, which require investigation. Another method is to consider an infinite metal divided up into cubes of side L and to suppose that each of these cubes has exactly the same propertie . For this 'cyclical metal' the boundary condition is that ψ must be triply periodic with period L. The solutions of equation $(1\cdot8\cdot1)$ are then

$$\psi = e^{2\pi i(\kappa_1 x + \kappa_2 y + \kappa_3 z)/L}, \qquad (1\cdot8\cdot2)$$

where κ_1, κ_2, κ_3 are integers, with the corresponding energies

$$E = -W + \tfrac{1}{2}h^2(\kappa_1^2 + \kappa_2^2 + \kappa_3^2)/(mL^2). \qquad (1\cdot8\cdot3)$$

The advantages of using a cyclical metal are that convergence difficulties are avoided, and that it is possible to consider a steady current. In dealing with a theory of conduction it is necessary to consider either an infinite metal, or a ring, or a cyclical metal, and the last is the simplest. We shall, however, use an infinite metal as far as possible.

If we consider a finite metal in the form of a cube of side L and use the condition that ψ shall vanish on the boundary, the wave functions are

$$\sin \pi\kappa_1 x/L \sin \pi\kappa_2 y/L \sin \pi\kappa_3 z/L, \qquad (1\cdot8\cdot4)$$

where κ_1, κ_2, κ_3 are positive integers, and the corresponding energies are

$$E = -W + \tfrac{1}{8}h^2(\kappa_1^2 + \kappa_2^2 + \kappa_3^2)/(mL^2). \qquad (1\cdot8\cdot5)$$

It might be thought at first sight that there was a paradox here, since, to each energy level given by $(1\cdot8\cdot3)$ there corresponds eight wave functions with values of κ_1, κ_2, κ_3 which merely differ in sign, whereas only one wave function is associated with each energy level given by (1·8·5). There is, however, a compensating factor in that the density of the states given by (1·8·5) is eight times that of the states given by (1·8·3), so that the total number of wave functions associated with the states whose energies are less than a given quantity is the same in both cases. Another way of looking at this question is the following. When we have a cyclical metal, the wave functions $e^{2\pi i(\pm \kappa_1 x + \kappa_2 y + \kappa_3 z)/L}$ are degenerate with each other, and represent electrons moving in opposite directions. If we introduce boundaries at $x = 0$ and $x = L$ these electron streams are no longer independent, and each degenerate state splits up into two non-degenerate ones, separated by a small energy difference, which tends to zero as L tends to infinity. The total number of states remains the same. We have discussed this rather trivial matter in detail, because it is a question which keeps recurring, and it is not always immediately obvious that the choice of boundary conditions is unimportant.

When the energy levels are given in terms of the integers κ_1, κ_2, κ_3 the average number of electrons with quantum numbers lying in the small range $\Delta\kappa_1$, $\Delta\kappa_2$, $\Delta\kappa_3$ is $\qquad 2f_0(E)\,\Delta\kappa_1\Delta\kappa_2\Delta\kappa_3.$

If the metal is infinite, it is more convenient to use quantum numbers k_1, k_2, k_3, which have the dimensions of a reciprocal length, defined by writing ψ as

$$\psi = e^{i(k_1 x + k_2 y + k_3 z)}. \tag{1·8·6}$$

These quantum numbers can take all real values. The number of electrons per unit volume with quantum numbers in the range (dk_1, dk_2, dk_3) is then

$$(1/4\pi^3)f_0\,dk_1\,dk_2\,dk_3. \tag{1·8·7}$$

We shall be continually replacing sums with respect to k_1, k_2, k_3 by integrals per unit volume, and each time a factor $1/8\pi^3$ is introduced.

It is sometimes convenient to use the velocity components u, v, w of the electrons. Comparing $E = -W + \frac{1}{2}m(u^2 + v^2 + w^2)$ with (1·8·3) we see that

$$(u, v, w) = h(\kappa_1, \kappa_2, \kappa_3)/(mL), \tag{1·8·8}$$

and so the number of electrons per unit volume with velocities lying in the range du, dv, dw is $\qquad 2(m/h)^3 f_0(E)\,du\,dv\,dw.$ $\qquad$ (1·8·9)

1·81. *The parameter ζ.* The parameter ζ is determined by n the number of free electrons per unit volume. It is convenient to use (1·8·9) and to take the energy zero to be $u = v = w = 0$. Then

$$n = 2\left(\frac{m}{h}\right)^3 \iiint f_0(E)\,du\,dv\,dw = \frac{8\sqrt{2}\,\pi m^{\frac{3}{2}}}{h^3}\int_0^\infty E^{\frac{1}{2}}f_0(E)\,dE$$

$$= -\frac{16\sqrt{2}\,\pi m^{\frac{3}{2}}}{3h^3}\int_0^\infty E^{\frac{3}{2}}\frac{\partial f_0}{\partial E}\,dE.$$

This can be evaluated by use of the asymptotic formula (1.7.2), which gives

$$n = \frac{16\sqrt{2}\,\pi m^{\frac{3}{2}}}{3h^3}\left(\zeta^{\frac{3}{2}} + \frac{\pi^2 k^2 T^2}{8\zeta^{\frac{1}{2}}} + \ldots\right). \qquad (1\cdot81\cdot1)$$

The first approximation to ζ is therefore

$$\zeta_0 = \frac{h^2}{8m}\left(\frac{3n}{\pi}\right)^{\frac{2}{3}}, \qquad (1\cdot81\cdot2)$$

and is the first term in the series

$$\zeta = \zeta_0 - \frac{\pi^2 k^2 T^2}{12\zeta_0} - \ldots \qquad (1\cdot81\cdot3)$$

For the above considerations to be valid it is necessary that $\zeta \gg kT$, i.e. that $T_0 \gg T$, where T_0 is defined by $kT_0 = \zeta_0$. The temperature T_0 is called the degeneracy temperature, since for temperatures much lower than this the majority of the electrons are in levels for which $f_0(E)$ is effectively unity, only a few electrons occupying levels situated in the region corresponding to the tail of the Fermi function. If we assume that the number of free electrons is equal to the number of valency electrons, then $n = 5\cdot9 \times 10^{22}$ for silver and we find that $T_0 = 6 \times 10^4$, so that at all normal temperatures the electron gas is highly degenerate, and the calculations of the present section are justified.

1·82. *The specific heat of the electrons.* One of Sommerfeld's main contributions was to show that the specific heat of the free electrons was very small in comparison with the normal specific heat at ordinary temperatures. The internal energy U per unit volume of the electrons is

$$U = 2\left(\frac{m}{h}\right)^3 \iiint E f_0(E)\,du\,dv\,dw = -\frac{16\sqrt{2}\,\pi m^{\frac{3}{2}}}{5h^3}\int_0^\infty E^{\frac{5}{2}}\frac{\partial f_0}{\partial E}dE. \quad (1\cdot82\cdot1)$$

Evaluation of this by means of (1·7·2) gives

$$U = \frac{16\sqrt{2}\,\pi m^{\frac{3}{2}}}{5h^3}(\zeta^{\frac{5}{2}} + \tfrac{5}{8}\pi^2 k^2 T^2\zeta^{\frac{1}{2}} + \ldots)$$

$$= \frac{16\sqrt{2}\,\pi m^{\frac{3}{2}}}{5h^3}(\zeta_0^{\frac{5}{2}} + \tfrac{5}{12}\pi^2 k^2 T^2\zeta_0^{\frac{1}{2}} + \ldots) \qquad (1\cdot82\cdot2)$$

on substituting for ζ from (1·81·3). The specific heat per unit volume is, to the first order in kT/ζ_0,

$$C_v = \frac{\partial U}{\partial T} = \frac{8\sqrt{2}\,\pi^3 m^{\frac{3}{2}}}{3h^3}k^2 T\zeta_0^{\frac{1}{2}} = \frac{4\pi^3 mk^2}{3h^2}\left(\frac{3n}{\pi}\right)^{\frac{1}{3}}T, \qquad (1\cdot82\cdot3)$$

and the ratio of this to the classical value of the specific heat of the free electrons is

$$\frac{C_v}{\frac{3}{2}nk} = \frac{8\pi^3 mk}{9h^2} \left(\frac{3}{\pi n^2}\right)^{\frac{1}{3}} T.$$

If we put $n = 5\cdot 9 \times 10^{22}$, the value assumed above for silver, this ratio is of the order of $5 \times 10^{-5}T$ and is negligibly small. The specific heat of the electrons can therefore be disregarded compared with the specific heat of the lattice vibrations at normal temperature where the specific heat of the latter is $3nk$. At sufficiently low temperatures, however, the specific heat due to the vibration of the atoms varies as T^3, and the electronic specific heat, being proportional to T, must eventually predominate (see § 6·2).

SOMMERFELD'S THEORY OF THE CONDUCTIVITIES

1·9. Having shown that the evidence afforded by the specific heat did not preclude there being free electrons equal in number to (or at least of the same order of magnitude as) the number of atoms, Sommerfeld proceeded to recalculate the conductivities using the Fermi-Dirac distribution function for the free electrons instead of the Maxwellian function. He did not attempt to consider the collision mechanism and took over unchanged the formulae (1·4·1) to (1·4·4) from Lorentz's theory, assuming, in addition, that the time of relaxation τ is a function of the energy E only. Insertion of the Fermi function for f_0 in (1·4·3) and (1·4·4) gives rise to the equations

$$j = \left(\epsilon^2 \mathscr{E} + \epsilon T \frac{\partial}{\partial x} \frac{\zeta}{T}\right)\mathscr{K}_1 + \frac{\epsilon}{T}\frac{\partial T}{\partial x}\mathscr{K}_2, \qquad (1\cdot9\cdot1)$$

$$w = \left(-\epsilon\mathscr{E} - T\frac{\partial}{\partial x}\frac{\zeta}{T}\right)\mathscr{K}_2 - \frac{1}{T}\frac{\partial T}{\partial x}\mathscr{K}_3, \qquad (1\cdot9\cdot2)$$

where

$$\mathscr{K}_n = -2\left(\quad\right)^3 \iiint \tau u^2 E^{n-1}\frac{\partial f_0}{\partial E} du\,dv\,dw, \qquad (1\cdot9\cdot3)$$

in analogy to equations (1·4·7) to (1·4·9). An alternative form for $\mathscr{K}_n$ is obtained by replacing u^2 by $\frac{1}{3}(u^2 + v^2 + w^2)$ and changing to E as the variable of integration. Then

$$\mathscr{K}_n = -\frac{16\sqrt{2}\,\pi m^{\frac{1}{2}}}{3h^3}\int \tau E^{n+\frac{1}{2}}\frac{\partial f_0}{\partial E}dE, \qquad (1\cdot9\cdot4)$$

which, by (1·7·2), is, to the second order in kT/ζ,

$$\mathscr{K}_n = \frac{16\sqrt{2}\,\pi m^{\frac{1}{2}}}{3h^3}\left[\tau(\zeta)\,\zeta^{n+\frac{1}{2}} + \tfrac{1}{6}\pi^2 k^2 T^2 \frac{d^2}{d\zeta^2}\{\tau(\zeta)\,\zeta^{n+\frac{1}{2}}\}\right]. \qquad (1\cdot9\cdot5)$$

As in § 1·4 the electrical conductivity is

$$\sigma = \epsilon^2 \mathscr{K}_1 = \tfrac{16}{3}\sqrt{2}\,\pi\epsilon^2 m^{\frac{1}{2}}\tau(\zeta)\,\zeta^{\frac{3}{2}}/h^3 = n\epsilon^2\tau(\zeta)/m, \qquad (1\cdot9\cdot6)$$

retaining only the first term in $\mathscr{K}_1$.

The calculation of the thermal conductivity is slightly more complicated. As in § 1·4 it is given by

$$\kappa = \frac{\mathscr{K}_1 \mathscr{K}_3 - \mathscr{K}_2^2}{\mathscr{K}_1 T},$$

but the use of the first approximation to $\mathscr{K}_n$ gives a zero result, and it is necessary to use the full expression (1·9·5) for $\mathscr{K}_n$. An elementary calculation gives as the lowest non-vanishing term

$$\kappa = \tfrac{16}{3} \sqrt{2}\,\pi \frac{m^{\frac{1}{2}}}{h^3} \tau(\zeta)\,\zeta^{\frac{3}{2}} \frac{\pi^2 k^2 T}{3} = \frac{\pi^2}{3} k^2 T \frac{n\tau(\zeta)}{m}. \tag{1·9·7}$$

Hence the Lorenz number is

$$L \equiv \frac{\kappa}{\sigma T} = \frac{\pi^2}{3} \left(\frac{k}{\epsilon}\right)^2. \tag{1·9·8}$$

1·91. *Criticism of Sommerfeld's theory.* Comparison of the formulae given by Sommerfeld's theory with those of Drude and Lorentz shows a considerable formal similarity between them, the Sommerfeld formulae resembling the Drude formulae more closely than the Lorentz formulae. The most striking success in all the theories is the prediction of the Wiede-mann-Franz law, but the numerical agreement is now much better, the theoretical value of the Lorenz number being $2·71 \times 10^{-13}$ e.s.u. as compared with the observed average value of $2·72 \times 10^{-13}$ e.s.u. at room temperature. The expressions for the individual conductivities are, however, not so satisfactory, as they involve the mean free path about which the theory has nothing to say. The best that can be done is to determine τ by comparing (1·9·6) and (1·9·7) with the experimental results. In order to obtain the correct order of magnitude for the conductivities and the correct temperature variations, it is necessary to assume that l varies as $1/T$ and at ordinary temperatures is of the order of 100 interatomic distances, being $5·2 \times 10^{-6}$ cm. for silver. In view of the fact that such long free paths are inexplicable on the basis of classical mechanics and that the fundamental formula (1·4·1) was taken over directly from the Drude-Lorentz theory, one cannot regard the Sommerfeld theory as satisfactory, but the main objection to the electron gas theory, the difficulty of the specific heats, has been removed. In addition, if once such mean free paths are admitted, the theoretical values of the second-order galvanomagnetic and thermomagnetic effects come into excellent alinement with the experimental values, whereas there was previously a wide divergence. Perhaps the most important feature of Sommerfeld's theory is that it destroys the notion that a free electron is necessarily a conduction electron. Since $f - f_0$ is proportional to $\partial f_0/\partial E$ in the presence of an electric field, we see that only those electrons whose energies lie in the range in which $\partial f_0/\partial E$ is not zero contribute to the conduction. For the Fermi distribution this range is small, extending over a range of order kT round the energy ζ_0. Only these electrons are strictly to be called conduction electrons, and this is in sharp contradistinction to the situation in the classical theory where conduction electrons and free electrons are necessarily the same.

In order to improve the theory, it is necessary to give precise meanings to the two fundamental ideas of a 'free electron' and of the 'mean free path'. It was shown by Bloch (1928) that an electron can move freely through a perfect crystal lattice without resistance, and that a finite free path can only be due to imperfections in the lattice. In general, the imperfections are caused predominantly by the thermal motions of the atoms and are strongly temperature dependent, increasing with increasing temperature. Impurities, however, also scatter the electrons, but the free path will then be determined by the numbering and scattering power of the impurities and will not vary appreciably with temperature. The resistance therefore consists of two parts: the 'impurity resistance' which is independent of the temperature and is small for metals of normal purity, and the resistance due to the thermal motions of the atoms. The free path for scattering by the thermal vibrations will increase rapidly as the temperature is lowered. For commercially pure metals, the impurity resistance is so small that it can be neglected at ordinary temperatures, but it becomes dominant at very low temperatures where the scattering by the thermal vibrations is very small. For this reason it is usually called the residual resistance, as it is the limit to which the resistance tends as the temperature tends to zero.

According to Bloch's analysis of the motion of an electron in a perfect lattice, all the electrons in a metal can be considered to be 'free', but it does not necessarily follow that they are all conduction electrons. Actually the free electrons in a solid form open and closed groups in much the same way as do electrons in an atom, and it is only when there are open groups that conduction electrons exist (Wilson, 1931). In this way we arrive at a theory which embraces metals, semi-conductors and insulators.

The theory of the free electrons is much simpler than that of the mean free path, since it is based upon the mechanics of a perfect lattice free from heat vibrations. In setting up such a theory we find that it enables us to understand many of the equilibrium properties of metals, such as cohesion and magnetic susceptibility. We shall therefore leave aside at the beginning all the more difficult phenomena which depend on the mean free path, and consider only the properties of an ideal metal.

REFERENCES

Bloch, F. (1928). The quantum mechanics of electrons in crystal lattices. *Z. Phys.* **52**, 555.

Bohr, N. (1911). *Studier over Metallernes Elektrontheori* (Copenhagen).

Chapman, S. and Cowling, T. G. (1939). *The mathematical theory of non-uniform gases* (Cambridge).

Debye, P. (1912). The theory of specific heats. *Ann. Phys., Lpz.* (4), **39**, 789.

Dirac, P. A. M. (1926). The theory of quantum mechanics. *Proc. Roy. Soc.* A, **112**, 661.

Drude, P. (1900). The electron theory of metals. *Ann. Phys., Lpz.* (4), **1**, 566.

Fermi, E. (1926). The quantization of the ideal monatomic gas. *Z. Phys.* **36**, 902.

Jeans, J. H. (1925). *The dynamical theory of gases* (Cambridge).

Königsberger, J. (1914). The electrical behaviour of variable conductors and their relation to the electron theory. *Jb. Radioakt.* **11**, 84.

Lorentz, H. A. (1904–5). The motion of electrons in metallic bodies. *Proc. Acad. Sci. Amst.* **7**, 438, 585, 684.

Lorenz, L. (1872). The determination of an absolute unit of heat. *Ann. Phys., Lpz.* (2), **147**, 429.

Lorenz, L. (1881). The conductivity of metals for heat and electricity. *Ann. Phys., Lpz.* (3), **13**, 422.

Pauli, W. (1927). Gas degeneration and paramagnetism. *Z. Phys.* **41**, 81.

Riecke, E. (1898). The theory of galvanic and thermal phenomena. *Ann. Phys., Lpz.* (3), **66**, 353.

Sommerfeld, A. (1928). The electron theory of metals based on Fermi statistics. *Z. Phys.* **47**, 1.

Weber, W. (1875). The motion of electricity in bodies of molecular constitution. *Ann. Phys., Lpz.* (2), **156**, 1.

Wiedemann, G. and Franz, R. (1853). The thermal conductivity of metals. *Ann. Phys., Lpz.* (2), **89**, 497.

Wilson, A. H. (1931). The theory of electronic semi-conductors. *Proc. Roy. Soc.* A, **133**, 458.

Chapter II

THE MOTION OF AN ELECTRON IN A PERFECT CRYSTAL LATTICE

ORDINARY DIFFERENTIAL EQUATIONS WITH PERIODIC COEFFICIENTS

2·1. As an introduction to the solution of the Schrödinger equation for an electron moving in the three-dimensional field of a crystal lattice, we consider the properties of the differential equation

$$\frac{d^2\psi}{dx^2} + \{\lambda + \phi(x)\}\,\psi = 0, \tag{2·1·1}$$

where λ is an arbitrary real parameter and $\phi(x) = \phi(x + 2\pi)$. The general properties of the solutions of an equation of this type are summed up in the following statement:

For a given value of λ, equation (2·1·1) has a solution of the form $\psi(x) = e^{\mu x} u(x)$, where the exponent μ is either complex or purely imaginary and where $u(x) = u(x + 2\pi)$. The solutions are said to be stable if μ is imaginary and unstable otherwise, and the solutions are unstable for sufficiently large negative values of λ if $\phi(x)$ is bounded above. As λ varies from $-\infty$ to ∞ regions of unstable and stable solutions alternate, there being an infinite number of regions of both types.

To prove and make this statement more precise we split it up into a number of theorems.

FLOQUET'S THEOREM. *The equation*

$$d^2\psi/dx^2 + \{\lambda + \phi(x)\}\,\psi = 0, \quad \text{where} \quad \phi(x) = \phi(x + 2\pi),$$

has a solution of the form

$$\psi = e^{\mu x} u(x), \quad \text{where} \quad u(x) = u(x + 2\pi).$$

Let $\psi_1(x)$, $\psi_2(x)$ be two fundamental solutions of the differential equation. Then the general solution is $\psi(x) = A\psi_1(x) + B\psi_2(x)$, where A and B are arbitrary constants. Now $\psi_1(x + 2\pi)$ and $\psi_2(x + 2\pi)$ are solutions of the differential equation, and they must therefore be expressible as linear combinations of the continuations of $\psi_1(x)$ and $\psi_2(x)$, so that equations exist of the type

$$\psi_1(x + 2\pi) = \alpha_1\psi_1(x) + \alpha_2\psi_2(x), \quad \psi_2(x + 2\pi) = \beta_1\psi_1(x) + \beta_2\psi_2(x), \tag{2·1·2}$$

where α_1, α_2, β_1, β_2 are definite constants. Thus, if

$$\psi(x) = A\psi_1(x) + B\psi_2(x),$$

then
$$\psi(x+2\pi) = (A\alpha_1 + B\beta_1)\,\psi_1(x) + (A\alpha_2 + B\beta_2)\,\psi_2(x).$$

If we now choose A, B and σ such that

$$A\alpha_1 + B\beta_1 = \sigma A, \quad A\alpha_2 + B\beta_2 = \sigma B, \tag{2·1·3}$$

then the solution $\psi(x)$ has the property

$$\psi(x+2\pi) = \sigma\psi(x). \tag{2·1·4}$$

The equations (2·1·3) for A and B have a non-zero solution if and only if

$$\begin{vmatrix} \alpha_1 - \sigma & \beta_1 \\ \alpha_2 & \beta_2 - \sigma \end{vmatrix} = 0, \tag{2·1·5}$$

so that (2·1·4) will hold if σ is taken to be either root of this quadratic. Now define $\sigma = e^{2\pi\mu}$ and $u(x) = e^{-\mu x}\psi(x)$. Then

$$u(x+2\pi) = e^{-\mu(x+2\pi)}\psi(x+2\pi) = e^{-\mu x}\psi(x) = u(x).$$

Thus the differential equation has a particular solution of the form $e^{\mu x}u(x)$, where $u(x+2\pi) = u(x)$.

2·11. *The choice of particular solutions.* It is convenient to choose the particular solutions to be such that

$$\psi_1(0) = 1, \quad \psi_1'(0) = 0; \quad \psi_2(0) = 0, \quad \psi_2'(0) = 1. \tag{2·11·1}$$

Then if we put $x = 0$ in equations (2·1·2) we find

$$\alpha_1 = \psi_1(2\pi), \quad \beta_1 = \psi_2(2\pi),$$

and if we differentiate the same equations and put $x = 0$ we find

$$\alpha_2 - \psi_1'(2\pi), \quad \beta_2 = \psi_2'(2\pi).$$

Equation (2·1·5) therefore becomes

$$\sigma^2 - \{\psi_1(2\pi) + \psi_2'(2\pi)\}\sigma + \psi_1(2\pi)\psi_2'(2\pi) - \psi_1'(2\pi)\psi_2(2\pi) = 0.$$

Now the Wronskian $\psi_1(x)\psi_2'(x) - \psi_1'(x)\psi_2(x)$ is constant,† so that

$$\psi_1(2\pi)\psi_2'(2\pi) - \psi_1'(2\pi)\psi_2(2\pi) = \psi_1(0)\psi_2'(0) - \psi_1'(0)\psi_2(0) = 1.$$

† Multiply the differential equation for ψ_2 by ψ_1, the equation for ψ_1 by ψ_2 and subtract. Then

$$0 = \psi_1\frac{d^2\psi_2}{dx^2} - \psi_2\frac{d^2\psi_1}{dx^2} = \frac{d}{dx}\left(\psi_1\frac{d\psi_2}{dx} - \psi_2\frac{d\psi_1}{dx}\right).$$

Therefore the equation for σ becomes

$$\sigma^2 - \{\psi_1(2\pi) + \psi_2'(2\pi)\}\sigma + 1 = 0, \qquad (2\cdot11\cdot2)$$

i.e.
$$\cosh 2\pi\mu = \tfrac{1}{2}\{\psi_1(2\pi) + \psi_2'(2\pi)\}. \qquad (2\cdot11\cdot3)$$

Now $\psi_1(x)$ and $\psi_2(x)$ are necessarily real for real x, and so $\cosh 2\pi\mu$ is real. It therefore follows that μ is purely imaginary if $|\cosh 2\pi\mu| \leqslant 1$ and complex if $|\cosh 2\pi\mu| > 1$. (If $\cosh 2\pi\mu > 1$, μ has an imaginary part ni, where n is an integer, but the factor e^{nix} can then be absorbed into $u(x)$, and thus μ can be reduced to a real quantity.) The end-points of the regions of λ for which $(2\cdot11\cdot2)$ has complex roots correspond to $\sigma = \pm 1$. We therefore investigate the distribution of those values of λ for which

$$\psi(\lambda, 2\pi) = \sigma\psi(\lambda, 0), \quad \psi'(\lambda, 2\pi) = \sigma\psi'(\lambda, 0), \quad \sigma^2 = 1. \qquad (2\cdot11\cdot4)$$

2·12. HAUPT'S THEOREM. *Let $\lambda_0, \lambda_1, \lambda_2, \ldots$ be the values of λ (the periodic eigenvalues) in ascending order of magnitude for which periodic solutions of equation $(2\cdot1\cdot1)$ exist, i.e. $\psi(\lambda_n, 0) = \psi(\lambda_n, 2\pi)$, $\psi'(\lambda_n, 0) = \psi'(\lambda_n, 2\pi)$. Also let $\bar{\lambda}_1, \bar{\lambda}_2, \ldots$ be the values of λ in ascending order of magnitude for which half-periodic solutions exist, i.e. $\psi(\bar{\lambda}_n, 0) = -\psi(\bar{\lambda}_n, 2\pi)$, $\psi'(\bar{\lambda}_n, 0) = -\psi'(\bar{\lambda}_n, 2\pi)$, so that $\psi(\bar{\lambda}_n, x)$ has period 4π. Then*

$$\lambda_0 < \bar{\lambda}_1 \leqslant \bar{\lambda}_2 < \lambda_1 \leqslant \lambda_2 < \ldots < \bar{\lambda}_{2n-1} \leqslant \bar{\lambda}_{2n} < \lambda_{2n-1} \leqslant \lambda_{2n} < \ldots.$$

The proof of this theorem depends upon the following properties of the solutions of differential equations subject to homogeneous boundary conditions (Haupt, 1914, 1919; Courant and Hilbert, 1924, pp. 338 and 366–7).

(a) If the coefficients in a differential equation are varied continuously, and if there are fixed two-point boundary conditions at $x = 0$ and $x = 2\pi$, the eigenvalues vary continuously and the number of zeros of any eigenfunction in $0 \leqslant x < 2\pi$ remains invariant.

(b) If the solutions $\psi(\lambda_m, x)$ and $\psi(\lambda_n, x)$ of $(2\cdot1\cdot1)$ for general values of λ_m and λ_n are such that they have m and n zeros respectively in $0 \leqslant x < 2\pi$, where $n > m$, then $\lambda_n > \lambda_m$.

(Haupt's proof of (a) applies to the general case in which the eigenfunctions for the interval $(0, 2\pi)$ satisfy the conditions

$$\psi(2\pi) = A\psi(0) + B\psi'(0), \quad \psi'(2\pi) = C\psi(0) + D\psi'(0),$$

where A, B, C, D are fixed real constants such that $AD - BC = 1$. Courant and Hilbert prove the theorem under less general conditions. Statement (b) is Sturm's oscillation theorem for a single-point boundary problem.)

To apply these properties, let the coefficients in $\phi(x)$ tend to zero, so that equation (2·1·1) becomes

$$d^2\psi/dx^2 + \lambda\psi = 0,$$

of which the solutions are $\psi = \sin\sqrt{\lambda}\,x$ and $\psi = \cos\sqrt{\lambda}\,x$. The solutions periodic in 2π are given by $\sqrt{\lambda} = n$, while those periodic in 4π are given by $\sqrt{\lambda} = n + \frac{1}{2}$, where n is an integer, and each eigenvalue is double with the exception of λ_0. We therefore see that the eigenvalues and eigenfunctions giving the periodic solutions are

$$\lambda_0 = 0; \quad \lambda_1 = \lambda_2 = 1; \quad \ldots; \quad \lambda_{2n-1} = \lambda_{2n} = n^2; \quad \ldots;$$

$$\text{constant}; \quad \sin x, \cos x; \quad \ldots; \quad \sin nx, \cos nx; \quad \ldots.$$

Now $\sin nx$ and $\cos nx$ have exactly $2n$ zeros in $0 \leqslant x < 2\pi$, and hence in the general case of $\phi(x) \neq 0$, by property (a), the periodic eigenfunctions $\psi(\lambda_{2n-1}, x)$ and $\psi(\lambda_{2n}, x)$ of equation (2·1·1) also have exactly $2n$ zeros in $0 \leqslant x < 2\pi$.

Similarly, the eigenvalues and eigenfunctions giving the half-periodic solutions are

$$\overline{\lambda}_1 = \overline{\lambda}_2 = \tfrac{1}{4}; \quad \overline{\lambda}_3 = \overline{\lambda}_4 = \tfrac{9}{4}; \quad \ldots; \quad \overline{\lambda}_{2n-1} = \overline{\lambda}_{2n} = (n-\tfrac{1}{2})^2; \quad \ldots;$$

$$\sin\tfrac{1}{2}x, \cos\tfrac{1}{2}x; \quad \sin\tfrac{3}{2}x, \cos\tfrac{3}{2}x; \quad \ldots; \quad \sin(n-\tfrac{1}{2})x, \cos(n-\tfrac{1}{2})x; \quad \ldots.$$

Also $\sin(n-\tfrac{1}{2})x$ and $\cos(n-\tfrac{1}{2})x$ have exactly $2n-1$ zeros in $0 \leqslant x < 2\pi$ and, as above, the half-periodic eigenfunctions $\psi(\overline{\lambda}_{2n-1}, x)$ and $\psi(\overline{\lambda}_{2n}, x)$ of the general equation (2·1·1) have exactly $2n-1$ zeros in $0 \leqslant x < 2\pi$.

We have now shown that $\psi(\overline{\lambda}_{2n-1}, x)$ and $\psi(\overline{\lambda}_{2n}, x)$ have exactly $2n-1$ zeros in $0 \leqslant x < 2\pi$, while $\psi(\lambda_{2n-1}, x)$ and $\psi(\lambda_{2n}, x)$ have exactly $2n$ zeros. It therefore follows from property (b) that $\lambda_{2n-1}, \lambda_{2n}$ are greater than $\overline{\lambda}_{2n-1}, \overline{\lambda}_{2n}$, which in turn are greater than $\lambda_{2n-3}, \lambda_{2n-2}$, since their eigenfunctions have $2n-2$ zeros. This proves Haupt's theorem.

In the particular case $\phi(x) = 0$ the eigenvalues are double, but this will not be so in general. In order to obtain the eigenvalues in ascending order of magnitude we associate λ_{2n-1} with $\sin nx$ and λ_{2n} with $\cos nx$ (and similarly for the $\overline{\lambda}$'s).

COROLLARY. The λ_n's are the values of λ for which $\sigma = 1$, and are therefore the roots of

$$F_1(\lambda) \equiv \psi_1(\lambda, 2\pi) + \psi_2'(\lambda, 2\pi) - 2 = 0. \tag{2·12·1}$$

Similarly the $\overline{\lambda}_n$'s are the roots of

$$F_2(\lambda) \equiv \psi_1(\lambda, 2\pi) + \psi_2'(\lambda, 2\pi) + 2 = 0. \tag{2·12·2}$$

Now σ is real or complex according as the discriminant of (2·11·2), namely, $F_1(\lambda)\,F_2(\lambda)$, is positive or negative. Also, by one of the results of Haupt's theorem, $F_1(\lambda)$ and $F_2(\lambda)$ have at most double zeros (these occur when

$\lambda_{2m-1} = \lambda_{2m}$ or $\bar{\lambda}_{2n-1} = \bar{\lambda}_{2n}$). Further, if $\phi(x)$ is bounded above, then for sufficiently large negative values of λ we have $\psi_1(x) = \cosh\sqrt{(-\lambda)}\,x$ and $\sqrt{(-\lambda)}\,\psi_2(x) = \sinh\sqrt{(-\lambda)}\,x$, so that $F_1(\lambda)$ and $F_2(\lambda)$ are both positive for $\lambda = -\infty$. The signs of $F_1(\lambda)$ and $F_2(\lambda)$ are therefore as follows:

$$F_1(\lambda) > 0 \quad \text{for} \quad \lambda < \lambda_0,$$

$$\left. \begin{aligned} F_1(\lambda) < 0 \quad &\text{for} \quad \lambda_{2n} < \lambda < \lambda_{2n+1} \\ F_1(\lambda) > 0 \quad &\text{for} \quad \lambda_{2n+1} < \lambda < \lambda_{2n+2} \end{aligned} \right\} \quad (n = 0, 1, 2, \ldots).$$

and

$$F_2(\lambda) > 0 \quad \text{for} \quad \lambda < \bar{\lambda}_1,$$

$$\left. \begin{aligned} F_2(\lambda) < 0 \quad &\text{for} \quad \bar{\lambda}_{2n-1} < \lambda < \bar{\lambda}_{2n} \\ F_2(\lambda) > 0 \quad &\text{for} \quad \bar{\lambda}_{2n} < \lambda < \bar{\lambda}_{2n+1} \end{aligned} \right\} \quad (n = 1, 2, 3, \ldots).$$

If we have a double eigenvalue at say $\lambda_{2n-1} = \lambda_{2n}$, then $F_1(\lambda)$ does not change sign there and the corresponding inequality is missing. Now μ is a pure imaginary when $F_1(\lambda)$ and $F_2(\lambda)$ have opposite signs, i.e. for those intervals of λ whose end-points are of different type. The stable solutions therefore have values of λ lying in the ranges $(\lambda_0, \bar{\lambda}_1)$, $(\bar{\lambda}_2, \lambda_1)$, $(\lambda_2, \bar{\lambda}_3)$, $\ldots$, $(\lambda_{2n-2}, \bar{\lambda}_{2n-1})$, $(\bar{\lambda}_{2n}, \lambda_{2n-1})$, $\ldots$, (see fig. II 1).

Fig. II 1. The continuous spectrum of a differential equation with periodic coefficients. The thick lines show the regions of λ in which stable solutions exist.

We have now proved all the theorems summarized at the beginning of §2·1, but before considering the extension to three-dimensional problems we shall illustrate the foregoing theory by solving a particular example. In physical problems we are normally interested in determining the regions of λ for which stable solutions exist, and the relation between $\mu = ik$ (k real) and λ in these regions.

2·13. *Application to a one-dimensional crystal.* The Schrödinger equation for an electron moving in a one-dimensional field is

$$\frac{d^2\psi}{dx^2} + \frac{8\pi^2 m}{h^2}\{E - V(x)\}\psi = 0. \tag{2·13·1}$$

If $V(x)$ is constant, the solutions are $\psi = e^{ikx}$, k being the wave number. If $V(x)$ is periodic the results of the preceding section show that the solutions are of the form $\psi = e^{ikx} u_k(x)$, where $u_k(x)$ has the same periodicity as $V(x)$. For the wave function to correspond to a state of motion in an infinite crystal, it is necessary for k to be real and ψ is then a modulated plane wave. There are, however, (an infinite number of) ranges of the energy E for which

k is imaginary, and it is impossible for electrons to exist in an infinite crystal with energies lying in these ranges. (Wave functions with imaginary wave numbers can exist in finite or semi-infinite crystals. They then correspond to electrons incident from outside and being reflected and transmitted by the crystal. If the crystal is semi-infinite, the electrons must be totally reflected.)

Kronig and Penney (1931) have considered a potential field for which the Schrödinger equation can be solved in terms of elementary functions. Their field shown in fig. II 2 and consists of rectangular humps of height V_0 and breadth b separated by regions of zero potential and of breadth a. If we put

$$\beta^2 = 8\pi^2 mE/h^2, \quad \gamma^2 = 8\pi^2 m(V_0 - E)/h^2, \qquad (2\cdot13\cdot2)$$

the Schrödinger equation is

$$d^2\psi/dx^2 + \beta^2\psi = 0 \quad (0 \leqslant x \leqslant a), \qquad d^2\psi/dx^2 - \gamma^2\psi = 0 \quad (a \leqslant x \leqslant a+b).$$

The particular solution ψ_1 satisfying $\psi_1(0) = 1$, $\psi_1'(0) = 0$ is

$$\psi_1(x) = \cos \beta x \quad (0 \leqslant x \leqslant a),$$
$$\psi_1(x) = \cos \beta a \cosh \gamma(x-a) - (\beta/\gamma) \sin \beta a \sinh \gamma(x-a) \quad (a \leqslant x \leqslant a+b), \Bigg\}$$
$$(2\cdot13\cdot3)$$

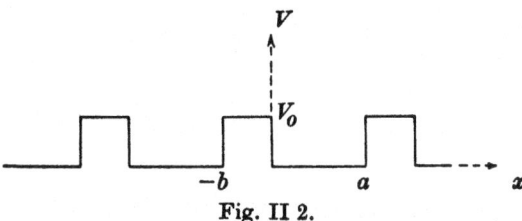

Fig. II 2.

and the particular solution ψ_2 satisfying $\psi_2(0) = 0$, $\psi_2'(0) = 1$ is

$$\psi_2(x) = \frac{1}{\beta}\sin \beta x \quad (0 \leqslant x \leqslant a),$$
$$\psi_2(x) = \frac{1}{\beta}\sin \beta a \cosh \gamma(x-a) + \frac{1}{\gamma}\cos \beta a \sinh \gamma(x-a) \quad (a \leqslant x \leqslant a+b). \Bigg\}$$
$$(2\cdot13\cdot4)$$

Therefore, from equation (2·11·3), we find, replacing the period 2π by $a+b$,

$$\cos k(a+b) = \tfrac{1}{2}\{\psi_1(a+b) + \psi_2'(a+b)\}$$

$$= \cos \beta a \cosh \gamma b + \frac{\gamma^2 - \beta^2}{2\beta\gamma} \sin \beta a \sinh \gamma b. \qquad (2\cdot13\cdot5)$$

The problem can be simplified still further by making b tend to zero and V_0 to infinity in such a way that bV_0 remains finite. The quantities β and γ are then always real. If we write

$$\lim 4\pi^2 mabV_0/h^2 = P, \qquad (2\cdot13\cdot6)$$

then $\gamma b \to 0$, $\gamma^2 b \to 2P/a$ and

$$\frac{\gamma^2-\beta^2}{2\beta\gamma}\sinh\gamma b \to \frac{\gamma^2-\beta^2}{2\beta}b \to \frac{P}{a\beta}.$$

The limiting form of the equation connecting k and β is therefore

$$\cos ka = P\frac{\sin\beta a}{\beta a} + \cos\beta a, \tag{2·13·7}$$

which can easily be solved graphically.

When $\beta a = n\pi + \eta$ $(n = 1, 2, 3, \ldots)$ the right-hand side of (2·13·7) is

$$(-1)^n\{1 + P\eta/(n\pi) - O(\eta^2)\}, \tag{2·13·8}$$

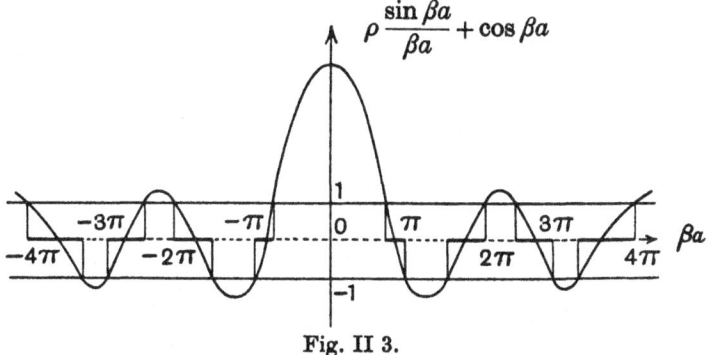

Fig. II 3.

which is numerically greater than unity for sufficiently small values of η. Now the left-hand side of (2·13·7) lies between -1 and 1, and so for all values of P there are values of βa in the neighbourhood of $n\pi$ for which k must be imaginary. The main features of equation (2·13·7) are illustrated in fig. II 3, which gives the graph of the function on the right-hand side with $P = \frac{3}{2}\pi$. The values of βa satisfying (2·13·7) are obtained by drawing a line parallel to the βa axis at a distance $\cos ka$ from it, and if we vary ka continuously from 0 to π we obtain all the possible values of βa and hence of the energy. These possible values of βa are the heavily drawn portions of the axis. The energy spectrum therefore consists of an infinite number of continuous bands separated by intervals in which there are no energy levels.

The quantity P represents in some manner the binding of the electrons, and it is a simple matter to see how the energy spectrum depends upon it. When $P = 0$, $\beta a = 2n\pi \pm ka$ (n an integer), and all values of the energy are allowed from zero to infinity. The electrons are then free, as in Sommerfeld's model of a metal, and there are no forbidden energy ranges. When P is infinite, the energy levels are independent of k, being given by $\beta a = n\pi$

$(n \geqslant 1)$, i.e. $E = n^2h^2/(8ma^2)$. The energy levels are discrete, and the electron is completely bound. It is caught between the potential walls and moves only in one cell of width a. The form of the energy spectrum for other values of P is shown in fig. II 4, the shaded portions representing allowed values of $(\beta a/\pi)^2$ and the unshaded portions disallowed values. The width of the disallowed regions, when they are fairly small, can be roughly estimated from the range for which (2·13·8) is numerically greater than unity.

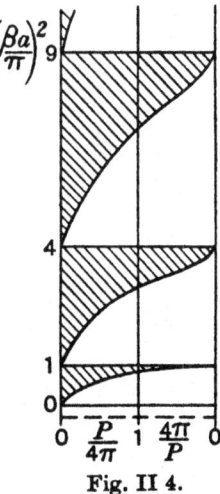

Fig. II 4.

2·14. The above results are particular examples of Floquet's and Haupt's theorems. There are two other general features to be noted. Rewriting (2·11·3) to refer to a crystal with lattice constant a, we see that the energy spectrum is determined by the equation

$$\cos ka = \tfrac{1}{2}\{\psi_1(E,a) + \psi_2'(E,a)\}.$$

The extreme limits of the left-hand side occur when

$$ka = \pm n\pi \quad (n = 0, 1, 2, \ldots).$$

The end-points of the allowed and forbidden energy regions are therefore given by

$$ka = (\ldots, -3\pi, -2\pi, -\pi, 0, \pi, 2\pi, 3\pi, \ldots).$$

Also the energy is an even function of ka, and we can enumerate the various allowed energy regions as belonging uniquely to the various intervals of ka. Thus for $E_n(ka)$, the energy in the nth allowed region, ka can be considered to be restricted to the intervals $\{-(n+1)\pi, -n\pi\}$ and $\{n\pi, (n+1)\pi\}$. This method of enumeration makes E a single-valued function of ka, and is convenient for many purposes. In certain cases, however, it is simpler to restrict ka to the interval $(-\pi, \pi)$, and E must then be treated as a multi-valued function of ka, it being necessary to specify which branch of $E(ka)$ is being considered. In this case, k is called the reduced wave vector.

If we consider a cyclical lattice which is such that all properties repeat in period $(2G+1)a$, it is necessary to impose the condition that ψ is periodic with period $(2G+1)a$. The values of the reduced wave vector k now become discrete and are $2\pi(0, \pm 1, \pm 2, \ldots, \pm G)/\{(2G+1)a\}$. To each value of k there corresponds one and only one energy level in each band, so that the number of energy levels in any band is $2G+1$. In other words, we have the important result that the number of energy levels in any band in a cyclical lattice (and therefore, on passing to the limit $G=\infty$, in an infinite lattice) is exactly equal to the number of cells in the lattice.

The Motion of an Electron in a Three-dimensional Lattice

2·2. We characterize the unit cell of a three-dimensional crystal by the three vectors a_1, a_2, a_3, which represent three concurrent sides of the parallelepiped forming the unit cell. A corner of any cell in the crystal can be denoted by the vector $g_1 a_1 + g_2 a_2 + g_3 a_3$, where g_1, g_2, g_3 are any positive or negative integers, including zero. For brevity we write

$$g_a = g_1 a_1 + g_2 a_2 + g_3 a_3. \tag{2·2·1}$$

Then, if $V(\mathbf{r})$ is the potential energy of an electron moving in the crystal, the periodicity of the potential is expressed by the relation

$$V(\mathbf{r} + g_a) = V(\mathbf{r}). \tag{2·2·2}$$

We now prove the analogue of Floquet's theorem for the Schrödinger equation of an electron moving in a field which satisfies (2·2·2).

Let E be the energy of a stationary state, which we suppose to be degenerate and to possess wave functions $\psi_1(\mathbf{r}, E), \ldots, \psi_l(\mathbf{r}, E)$. Now $\psi_\kappa(\mathbf{r} + a_i, E)$ is a solution of the Schrödinger equation with the same energy, and it is therefore expressible as a linear combination of the functions given above:

$$\psi_\kappa(\mathbf{r} + a_i, E) = \sum_{\lambda=1}^{l} P_{\kappa\lambda}^{(i)} \psi_\lambda(\mathbf{r}, E) \quad (i = 1, 2, 3).$$

The matrices $P^{(i)}$ all commute with each other, since the displacement operators, to which they correspond, commute, and it is therefore possible to reduce them simultaneously to the diagonal form by a canonical transformation

$$\psi_\lambda' = \sum_{\kappa=1}^{l} T_{\lambda\kappa} \psi_\kappa,$$

where the matrix $\mathbf{T}$ is such that $\mathbf{T} P^{(i)} \mathbf{T}^{-1}$ ($i = 1, 2, 3$) is a diagonal matrix, i.e. (on multiplying on the right by $\mathbf{T}$)

$$\sum_\lambda T_{\kappa\lambda} P_{\lambda\mu}^{(i)} = e^{ik_i} T_{\kappa\mu} \quad (i = 1, 2, 3).$$

Then
$$\psi_\lambda'(\mathbf{r} + a_i, E) = \sum_\kappa T_{\lambda\kappa} \psi_\kappa(\mathbf{r} + a_i, E) = \sum_{\kappa, \mu} T_{\lambda\kappa} P_{\kappa\mu}^{(i)} \psi_\mu(\mathbf{r}, E)$$

$$= e^{ik_i} \sum_\mu T_{\lambda\mu} \psi_\mu(\mathbf{r}, E) = e^{ik_i} \psi_\lambda'(\mathbf{r}, E).$$

This shows that any wave function can be written in the form

$$\psi_\mathbf{k}(\mathbf{r}) = e^{i\mathbf{k}\cdot\mathbf{r}} u_\mathbf{k}(\mathbf{r}), \tag{2·2·3}$$

where $\mathbf{k}$ is the wave vector (k_1, k_2, k_3), and $u_\mathbf{k}(\mathbf{r})$ has the same periodicity as the potential, so that
$$u_\mathbf{k}(\mathbf{r} + g_a) = u_\mathbf{k}(\mathbf{r}). \tag{2·2·4}$$

All the results proved for a one-dimensional periodic field apply equally to three-dimensional fields if we keep two of the coordinates fixed (the axes being supposed to be parallel to the axes of the unit cell). Three-dimensional crystals may, however, have properties which are absent in the one-dimensional case, and we proceed to investigate these by special methods of approximation, but before doing so it is necessary to describe the notation used and to establish some elementary properties of crystal lattices.

The Reciprocal Lattice

2·3. The most obvious and intuitive way of describing a crystal lattice is by enumerating the lattice points. It can, however, equally well be described by enumerating the planes on which the lattice points lie, and in actual practice it is the planes that are determined by, for example, the X-ray reflexions. Both descriptions must of course be equivalent, since three lattice points determine a plane and three lattice planes intersect in a lattice point. The simplest way of describing a set of parallel crystal planes, i.e. with a given direction of the normal, is by giving the ratios of the intercepts on the crystal axes. It is usual to define these by the Miller indices (h_1, h_2, h_3) of a crystal plane, which are such that the intercepts on the crystal axes are

$$pa_1/h_1, \quad pa_2/h_2, \quad pa_3/h_3, \tag{2·3·1}$$

where the proportionality factor p is chosen so that the h's are integers with no common factor.

In general a_1, a_2, a_3 form an oblique system of cartesian axes, and the normal definition of the oblique coordinates (x_1, x_2, x_3) of a point P is as follows. Let the planes through P parallel to the coordinate planes (the $a_2 a_3$, $a_3 a_1$, $a_1 a_2$ planes) cut the three axes in A_1, A_2, A_3 respectively. Then, if O is the origin, the coordinates are defined by the magnitudes and signs of OA_1, OA_2, OA_3. The coordinates defined in this way are called the contravariant coordinates (i.e. contravariant with respect to transformation of the axes). An alternative definition employs the covariant coordinates which are most simply defined in terms of the reciprocal vectors b_1, b_2, b_3, which are perpendicular to the coordinate planes and satisfy the relations

$$b_i . a_j = \delta_{ij} \quad (i, j = 1, 2, 3). \tag{2·3·2}$$

The explicit expressions for the b's are

$$b_1 = \frac{a_2 \times a_3}{a_1 . (a_2 \times a_3)}, \quad b_2 = \frac{a_3 \times a_1}{a_1 . (a_2 \times a_3)}, \quad b_3 = \frac{a_1 \times a_2}{a_1 . (a_2 \times a_3)}, \tag{2·3·3}$$

and the same formulae hold with the a's and b's interchanged. Also

$$b_1 . (b_2 \times b_3) = (a_2 \times a_3) . \{(a_3 \times a_1) \times (a_1 \times a_2)\} / \{a_1 . (a_2 \times a_3)\}^3.$$

But $$(\mathbf{a_3} \times \mathbf{a_1}) \times (\mathbf{a_1} \times \mathbf{a_2}) = \{\mathbf{a_1} . (\mathbf{a_2} \times \mathbf{a_3})\} \mathbf{a_1},$$

and so $$\mathbf{b_1} . (\mathbf{b_2} \times \mathbf{b_3}) = 1/\{\mathbf{a_1} . (\mathbf{a_2} \times \mathbf{a_3})\}. \tag{2·3·4}$$

The vector $\mathbf{b_1}$ is therefore perpendicular to the plane of $\mathbf{a_2}$ and $\mathbf{a_3}$, and the volume of the parallelepiped formed by $\mathbf{b_1}$, $\mathbf{b_2}$, $\mathbf{b_3}$ is the reciprocal of that formed by $\mathbf{a_1}$, $\mathbf{a_2}$, $\mathbf{a_3}$.

To resolve a vector $\mathbf{r}$ into its contravariant components r_i along the axes $\mathbf{a_i}$ write $\mathbf{r} = r_1\mathbf{a_1} + r_2\mathbf{a_2} + r_3\mathbf{a_3}$ and form the scalar product with $\mathbf{b_1}$, which gives $r_1 = \mathbf{b_1} . \mathbf{r}$. Hence

$$\mathbf{r} = \mathbf{a_1}(\mathbf{b_1} . \mathbf{r}) + \mathbf{a_2}(\mathbf{b_2} . \mathbf{r}) + \mathbf{a_3}(\mathbf{b_3} . \mathbf{r}). \tag{2·3·5}$$

Similarly, the covariant resolution of $\mathbf{r}$ is given by

$$\mathbf{r} = \mathbf{b_1}(\mathbf{a_1} . \mathbf{r}) + \mathbf{b_2}(\mathbf{a_2} . \mathbf{r}) + \mathbf{b_3}(\mathbf{a_3} . \mathbf{r}), \tag{2·3·6}$$

and so the covariant components of $\mathbf{r}$ are $\mathbf{a_i} . \mathbf{r}$, while the contravariant components are $\mathbf{b_i} . \mathbf{r}$.

2·31. If we construct all vectors $\mathbf{h_b} \equiv h_1\mathbf{b_1} + h_2\mathbf{b_2} + h_3\mathbf{b_3}$, where the h's are integers, the end-points of these vectors relative to the origin form a lattice which is known as the reciprocal lattice and which is intimately connected with the crystal planes. The vector $\mathbf{h_b}$ is in fact perpendicular to the crystal plane with Miller indices (h_1, h_2, h_3). For the vectors $\mathbf{a_1}/h_1 - \mathbf{a_2}/h_2$ and $\mathbf{a_1}/h_1 - \mathbf{a_3}/h_3$ are the vectors joining the points where the plane meets the $\mathbf{a_1}$, $\mathbf{a_2}$ and the $\mathbf{a_1}$, $\mathbf{a_3}$ axes respectively, and they therefore lie in the plane. Now

$$\mathbf{h_b} . (\mathbf{a_1}/h_1 - \mathbf{a_2}/h_2) = \mathbf{h_b} . (\mathbf{a_1}/h_1 - \mathbf{a_3}/h_3) = 0,$$

so that $\mathbf{h_b}$ is at right angles to the plane (h_1, h_2, h_3). We can therefore describe the crystal planes either by the Miller indices in the lattice space (the **a** space) or by a vector in the reciprocal space (the **b** space), and since an assembly of points is more easily visualized than an assembly of planes, the latter description is the more convenient. (Note that two vectors in the reciprocal space which have the same ratios $h_1 : h_2 : h_3$ define the same set of crystal planes.)

If we write the equations of the planes defined by (2·3·1) as

$$\frac{x_1 h_1}{a_1} + \frac{x_2 h_2}{a_2} + \frac{x_3 h_3}{a_3} = n \quad (n = 0, \pm 1, ...), \tag{2·31·1}$$

then, for given h_1, h_2, h_3 every lattice point lies on one of these planes. For the condition that the lattice point $g_1\mathbf{a_1} + g_2\mathbf{a_2} + g_3\mathbf{a_3}$ lies on the nth plane is that $g_1 h_1 + g_2 h_2 + g_3 h_3 = n$, and, since the g's and h's are integers, n can always be chosen to satisfy this relation.

If d_h is the distance apart of the planes (the distance from the origin to the first plane), the density of lattice points on the planes is proportional to d_h. For the number of lattice points per unit volume must be equal to the product of the density of lattice points on a plane and the density of planes, and the latter is equal to the reciprocal of the distance of the planes apart.

The vector $\mathbf{h}_b$ not only determines the orientation of a set of lattice planes but also the density of lattice points on the planes. To prove this, we have merely to calculate the distance of the first of the planes (2·31·1) from the origin. Since this plane goes through the points $\mathbf{a}_1/h_1$, $\mathbf{a}_2/h_2$, $\mathbf{a}_3/h_3$, the position vector $\mathbf{r}$ of any point on it relative to the origin can be considered as defined by the centroid of masses λ_1, λ_2, λ_3 placed at these points, i.e.

$$\mathbf{r} = \frac{1}{\lambda_1 + \lambda_2 + \lambda_3} \left(\frac{\lambda_1 \mathbf{a}_1}{h_1} + \frac{\lambda_2 \mathbf{a}_2}{h_2} + \frac{\lambda_3 \mathbf{a}_3}{h_3} \right).$$

Now since the direction of the normal to the plane is given by $\mathbf{h}_b$, the distance d_h of the plane from the origin is given by

$$d_h = \mathbf{r} \cdot \mathbf{h}_b / |\mathbf{h}_b|.$$

But $\mathbf{h}_b \cdot \mathbf{a}_i = h_i$, and so $\mathbf{r} \cdot \mathbf{h}_b = 1$, which gives

$$d_h = 1/|\mathbf{h}_b|.$$

Note that, in order to obtain the distance between consecutive planes, it is essential to choose the h's to have no common factor. Where it is desired to emphasize this, it is usual to write $\mathbf{h}_b^*$ and

$$d_h = 1/|\mathbf{h}_b^*|. \tag{2·31·2}$$

X-ray Reflexions in a Crystal

2·4. The motion of electrons in a crystal has many similarities to the diffraction of X-rays, and as the latter phenomenon is easier to understand it is convenient to give a brief résumé of the simpler facts before considering the case of electrons.

Consider a beam of X-rays traversing a crystal lattice in the direction $\mathbf{s}_0$ and being scattered in the direction $\mathbf{s}$, $\mathbf{s}_0$ and $\mathbf{s}$ being unit vectors. Let P and Q be any two atoms in the crystal, and let QA and QB be drawn at right angles to the rays incident on and reflected from the atom P (see fig. II 5). The path difference between the rays reflected by P and Q is $AP + PB$. Now, if the vector PQ is $\mathbf{r}$, then $AP = -\mathbf{r} \cdot \mathbf{s}_0$ and $PB = \mathbf{r} \cdot \mathbf{s}$, and if the rays are to reinforce one another the path difference must be an integral multiple m of the wave length λ. Hence

$$\mathbf{r} \cdot (\mathbf{s} - \mathbf{s}_0) = m\lambda,$$

and this must be true for all the lattice vectors $\mathbf{r} = \mathbf{g}_a$ if the waves scattered by all the atoms are not to suffer destructive interference. We must therefore have

$$\mathbf{a}_i \cdot (\mathbf{s} - \mathbf{s}_0) = h_i \lambda \quad (i = 1, 2, 3), \tag{2·4·1}$$

where the h's are (positive or negative) integers. These three equations can be combined into one:

$$\mathbf{s} - \mathbf{s}_0 = \lambda(h_1 \mathbf{b}_1 + h_2 \mathbf{b}_2 + h_3 \mathbf{b}_3) = \lambda \mathbf{h}_b, \tag{2·4·2}$$

which means that $(\mathbf{s} - \mathbf{s}_0)/\lambda$ must be a vector of the reciprocal lattice.

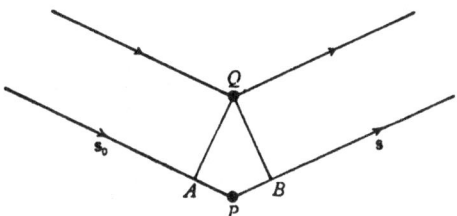

Fig. II 5. The scattering of X-rays by two atoms in a crystal.

The relation (2·4·2) is Laue's condition for an X-ray reflexion to occur. It can be represented diagrammatically as follows. From the origin O of the reciprocal lattice draw the vector $-\mathbf{s}_0/\lambda$ to the point R (which in general is not a point of the lattice). From R draw a sphere of radius $1/\lambda$. If this sphere goes through a reciprocal lattice point S, the vector OS is a vector of the reciprocal lattice satisfying the reflexion condition (2·4·2) (see fig. II 6).

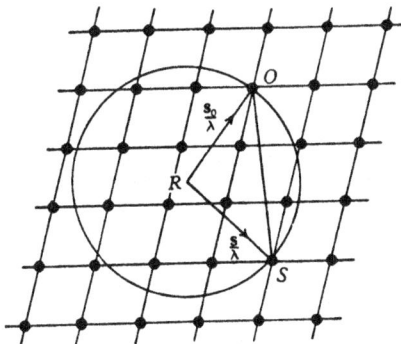

Fig. II 6. The reflexion condition in the reciprocal lattice.

2·41. The reflexion condition can also be formulated in terms of lattice planes as follows. Since $\mathbf{h}_b$ is at right angles to the plane (h_1, h_2, h_3), this plane must be parallel to the bisector of the angle ORS, and the secondary ray can be considered as being derived from the primary ray by being reflected in the plane (h_1, h_2, h_3). If θ is the angle of reflexion, then ORS is 2θ and $OS = |\mathbf{h}_b| = 2\lambda^{-1}\sin\theta$.

If the integers h_1, h_2, h_3 have a common factor n, we write

$$h_1, h_2, h_3 = n(h_1^*, h_2^*, h_3^*), \tag{2·41·1}$$

and call the reflexion the nth order reflexion at the (h_1^*, h_2^*, h_3^*) plane. Then $|\mathbf{h}_b| = n|\mathbf{h}_b^*| = 2\lambda^{-1}\sin\theta$, which, by (2·31·2), can be written in the form

$$n\lambda = 2d_h \sin\theta. \tag{2·41·2}$$

This is Bragg's form of the reflexion condition.

THE MOTION OF NEARLY FREE ELECTRONS IN A THREE-DIMENSIONAL LATTICE

2·5. We can obtain a solution of the Schrödinger equation in the form of a triple Fourier series, but it must be borne in mind that such series are usually only slowly convergent. We first expand the potential V as a triple Fourier series

$$V = \sum_{g_1, g_2, g_3 = -\infty}^{\infty} V_g e^{2\pi i \mathbf{g}_b \cdot \mathbf{r}}. \tag{2·5·1}$$

To verify this we have merely to note that $e^{2\pi i \mathbf{g}_b \cdot \mathbf{r}}$ is triply periodic with the unit cell as its period. (The expression $\mathbf{g}_b \cdot \mathbf{r}$ denotes $\Sigma g_i(\mathbf{b}_i \cdot \mathbf{r})$ and, if we replace $\mathbf{r}$ by $\mathbf{r} + \Sigma h_j \mathbf{a}_j$, $\mathbf{g}_b \cdot \mathbf{r}$ becomes

$$\mathbf{g}_b \cdot \mathbf{r} + \Sigma\Sigma g_i(\mathbf{b}_i \cdot h_j \mathbf{a}_j) = \mathbf{g}_b \cdot \mathbf{r} + \Sigma\Sigma g_i h_j \delta_{ij} = \mathbf{g}_b \cdot \mathbf{r} + \text{an integer},$$

so that $e^{2\pi i \mathbf{g}_b \cdot \mathbf{r}}$ is unaltered.) Similarly, ψ can be written

$$\psi_k(\mathbf{r}) = e^{i\mathbf{k} \cdot \mathbf{r}} \sum_g c_g e^{2\pi i \mathbf{g}_b \cdot \mathbf{r}}. \tag{2·5·2}$$

On substituting these expressions into the Schrödinger equation, we obtain an infinite set of linear equations for the c_g's:

$$\{E - V_{000} - \tfrac{1}{2}\hbar^2(\mathbf{k} + 2\pi\mathbf{g}_b)^2/m\} c_g = \sum_{h \neq 0} c_{g-h} V_h. \tag{2·5·3}$$

These equations are only soluble provided the infinite determinant of the set vanishes, and the roots of this determinant are the energy levels. In general we have to find the roots by approximate methods. The features present in the one-dimensional problem recur here, and we shall find that the spectrum consists of bands separated by regions of disallowed energy values.

If all the Fourier coefficients of V were zero except V_{000}, we should have perfectly free electrons. The energy would then be

$$V_{000} + \tfrac{1}{2}\hbar^2 |\mathbf{k}|^2/m, \tag{2·5·4}$$

and all the coefficients in (2·5·2) would be zero except c_{000}, which we can put equal to unity. Therefore provided the Fourier coefficients of V are small we may hope to obtain an approximate solution of (2·5·3) by neglecting products of these coefficients. If this is so, we shall say that the electrons are nearly free. On the right hand of (2·5·3) put all the quantities c_{g-h} zero except c_{000}, and put for E the zero approximation given by (2·5·4). We then obtain

$$c_g = -\frac{2\pi m}{h^2}\frac{V_g}{\pi\,|\,\mathbf{g}_b\,|^2 + \mathbf{g}_b . \mathbf{k}}. \qquad (2·5·5)$$

If we now substitute this value of c_g back into (2·5·3) we obtain a second approximation for the energy

$$E = V_{000} + \frac{\hbar^2\,|\,\mathbf{k}\,|^2}{2m} - \frac{m}{2\pi\hbar^2}\sum_{g\neq0}\frac{|\,V_g\,|^2}{\pi\,|\,\mathbf{g}_b\,|^2 + \mathbf{g}_b . \mathbf{k}}, \qquad (2·5·6)$$

since V_g and V_{-g} are complex conjugates.

2·51. The foregoing method is obviously inappropriate when any term in the expression (2·5·6) becomes infinite, i.e. when

$$\pi\,|\,\mathbf{g}_b\,|^2 + \mathbf{g}_b . \mathbf{k} = 0 \quad (V_g\neq0), \qquad (2·51·1)$$

which is the condition for the electron wave to suffer a Bragg reflexion by the (g_1, g_2, g_3) plane. For (2·51·1) can be written as $|\,\mathbf{k}\,|^2 = |\,\mathbf{k} + 2\pi\mathbf{g}_b\,|^2$, and this is equivalent to the Laue condition (2·4·2) for the total reflexion of an electron wave with 'wave-length' $2\pi/k$. The electron wave with wave vector $\mathbf{k}$ is reflected by the (g_1, g_2, g_3) plane and acquires a wave vector $\mathbf{k} + 2\pi\mathbf{g}_b$. When these are numerically equal the electron wave is totally reflected by the lattice planes and it cannot be propagated through the crystal. When the condition (2·51·1) is fulfilled, or nearly fulfilled, we must adopt another method of approximation. We see from (2·5·5) that the corresponding coefficient in the expansion becomes large, and so we approximate to the infinite set of equations (2·5·3) by neglecting all the coefficients except c_{000} and c_g (Peierls, 1930). We then have

$$\begin{aligned}(E - V_{000} - \tfrac{1}{2}\hbar^2\,|\,\mathbf{k}\,|^2/m)\,c_{000} &= c_g V_{-g},\\(E - V_{000} - \tfrac{1}{2}\hbar^2\,|\,\mathbf{k} + 2\pi\mathbf{g}_b\,|^2/m)\,c_g &= c_{000}V_g,\end{aligned}\Bigg\} \qquad (2·51·2)$$

giving two values of E for each value of $\mathbf{k}$:

$$E = V_{000} + \frac{\hbar^2}{4m}(|\,\mathbf{k}\,|^2 + |\,\mathbf{k} + 2\pi\mathbf{g}_b\,|^2) \pm \left\{|\,V_g\,|^2 + \frac{\pi^2\hbar^4}{m^2}(\pi\,|\,\mathbf{g}_b\,|^2 + \mathbf{g}_b . \mathbf{k})^2\right\}^{\frac{1}{2}}. \qquad (2·51·3)$$

When the condition (2·51·1) for Bragg reflexion is exactly fulfilled there are two values of the energy:

$$E = V_{000} + \tfrac{1}{2}h^2\,|\,\mathbf{k}\,|^2/m \pm |\,V_g\,|,$$

and the energy has a discontinuity. If $\mathbf{k}$ is such that more than two coefficients are large simultaneously, we have more than two equations of the form (2·51·2), and the energy is correspondingly more complicated.

If we choose all possible integral values for g_1, g_2, g_3, we obtain from (2·51·1) a series of planes drawn in a space in which k_1, k_2, k_3 are rectangular axes, and the energy is discontinuous across these planes. The planes are perpendicular to the direction of $\mathbf{g}_b$ and are at a distance $\pi \,|\, \mathbf{g}_b \,|$ from the origin O. They divide the $\mathbf{k}$ space into a number of zones, which are called Brillouin zones. As we have already seen, the energy is not necessarily a single-valued function of $\mathbf{k}$. The most convenient convention to make the energy single-valued is to demand that the energy should be a continuous function of $\mathbf{k}$ throughout any one Brillouin zone, and that the jump in the energy when passing outwards across a zone boundary should be positive. With this convention the wave functions pass over continuously into the wave functions of a free electron when we make all the Fourier coefficients of V tend to zero. It must, however, be borne in mind that there are various equivalent ways of arranging the surfaces so as to make E a single-valued function of $\mathbf{k}$ (see e.g. Seitz, 1940, pp. 292ff.).

For values of $\mathbf{k}$ far from the zone boundaries the energy is given approximately by (2·5·6), while near the zone boundaries we must use the expression (2·51·3). We may obtain the energy in a simpler form if we consider its variation in a direction perpendicular to the zone boundary. We change the axes in the $\mathbf{k}$ space, taking new rectangular coordinates $(\mathfrak{k}_1', \mathfrak{k}_2, \mathfrak{k}_3)$ where $O\mathfrak{k}_1'$ is parallel to $\mathbf{g}_b$, and $O\mathfrak{k}_2$, $O\mathfrak{k}_3$ are parallel to the zone boundary. With these coordinates the plane (2·51·1) is simply

$$\mathfrak{k}_1' + \pi \,|\, \mathbf{g}_b \,| = 0,$$

and putting $\mathfrak{k}_1' = -\pi \,|\, \mathbf{g}_b \,| + \mathfrak{k}_1$, we have for (2·51·3)

$$E = V_{000} + \frac{\hbar^2}{2m}(\mathfrak{k}_1^2 + \mathfrak{k}_2^2 + \mathfrak{k}_3^2 + \pi^2 \,|\, \mathbf{g}_b \,|^2) \pm \left\{ |\, V_g \,|^2 + \frac{\pi^2 \hbar^4}{m^2} \,|\, \mathbf{g}_b \,|^2 \mathfrak{k}_1^2 \right\}^{\frac{1}{2}}. \quad (2\cdot51\cdot4)$$

Expanding the square root we have approximately

$$E = V_{000} + \frac{\hbar^2}{2m}(\mathfrak{k}_1^2 + \mathfrak{k}_2^2 + \mathfrak{k}_3^2 + \pi^2 \,|\, \mathbf{g}_b \,|^2) - |\, V_g \,| - \frac{\pi^2 \hbar^4}{2m^2} \frac{|\, \mathbf{g}_b \,|^2 \mathfrak{k}_1^2}{|\, V_g \,|} \quad (2\cdot51\cdot5)$$

for $\mathfrak{k}_1 > 0$, and

$$E = V_{000} + \frac{\hbar^2}{2m}(\mathfrak{k}_1^2 + \mathfrak{k}_2^2 + \mathfrak{k}_3^2 + \pi^2 \,|\, \mathbf{g}_b \,|^2) + |\, V_g \,| + \frac{\pi^2 \hbar^4}{2m^2} \frac{|\, \mathbf{g}_b \,|^2 \mathfrak{k}_1^2}{|\, V_g \,|} \quad (2\cdot51\cdot6)$$

for $\mathfrak{k}_1 < 0$. If we plot the energy as a function of one variable k_1 we obtain a curve of the type shown in fig. II 7. The discontinuity in the energy is the only general feature, and occurs at every zone boundary. The broken curve shows how the energy in the second zone would look if expressed as a function of the reduced wave vector.

2·52. So far the properties of the three-dimensional problem are similar to those of the one-dimensional, though more complicated, but there is one very important difference. In a one-dimensional lattice there are always forbidden energy ranges. In a three-dimensional lattice there are always discontinuities in the energy, but there may be no forbidden energy ranges. This is most easily seen by expressing the energy as a function of the reduced wave vector. The energy surface then consists of a number of detached parts one above the other. These surfaces do not cut each other along the boundary of the **k** region, but there is nothing to prevent their cutting inside. In fact they certainly will do so provided the energy discontinuities are small enough, and for nearly free electrons we have an overlapping of the energy bands.

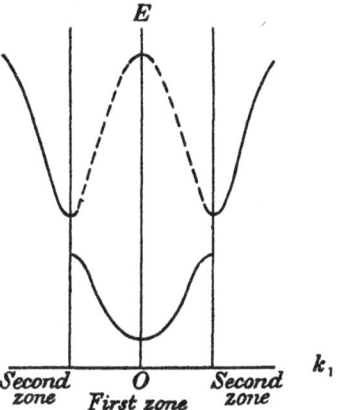

Fig. II 7. The energy as a function of the wave number. The broken curve gives the energy in terms of the reduced wave number.

To see the shape of the zones in a particular case, let us consider a simple cubic lattice with lattice constant a. The unit vectors of the reciprocal lattice are parallel to those of the crystal lattice and are of length $1/a$, so the equations of the zone boundaries are

$$a(g_1 k_1 + g_2 k_2 + g_3 k_3) + \pi(g_1^2 + g_2^2 + g_3^2) = 0.$$

The first zone is $-\pi \leqslant a(k_1, k_2, k_3) \leqslant \pi$. The second zone lies between the first zone and the dodecahedron formed by planes such as

$$a(k_1 \pm k_2) \pm 2\pi = 0. \tag{2·52·1}$$

The third zone lies outside the dodecahedron and is bounded by the planes (2·52·1) and the planes $ak_i = \pm \pi$. The fourth zone lies outside the third, and is bounded externally by the planes

$$a(k_1 \pm k_2 \pm k_3) \pm 3\pi = 0 \tag{2·52·2}$$

and by the planes $ak_i = \pm 2\pi$. Its surface consists of eight hexagons and six squares. The external surfaces of these zones are shown in fig. II 8. The projection of these zones on the central cube is complicated, except for the dodecahedron. Diagrams are given by Brillouin (1930), who also considers in detail the zones for a two-dimensional lattice.

THE MOTION OF TIGHTLY BOUND ELECTRONS
IN A THREE-DIMENSIONAL LATTICE

2·6. There is another method of approximation which is valid when the force attracting an electron to a lattice point is of very short range. Let us consider only cubic crystals with lattice constant a and expand the potential energy V as

$$V(\mathbf{r}) = \sum_{\mathbf{g}} U_{\mathbf{g}}(\mathbf{r}),$$

where the potential energy $U_{\mathbf{g}}$ is a function of the distance from the lattice point $(g_1 a, g_2 a, g_3 a)$ and is due partly to the ion at the lattice point and

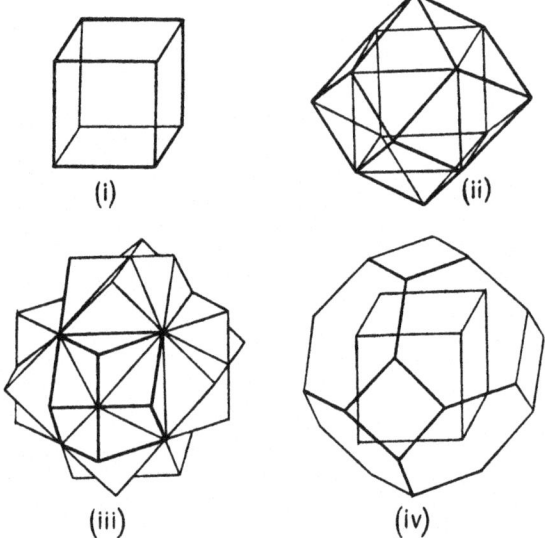

Fig. II 8. The first four Brillouin zones of a cubic lattice.

partly to the other electrons. We shall assume that $U_{\mathbf{g}}$ is large near the lattice point and falls off very rapidly with distance. Also we suppose that the Schrödinger equation

$$\nabla^2 \phi_{\mathbf{g}} + \frac{8\pi^2 m}{h^2} (E - U_{\mathbf{g}}) \phi_{\mathbf{g}} = 0 \tag{2·6·1}$$

can be solved. Then in order to determine the stationary states of an electron in the lattice we assume as zero approximation that the electron is in the neighbourhood of one particular lattice point, and neglect the influence of the other atoms. This state is degenerate, since placing the electron in a similar orbit round any other lattice point gives a state of the same energy, the order of the degeneracy being equal to the number of atoms in the crystal. When we take into consideration the forces due to the neighbouring atoms, the degeneracy is removed, and each energy level splits up into a number of levels forming a band. In order to obtain a simple solution of the Schrödinger equation it is necessary to assume that the presence of the neighbouring atoms causes an energy change small compared with the energy differences between the various levels of (2·6·1). This is a more exact way of saying that the electrons are tightly bound.

2·61. We shall first deal with s states (Bloch, 1928), ϕ_g being then spherically symmetrical. If E_1 is the corresponding energy value of (2·6·1), we obtain a solution of the Schrödinger equation for the whole lattice by putting

$$\psi = \sum_g c_g \phi_g, \quad E = E_1 + \eta_1. \tag{2·61·1}$$

The equations determining η_1 and c_g to the first order are easily found to be

$$\sum_g c_g \iiint (\eta_1 - U_g') \phi_g \phi_h \, dx \, dy \, dz = 0 \tag{2·61·2}$$

for all values of h_1, h_2, h_3, where $U_g' = V - U_g$. Since ϕ_g decreases rapidly as we go away from the lattice point g, we assume that

$$\int \phi_g \phi_h \, d\tau = \delta_g^h,$$

and we write $\quad -\int U_g' \phi_g^2 \, d\tau = \alpha_1, \quad -\int U_g' \phi_g \phi_h \, d\tau = J(g-h),$

where $J(g-h)$ depends on g and h only through their difference $|g-h|$. Since ϕ_g and U_g are spherically symmetrical, α_1 and J will in general be positive, provided ϕ_g has no nodes except inside the atomic core. The equations (2·61·2) then reduce to the infinite set

$$(\eta_1 + \alpha_1) c_g + \sum_h J(g-h) c_h = 0, \tag{2·61·3}$$

which are satisfied by $\quad c_g(\mathbf{k}) = e^{iak \cdot g} \tag{2·61·4}$

and $\quad \eta_1 = -\alpha_1 - 2 \sum_p' J(\mathbf{p}) \cos a(\mathbf{k} \cdot \mathbf{p}), \tag{2·61·5}$

where $\sum'$ means that the summation is to be restricted to the range $p_1 \geqslant 0$, $-\infty \leqslant p_2$, $p_3 \leqslant \infty$, and $p_1 = p_2 = p_3 = 0$ is to be omitted. The quantities $J(\mathbf{p})$ decrease rapidly with increasing $|\mathbf{p}|$, and we shall therefore only consider interactions between neighbouring atoms. The wave function has the correct periodicity, since we may write

$$\psi = \sum_g e^{iak \cdot g} \phi_g = e^{i\mathbf{k} \cdot \mathbf{r}} \sum_g e^{-i\mathbf{k} \cdot (\mathbf{r} - ga)} \phi(\mathbf{r} - ga), \tag{2·61·6}$$

and the sum in the last expression is obviously periodic in the lattice constant.

In a simple cubic lattice each atom has six nearest neighbours at a distance a away. If we put

$$J(1, 0, 0) = J(0, 1, 0) = J(0, 0, 1) = \beta_1$$

and neglect all other J's, we obtain for the energy

$$E = E_1 - \alpha_1 - 2\beta_1(\cos ak_1 + \cos ak_2 + \cos ak_3). \tag{2·61·7}$$

These expressions for the energy are periodic functions of $\mathbf{k}$. To obtain each energy level once and once only, we must restrict $\mathbf{k}$ to lie in the first Brillouin zone, i.e. we must use the reduced wave vector.

2·62. We next consider a p state. Here the level is triply degenerate, but it will not split up in a cubic crystal. For the independent wave functions of the p state can be chosen to be $xf(r)$, $yf(r)$ and $zf(r)$, where $f(r)$ is a function of r only, and these three functions will give the same energy in a field with cubic symmetry. We consider the levels derived from one of these wave functions, the others being obvious from symmetry.

Let $\chi_\mathbf{g} = x_\mathbf{g} f(r_\mathbf{g})$ be a wave function of the isolated atom corresponding to the energy value E_2. Then we solve the Schrödinger equation as before by

$$\psi = \sum_\mathbf{g} c_\mathbf{g} \chi_\mathbf{g}, \quad E = E_2 + \eta_2, \tag{2·62·1}$$

obtaining a set of equations similar to (2·61·2). Here, however, the equations involve more than two constants α and β, since χ is not spherically symmetrical. As before we put

$$-\int U'_\mathbf{g} \chi_\mathbf{g}^2 d\tau = \alpha_2.$$

Among the non-vanishing integrals we must distinguish two further types. First the type

$$-\int U'_\mathbf{g} \chi_\mathbf{g} \chi_{g_1, g_2+1, g_3} d\tau.$$

This we put equal to β_2, which is essentially positive, since the only factor which makes it differ in form from the integral for β_1 is $x_\mathbf{g}^2$ which is positive. Then there is the type

$$-\int U'_\mathbf{g} \chi_\mathbf{g} \chi_{g_1+1, g_2, g_3} d\tau.$$

This we put equal to $-\gamma_2$, where γ_2 is positive, since in the integrand there occurs a factor $x_{g_1} x_{g_1+1}$. Little can be said about β_2 and γ_2 except that they are both positive and of the same order of magnitude. The equations corresponding to (2·61·3) are

$$(\eta_2 + \alpha_2) c_\mathbf{g} - \gamma_2(c_{g_1+1, g_2, g_3} + c_{g_1-1, g_2, g_3})$$
$$+ \beta_2(c_{g_1, g_2+1, g_3} + c_{g_1, g_2-1, g_3} + c_{g_1, g_2, g_3+1} + c_{g_1, g_2, g_3-1}) = 0. \tag{2·62·2}$$

These are solved as before by $c_g(\mathbf{k}) = e^{ia\mathbf{k}\cdot\mathbf{g}}$, giving

$$E = E_2 - \alpha_2 + 2\gamma_2 \cos ak_1 - 2\beta_2(\cos ak_2 + \cos ak_3). \qquad (2\cdot62\cdot3)$$

There are also two similar expressions obtained by permuting k_1, k_2 and k_3.

2·63. The energy levels of tightly bound electrons bear some resemblances to those of nearly free electrons. If we expand the cosines in (2·61·7), we obtain for small values of $a\,|\,\mathbf{k}\,|$

$$E = E_1 - \alpha_1 - 6\beta_1 + \beta_1 a^2\,|\,\mathbf{k}\,|^2. \qquad (2\cdot63\cdot1)$$

To this approximation the electrons behave as if they were free electrons with an effective mass given by

$$m^* = \hbar^2/(2a^2\beta_1), \qquad (2\cdot63\cdot2)$$

as is seen by comparing (2·63·1) with (2·5·4). For the approximation of tightly bound electrons to be valid, β_1 must be small, and m^* must be greater than the ordinary mass m but of the same order of magnitude. If we put

$$ak_1' = \pi - ak_1, \quad ak_2' = \pi - ak_2, \quad ak_3' = \pi - ak_3$$

at the extreme limit of the top of the energy band, then, for small values of $a\,|\,\mathbf{k}'\,|$, (2·61·7) becomes

$$E = E_1 - \alpha_1 + 6\beta_1 - \beta_1 a^2\,|\,\mathbf{k}'\,|^2, \qquad (2\cdot63\cdot3)$$

which is again of the same form as for free electrons, with a negative sign in front of β_1.

For the perturbation method to be valid it is necessary that the unperturbed energy levels $E_1, E_2, \dots$ should be far apart. Therefore if (2·61·7) gives the states derived from a term ns, and (2·62·3) those derived from a term np, the energy difference between these two systems of states must be large. Now the greatest of the energies (2·61·7) is given by putting $k_1 = k_2 = k_3 = \pi/a$ and is $E_1 - \alpha_1 + 6\beta_1$, while the least of the energies (2·62·3) is given by putting $k_1 = \pi/a$, $k_2 = k_3 = 0$ and is $E_2 - \alpha_2 - 4\beta_2 - 2\gamma_2$. We must therefore have $E_2 - \alpha_2 - 4\beta_2 - 2\gamma_2 > E_1 - \alpha_1 + 6\beta_1$. All the energies of the second zone lie above those of the first zone, and there are forbidden ranges of the energy, whereas with nearly free electrons the energy can take all values between V_{000} and infinity, in spite of the discontinuities.

When the electrons are tightly bound it is most convenient to use the reduced wave numbers, and to distinguish the various zones by another quantum number. In fig. II 9 the energy curves for an s state and a p_x state are given as functions of k_1, and also of k_2, for fixed values of the other quantum numbers. In fig. II 10 the energy contours, as given by the free electron model, are drawn for a two-dimensional simple cubic lattice, while in fig. II 11 we show the general shape of a three-dimensional energy surface for tightly bound s-electrons when the energy is in the upper half of the energy band.

ENERGY BANDS OF STANDARD FORM

2·7. Since both models that we have discussed possess the same qualitative features, we are on safe ground in assuming that these features are general ones. It is, however, clear that neither model can give a good quantitative picture of the wave functions and energy levels, and more refined calculations are required which are discussed in § 3·3. The inner

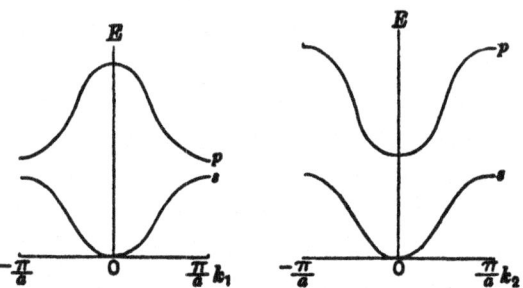

Fig. II 9. The energy levels for s and p states.

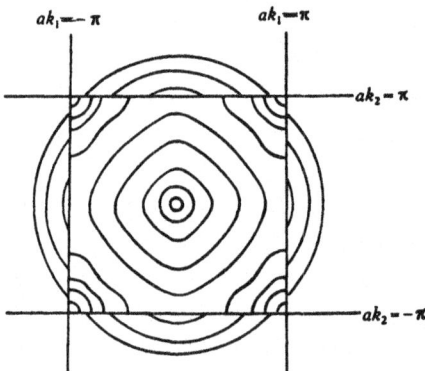

Fig. II 10. The energy contours in the plane $k_3 = 0$.

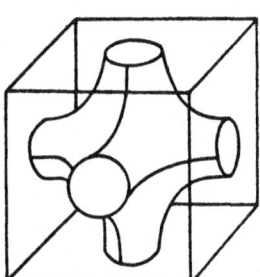

Fig. II 11. An energy surface for tightly bound s-electrons near the top of the energy band.

electrons can quite definitely be regarded as tightly bound, the values of the constants $\alpha, \beta, \gamma, \ldots$ being so small that in practice they can be taken to be zero. The valency electrons, however, cannot be considered to be either tightly bound or nearly free, and the crystal energies are of the same order of magnitude as the excitation energies of the free atoms. Nevertheless, it is desirable in many problems to assume some reasonable form for the energy levels in order to carry the calculations to completion, and the following simple cases have been found convenient for calculation.

A normal energy band of standard form. This is defined to be a band in which the energy is given by

$$E = \tfrac{1}{2}\hbar^2\,|\,\mathbf{k}\,|^2/m^*. \qquad (2\cdot7\cdot1)$$

It simulates the behaviour of nearly free electrons and also of tightly bound electrons near the bottom of an energy band (cf. equation (2·63·1)). In general we should expect the effective mass m^* to be of the same order as but greater than the electron mass m.

An inverted band of standard form. This is defined to be a band in which the energy is given by

$$E = A - \tfrac{1}{2}\hbar^2\,|\,\mathbf{k}\,|^2/m^*. \qquad (2\cdot7\cdot2)$$

This simulates the behaviour of tightly bound electrons near the top of an energy band (cf. equation (2·63·3)), and also of nearly free electrons with wave vectors near the boundary of a Brillouin zone (cf. equation (2·51·5)).

Overlapping bands of standard form. The simplest model in which two energy bands overlap is obtained if we assume that there is one band given by (2·7·1) and one given by (2·7·2) with $A > 0$, usually with different values of m^*. The same model with $A < 0$ would give a crystal with two non-overlapping bands.

2·71. *The density of states.* Although a knowledge of the energy levels and wave functions of a crystal is essential for a complete description of all its properties, and in particular for the discussion of transport phenomena, yet knowledge of a much less detailed nature is sufficient for a discussion of the simpler equilibrium phenomena. For example, the specific heat of the electrons at low temperatures can be found if the density of states $\mathfrak{n}(E)$ is known. We define $\mathfrak{n}(E)\,dE$ as being the number of energy levels per unit volume (for one direction of the electron spin) lying in the range $E, E + dE$. Hence

$$\mathfrak{n}(E)\,dE = \frac{1}{8\pi^3}\int d\mathbf{k}, \qquad (2\cdot71\cdot1)$$

where $d\mathbf{k} = dk_1\,dk_2\,dk_3$, the integral being taken over the part of the $\mathbf{k}$ space lying between the energy contours E and $E + dE$. The proportionality factor is fixed by the fact that, when the integral is taken through the fundamental cube $-\pi/a \leqslant k_1, k_2, k_3 \leqslant \pi/a$, the number of energy levels per unit volume must be equal to the number of atoms per unit volume, namely, $1/a^3$.

For perfectly free electrons the energy contours are spheres, and, since the volume in $\mathbf{k}$ space contained between two concentric spheres of radii k and $k + dk$ is $4\pi k^2 dk$, we have

$$\mathfrak{n}(E)\,dE = k^2\,dk/(2\pi^2). \qquad (2\cdot71\cdot2)$$

Further, $E = h^2 k^2/(8\pi^2 m)$, and so

$$\mathfrak{n}(E) = 2\pi (2m)^{\frac{3}{2}} h^{-3} E^{\frac{1}{2}}. \tag{2·71·3}$$

This expression for $\mathfrak{n}(E)$ is not correct when we take the zone structure into account, but it is still a good approximation when the energy contours do not approach the zone boundaries too closely. Even when the binding forces are fairly large, and the energy bands narrow, the energy is given by $E = h^2 k^2/(8\pi^2 m^*)$ for not too large values of k, where m^* is the effective mass of the electron, and we can therefore use (2·71·3) with m^* instead of m. The effect of the binding is, therefore, to increase $\mathfrak{n}(E)$ but to leave the variation with E practically unchanged.

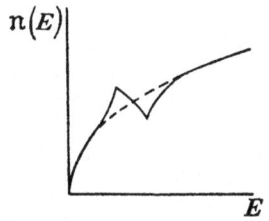

Fig. II 12. The influence of the energy discontinuities on the density of states.

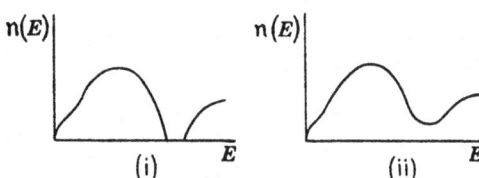

Fig. II 13. The density of levels for (i) an insulator, and (ii) a semi-metal.

The zone structure has a very marked effect on $\mathfrak{n}(E)$, which is usually exceedingly complicated. Fig. II 12 shows $\mathfrak{n}(E)$ when the energy contours are of the type given in fig. II 10, p. 42. For small values of E the energy contours are spheres, so that $\mathfrak{n}(E) \propto E^{\frac{1}{2}}$. As the contours deviate from the spherical shape, $\mathfrak{n}(E)$ increases more rapidly than $E^{\frac{1}{2}}$ and reaches a maximum when the contour just touches the zone boundaries. The density of states now rapidly decreases with E until the minimum energy of the second zone is reached; the energy levels of the second zone then start contributing, and $\mathfrak{n}(E)$ increases once more. If the energy zones do not overlap, the density of states in the first zone reaches a maximum and then drops to zero. There follows a region in which there are no energy levels, after which the energy levels of the second zone begin. This behaviour is shown in fig. II 13 (i). If the two zones only just overlap, as in a semi-metal like bismuth, the form of $\mathfrak{n}(E)$ is as shown in fig. II 13 (ii).

2·711. *The density of states for overlapping bands of standard form.* When a band is of standard form the density of states is always $\frac{1}{2}(k/\pi)^2 \, dk/dE$. When the band is inverted, the density of states is therefore

$$\mathfrak{n}(E) = 2\pi (2m^*)^{\frac{3}{2}} h^{-3} (A - E)^{\frac{1}{2}}. \tag{2·711·1}$$

Energy bands of standard form

When there are two overlapping bands, the effective mass for the normal band being m_1 and that for the other being m_2, the density of states is given by

$$n(E) = 2\pi h^{-3}\{(2m_1)^{\frac{3}{2}} E^{\frac{1}{2}} + (2m_2)^{\frac{3}{2}} (A - E)^{\frac{1}{2}}\}. \qquad (2\cdot711.2)$$

The first term must be omitted when $E < 0$ and the second when $E > A$. The energy levels and the density of states are shown in fig. II 14.

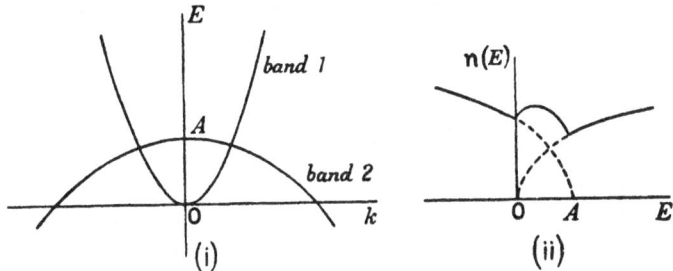

Fig. II 14. The energy levels for two overlapping bands of normal form and the density of states.

THE CURRENT

2·8. Although the determination of the current $\mathbf{j}$ and the velocity $\mathbf{v}$ associated with a stationary state necessitates an exact knowledge of the wave function, it is possible to find the mean value $\bar{\mathbf{v}}$ of the velocity over a unit cell provided that the energy is known as a function of $\mathbf{k}$.

We write the wave equation in the form

$$(L + E_\mathbf{k})[\psi_\mathbf{k}] \equiv (h^2/8\pi^2 m) \nabla^2 \psi_\mathbf{k} + (E_\mathbf{k} - V) \psi_\mathbf{k} = 0, \qquad (2\cdot8\cdot1)$$

and on differentiating this with respect to k_1 we obtain

$$(L + E_\mathbf{k})\left[e^{i\mathbf{k}\cdot\mathbf{r}}\frac{\partial u_\mathbf{k}}{\partial k_1}\right] + \frac{ih^2}{4\pi^2 m}\frac{\partial \psi_\mathbf{k}}{\partial x} + \frac{\partial E_\mathbf{k}}{\partial k_1}\psi_\mathbf{k} = 0. \qquad (2\cdot8\cdot2)$$

If we multiply this by $\psi_\mathbf{k}^*$, the resulting expression is periodic in the unit cell, and on integrating over the unit cell (such an integration being denoted by $\int d\tau_0$), we find

$$\frac{\partial E}{\partial k_1} + \frac{ih^2}{4\pi^2 m}\int \psi_\mathbf{k}^* \frac{\partial \psi_\mathbf{k}}{\partial x} d\tau_0 + \int \psi_\mathbf{k}^*(L + E_\mathbf{k})\left[e^{i\mathbf{k}\cdot\mathbf{r}}\frac{\partial u_\mathbf{k}}{\partial k_1}\right] d\tau_0 = 0.$$

The last term can be transformed by Green's theorem into

$$\int e^{i\mathbf{k}\cdot\mathbf{r}}\frac{\partial u_\mathbf{k}}{\partial k_1}(L + E_\mathbf{k})[\psi_\mathbf{k}^*] d\tau_0,$$

which is zero. (The surface integrals vanish owing to the periodicity of the integrand.) Hence

$$\frac{2\pi}{h}\frac{\partial E}{\partial k_1} = \frac{h}{2\pi m i}\int \psi_{\mathbf{k}}^* \frac{\partial \psi_{\mathbf{k}}}{\partial x}\, d\tau_0 = \frac{h}{4\pi m i}\int \left(\psi_{\mathbf{k}}^* \frac{\partial \psi_{\mathbf{k}}}{\partial x} - \psi_{\mathbf{k}}\frac{\partial \psi_{\mathbf{k}}^*}{\partial x}\right) d\tau_0, \qquad (2\cdot8\cdot3)$$

a surface integral again vanishing. This is the desired relation

$$h\bar{\mathbf{v}} = 2\pi \,\mathrm{grad}_{\mathbf{k}}\, E. \qquad (2\cdot8\cdot4)$$

We can obtain further relations by successive differentiation. We write (2·8·3) in the form

$$\frac{4\pi^2 m}{h^2}\frac{\partial E}{\partial k_1} = k_1 + \frac{1}{i}\int u_{\mathbf{k}}^* \frac{\partial u_{\mathbf{k}}}{\partial x}\, d\tau_0.$$

Then on differentiation with respect to k_1 and integration by parts we find

$$\frac{4\pi^2 m}{h^2}\frac{\partial^2 E}{\partial k_1^2} = 1 + i\int \left(\frac{\partial u_{\mathbf{k}}^*}{\partial x}\frac{\partial u_{\mathbf{k}}}{\partial k_1} - \frac{\partial u_{\mathbf{k}}}{\partial x}\frac{\partial u_{\mathbf{k}}^*}{\partial k_1}\right) d\tau_0. \qquad (2\cdot8\cdot5)$$

Similarly

$$\frac{4\pi^2 m}{h^2}\frac{\partial^2 E}{\partial k_1 \partial k_2} = i\int \left(\frac{\partial u_{\mathbf{k}}^*}{\partial x}\frac{\partial u_{\mathbf{k}}}{\partial k_2} - \frac{\partial u_{\mathbf{k}}}{\partial x}\frac{\partial u_{\mathbf{k}}^*}{\partial k_2}\right) d\tau_0$$

$$= i\int \left(\frac{\partial u_{\mathbf{k}}^*}{\partial y}\frac{\partial u_{\mathbf{k}}}{\partial k_1} - \frac{\partial u_{\mathbf{k}}}{\partial y}\frac{\partial u_{\mathbf{k}}^*}{\partial k_1}\right) d\tau_0. \qquad (2\cdot8\cdot6)$$

Equation (2·8·4) shows that even tightly bound electrons are free to move through the crystal, their mobility only becoming zero when the band width vanishes. The general form of v_1 as a function of k_1 is shown in fig. II 15. It will be seen that near the top of a band v_1 is a decreasing function of k_1, and the acceleration behaves anomalously, as discussed in § 2·84.

Fig. II 15. The velocity as a function of the wave number.

2·81. *The non-diagonal elements of the current.* The non-diagonal elements

$$\frac{h}{2\pi i}\int \psi_{\mathbf{k}''}^* \,\mathrm{grad}\, \psi_{\mathbf{k}'}\, d\tau \qquad (2\cdot81\cdot1)$$

of the momentum play an important part in several phenomena; in particular, they determine the optical transition probabilities. It is convenient to use the reduced wave vector $\mathbf{k}$ and a zone quantum number m' for enumerating the states, and (2·81·1) is then different from zero only if $\mathbf{k}' = \mathbf{k}'' = \mathbf{k}$. If we write

$$(m'' \mid p_x \mid m') = \frac{h}{2\pi i}\int \psi_{\mathbf{k}, m''}^* \frac{\partial \psi_{\mathbf{k}, m'}}{\partial x}\, d\tau, \quad f_{m''m', x} = \frac{2}{m}\frac{\mid (m'' \mid p_x \mid m')\mid^2}{E_{m''} - E_{m'}},$$

$$(2\cdot81\cdot2)$$

there is an important identity, known as the f-sum rule, connecting the f's. It is

$$\sum_{m''\neq m'} f_{m''m',x} = 1 - \frac{4\pi^2 m}{h^2} \frac{\partial^2 E_{m'}}{\partial k_1^2}. \tag{2·81·3}$$

The proof depends upon the fact that, for fixed $\mathbf{k}$, the functions $u_{\mathbf{k},m'}$ form a complete orthogonal system in the unit cell, so that

$$\int u_{\mathbf{k},m''}^* u_{\mathbf{k},m'} d\tau_0 = \delta_{m'}^{m''}, \quad \sum_{m'} u_{\mathbf{k},m'}(\mathbf{r}) \int u_{\mathbf{k},m'}^*(\mathbf{r}') \phi(\mathbf{r}') d\tau_0' = \phi(\mathbf{r}), \tag{2·81·4}$$

where $\phi(\mathbf{r})$ is any quadratically integrable function which is periodic in the unit cell.

We first of all obtain an alternative expression for $(m'' \mid p_x \mid m')$. We apply (2·8·2) to $\psi_{\mathbf{k},m'}$, multiply by $\psi_{\mathbf{k},m''}^*$, integrate over the unit cell, and use Green's theorem to transform the integrals. We obtain (omitting $\mathbf{k}$ where no ambiguity is possible)

$$\frac{ih^2}{4\pi^2 m} \int u_{m''}^* \frac{\partial u_{m'}}{\partial x} d\tau_0 = -\int \psi_{m''}^* (L + E_{m'}) \left[e^{i\mathbf{k}\cdot\mathbf{r}} \frac{\partial u_{m'}}{\partial k_1} \right] d\tau_0$$

$$= -\int e^{i\mathbf{k}\cdot\mathbf{r}} \frac{\partial u_{m'}}{\partial k_1} (L + E_{m'}) [\psi_{m''}^*] d\tau_0 = (E_{m''} - E_{m'}) \int u_{m''}^* \frac{\partial u_{m'}}{\partial k_1} d\tau_0, \tag{2·81·5}$$

since the surface integrals resulting from the use of Green's theorem are zero due to the integrand having the unit cell as its period. Hence

$$\sum_{m''\neq m'} \frac{2}{m} \frac{\mid (m'' \mid p_x \mid m') \mid^2}{E_{m''} - E_{m'}} = i \sum_{m''\neq m'} \left\{ \int u_{m''}^* \frac{\partial u_{m'}}{\partial x} d\tau_0 \int u_{m''} \frac{\partial u_{m''}^*}{\partial k_1} d\tau_0 \right.$$

$$\left. - \int u_{m''} \frac{\partial u_{m'}^*}{\partial x} d\tau_0 \int u_{m''}^* \frac{\partial u_{m'}}{\partial k_1} d\tau_0 \right\}. \tag{2·81·6}$$

We can add the term arising from $m'' = m'$, since it is zero because

$$\frac{\partial}{\partial k_1} \int \mid \psi_{m'} \mid^2 d\tau_0 = 0,$$

and we can then sum the resulting expression by using the completeness relation (the second of the equations (2·81·4)). We obtain

$$i \int \left(\frac{\partial u_{m'}}{\partial x} \frac{\partial u_{m'}^*}{\partial k_1} - \frac{\partial u_{m'}^*}{\partial x} \frac{\partial u_{m'}}{\partial k_1} \right) d\tau_0 \tag{2·81·7}$$

for the right-hand side of (2·81·6). Hence, by (2·8·5), we have

$$\frac{4\pi^2 m}{h^2} \frac{\partial^2 E_{m'}}{\partial k_1^2} = 1 - \sum_{m''\neq m'} f_{m''m',x},$$

which is the f-sum rule. By a similar argument it can be shown that

$$\frac{1}{m} \sum_{m''\neq m'} \frac{(m' \mid p_x \mid m'')(m'' \mid p_y \mid m') + (m' \mid p_y \mid m'')(m'' \mid p_x \mid m')}{E_{m''} - E_{m'}}$$

$$= -\frac{4\pi^2 m}{h^2} \frac{\partial^2 E_{m'}}{\partial k_1 \partial k_2}. \tag{2·81·8}$$

2·82. *The influence of uniform electric and magnetic fields.* Under the action of a uniform electric field and a uniform magnetic field, an electron in a crystal lattice will move through a succession of states in a band very much in the same way as if it were free. If, however, the electric field is sufficiently strong and the time for which it acts is sufficiently large, the electron will not stay in the band but will be excited into the next one. This second effect is only of interest for an insulator, since the field inside a metal is always small. We should, therefore, expect to find a close analogy between the classical and actual motions for small values of t. This is most easily shown by finding solutions of the Schrödinger equation of the form

$$\Psi'(\mathbf{r}, t) = \sum_{m'} \int \alpha_{m'}(\mathbf{k}, t) \, \psi_{\mathbf{k}, m'}(\mathbf{r}) \, d\mathbf{k}, \qquad (2\cdot82\cdot1)$$

which is almost equivalent to solving the equation by using momentum variables.

We require a number of lemmas, to prove which we require the following generalization of the Fourier integral theorem (Jones and Zener, 1934). Let $\phi(\mathbf{k}, \mathbf{r})$ be a continuous function of $\mathbf{k}$ and $\mathbf{r}$ such that

$$\phi(\mathbf{k}, \mathbf{r} + \mathbf{a}_i) = \phi(\mathbf{k}, \mathbf{r}) \quad (i = 1, 2, 3).$$

Then
$$\frac{1}{(2\pi)^3} \int d\tau \int e^{i(\mathbf{k}' - \mathbf{k}) \cdot \mathbf{r}} \, \phi(\mathbf{k}', \mathbf{r}) \, d\mathbf{k}' = \frac{1}{\Delta} \int \phi(\mathbf{k}, \mathbf{r}) \, d\tau_0, \qquad (2\cdot82\cdot2)$$

where $\Delta = \mathbf{a}_1 . (\mathbf{a}_2 \times \mathbf{a}_3)$ is the volume of the unit cell, and where the integration with respect to $\mathbf{k}'$ is taken over all values of the reduced wave vector, i.e. over the domain $-\pi \leqslant \mathbf{k}' . \mathbf{a}_i \leqslant \pi$. The space integral on the left is over the whole of space, while that on the right is over the unit cell.

Now let $\psi_{\mathbf{k}}(\mathbf{r}) = e^{i\mathbf{k} \cdot \mathbf{r}} u_{\mathbf{k}}(\mathbf{r})$ and $\phi_{\mathbf{k}}(\mathbf{r}) = e^{i\mathbf{k} \cdot \mathbf{r}} v_{\mathbf{k}}(\mathbf{r})$ be two wave functions expressed in terms of the reduced wave vector. They are periodic functions of the reduced wave vector with the fundamental domain as their period, and they are normalized in the unit cell. We then have the following identities, where $\alpha(\mathbf{k})$ is a twice differentiable periodic function of the reduced wave vector. (It is necessarily a periodic function of $\mathbf{k}$ in order that the functions should have a physical significance.)

(1)
$$\frac{\Delta}{(2\pi)^3} \int d\tau \int \alpha(\mathbf{k}') \, \phi_{\mathbf{k}}^* \, x \psi_{\mathbf{k}} \, d\mathbf{k}'$$

$$= -\frac{i\Delta}{(2\pi)^3} \int d\tau \int \alpha(\mathbf{k}') \, \phi_{\mathbf{k}}^* \, \frac{\partial e^{i\mathbf{k}' \cdot \mathbf{r}}}{\partial k_1'} \, u_{\mathbf{k}'} \, d\mathbf{k}'$$

$$= i \frac{\partial \alpha(\mathbf{k})}{\partial k_1} \int v_{\mathbf{k}}^* u_{\mathbf{k}} \, d\tau_0 + i\alpha(\mathbf{k}) \int v_{\mathbf{k}}^* \frac{\partial u_{\mathbf{k}}}{\partial k_1} \, d\tau_0. \qquad (2\cdot82\cdot3)$$

(2) $\quad \dfrac{\Delta}{(2\pi)^3}\displaystyle\int d\tau \int \alpha(\mathbf{k}')\,\phi_{\mathbf{k}}^*\,y\,\dfrac{\partial \psi_{\mathbf{k}'}}{\partial x}\,d\mathbf{k}'$

$\quad = \dfrac{\Delta}{(2\pi)^3}\displaystyle\int d\tau \int \alpha(\mathbf{k}')\,\phi_{\mathbf{k}}^*\,\dfrac{\partial e^{i\mathbf{k}'\cdot\mathbf{r}}}{\partial k_2'}\left(k_1' u_{\mathbf{k}'} - i\,\dfrac{\partial u_{\mathbf{k}'}}{\partial x}\right)d\mathbf{k}'$

$\quad = i\,\dfrac{\partial \alpha(\mathbf{k})}{\partial k_2}\displaystyle\int \phi_{\mathbf{k}}^*\,\dfrac{\partial \psi_{\mathbf{k}}}{\partial x}\,d\tau_0 - \alpha(\mathbf{k})\int v_{\mathbf{k}}^* k_1 \dfrac{\partial u_{\mathbf{k}}}{\partial k_2}\,d\tau_0 - i\alpha(\mathbf{k})\int \dfrac{\partial v_{\mathbf{k}}^*}{\partial x}\dfrac{\partial u_{\mathbf{k}}}{\partial k_2}\,d\tau_0.$ (2·82·4)

(3) $\quad \dfrac{\Delta}{(2\pi)^3}\displaystyle\int d\tau \int \alpha(\mathbf{k}')\,\phi_{\mathbf{k}}^*\,x^2 \psi_{\mathbf{k}'}\,d\mathbf{k}'$

$\quad = -\dfrac{\Delta}{(2\pi)^3}\displaystyle\int d\tau \int \alpha(\mathbf{k}')\,\phi_{\mathbf{k}}^*\,\dfrac{\partial^2 e^{i\mathbf{k}'\cdot\mathbf{r}}}{\partial k_1'^2}\,u_{\mathbf{k}'}\,d\mathbf{k}'$

$\quad = -\dfrac{\partial^2 \alpha(\mathbf{k})}{\partial k_1^2}\displaystyle\int v_{\mathbf{k}}^* u_{\mathbf{k}}\,d\tau_0 - 2\,\dfrac{\partial \alpha(\mathbf{k})}{\partial k_1}\int v_{\mathbf{k}}^*\dfrac{\partial u_{\mathbf{k}}}{\partial k_1}\,d\tau_0 - \alpha(\mathbf{k})\int v_{\mathbf{k}}^*\dfrac{\partial^2 u_{\mathbf{k}}}{\partial k_1^2}\,d\tau_0.$ (2·82·5)

The proofs of these formulae are straightforward. One or more integrations by parts with respect to k_1' reduce the integrals to a form in which they can be immediately evaluated by using (2·82·2), the surface integrals over the surface of the fundamental $\mathbf{k}$ domain vanishing since all the functions are periodic functions of the reduced wave vector.

In general these formulae are applied to solutions of the Schrödinger equation characterized by a zone number. The integrals which occur are then expressible in terms of the diagonal and non-diagonal elements of the current or of similar quantities.

2·83. When there is a uniform electric field $\mathscr{E}$, the Schrödinger equation is

$$\frac{h}{2\pi i}\frac{\partial \Psi}{\partial t} = \frac{h^2}{8\pi^2 m}\nabla^2 \Psi - V\Psi - e\mathscr{E}\cdot\mathbf{r}\Psi. \qquad (2\cdot83\cdot1)$$

Substituting the expression (2·82·1) into this, multiplying by $\psi_{\mathbf{k}',m''}^*$ and integrating over all space, we obtain on using (2·82·2) and (2·82·3)

$$\frac{h}{2\pi i}\frac{\partial \alpha_{m'}(\mathbf{k})}{\partial t} = -E_{\mathbf{k},m'}\alpha_{m'}(\mathbf{k}) - ie\mathscr{E}\cdot\left\{\mathrm{grad}_{\mathbf{k}}\,\alpha_{m'} + \sum_{m''}\alpha_{m''}\int u_{m'}^*\,\mathrm{grad}_{\mathbf{k}}\,u_{m''}\,d\tau_0\right\},$$
$$(2\cdot83\cdot2)$$

which gives

$$\left(\frac{h}{2\pi}\frac{\partial}{\partial t} - e\mathscr{E}\cdot\mathrm{grad}_{\mathbf{k}}\right)|\alpha_{m'}|^2$$

$$= e\mathscr{E}\cdot\left\{\alpha_{m'}^*\sum_{m''}\alpha_{m''}\int u_{m'}^*\,\mathrm{grad}_{\mathbf{k}}\,u_{m''}\,d\tau_0 + \alpha_{m'}\sum_{m''}\alpha_{m''}^*\int u_{m'}\,\mathrm{grad}_{\mathbf{k}}\,u_{m''}^*\,d\tau_0\right\}.$$

If we sum this equation with respect to m', the right-hand side gives zero, since

$$\int u_{m'}^{*} u_{m''} d\tau_0 = \delta_{m'}^{m''},$$

and so

$$\left(\frac{h}{2\pi}\frac{\partial}{\partial t} - \epsilon \mathscr{E}. \operatorname{grad}_{\mathbf{k}}\right) \sum_{m'} |\alpha_{m'}|^2 = 0.$$

If the electron is initially in the band m', we may assume that, for sufficiently small times, only $\alpha_{m'}$ is of importance, and then

$$\frac{h}{2\pi}\frac{\partial |\alpha_{m'}|^2}{\partial t} = \epsilon \mathscr{E}. \operatorname{grad}_{\mathbf{k}} |\alpha_{m'}|^2, \qquad (2\cdot83\cdot3)$$

which is the same relation as for free electrons.

2·84. *The acceleration.* Equation (2·83·3) has the general solution $|\alpha_{m'}|^2 = F(\mathbf{k} + 2\pi\epsilon\mathscr{E}t/h)$, where F is an arbitrary function. This shows that the rate of change of $\mathbf{k}$ due to an electric field $\mathscr{E}$ is given by

$$dk_i/dt = -(2\pi\epsilon/h)\,\mathscr{E}_i.$$

The acceleration is therefore given by

$$\frac{dv_i}{dt} = \sum_j \frac{\partial v_i}{\partial k_j}\frac{\partial k_j}{\partial t} = \frac{2\pi}{h}\sum_j \frac{\partial^2 E}{\partial k_i \partial k_j}\frac{dk_j}{dt} = -\frac{4\pi^2\epsilon}{h^2}\sum_j \frac{\partial^2 E}{\partial k_i \partial k_j}\mathscr{E}_j, \qquad (2\cdot84\cdot1)$$

which we can write as

$$dv_i/dt = -\epsilon \sum_j m_{ij}^{-1}\mathscr{E}_j, \qquad (2\cdot84\cdot2)$$

where the tensor defined by

$$m_{ij}^{-1} = \frac{4\pi^2}{h^2}\frac{\partial^2 E}{\partial k_i \partial k_j} \qquad (2\cdot84\cdot3)$$

is the reciprocal of the mass tensor. The acceleration is not in general in the direction of the field, and if the second derivatives of E are negative the acceleration can be in the opposite direction to $-\epsilon\mathscr{E}$. If E is spherically symmetrical, the mass tensor is a scalar. In particular, if $E = h^2 |\mathbf{k}|^2/(8\pi^2 m^*)$ we have $dv_i/dt = -\epsilon\mathscr{E}_i/m^*$, and the electron behaves as if it had an effective mass m^*. If, on the other hand, $E = \text{constant} - h^2|\mathbf{k}|^2/(8\pi^2 m^*)$, as occurs near the top of an energy band, we have $dv_i/dt = \epsilon\mathscr{E}_i/m^*$, and the electron behaves as if it had a negative mass, or equivalently as if it had positive charge ϵ and an effective mass m^*. Since positive charges are easier to comprehend than negative masses, we prefer to consider that the effective charge is positive when the second derivatives of E are negative, and this interpretation will be used throughout this book.

2·85. *The influence of a magnetic field.* When there is a magnetic field present, the vector potential $\mathbf{A}$ occurs explicitly in the Schrödinger equation, and the simple relation $m\mathbf{v} = \mathbf{p}$ no longer holds. The average velocity for a state Ψ is in fact given by

$$[\mathbf{v}] = \frac{h}{4\pi mi}\int(\Psi^*\operatorname{grad}\Psi - \Psi\operatorname{grad}\Psi^*)\,d\tau + \frac{\epsilon}{mc}\int\mathbf{A}\Psi^*\Psi\,d\tau. \quad (2\cdot85\cdot1)$$

Now $\mathbf{A}$ is not uniquely defined by the magnetic field and so there is a degree of arbitrariness in the definition of $[\mathbf{v}]$. For example, if the magnetic field is $(0,0,H)$, we can choose $\mathbf{A}$ to be either $\frac{1}{2}(-Hy, Hx, 0)$ or $(-Hy, 0, 0)$ or $(0, Hx, 0)$. These various choices of $\mathbf{A}$ cannot affect in any way the physical properties of the system, although the equations determining the motion of a wave packet are clearly different for the different choices of $\mathbf{A}$. It should be possible to carry out the calculations with an arbitrary choice for $\mathbf{A}$ and arrive finally at a set of equations in which all arbitrariness has disappeared. There seems, however, to be no simple and elegant way of carrying out the calculations in this manner, and it is easier to lay down an absolute gauge system for a particular wave packet by requiring the mean value of $\mathbf{A}$ to be zero (Jones and Zener, 1934). We then also obtain the usual relation between $\bar{\mathbf{v}}$ and E in the absence of a magnetic field, namely, $h\bar{\mathbf{v}} = 2\pi\operatorname{grad}_k E$. This can be done by making use of the fact that the Schrödinger equation is invariant under the gauge transformation

$$\psi \to \psi\, e^{2\pi i\epsilon\Lambda/hc}, \quad \mathbf{A} \to \mathbf{A} - \operatorname{grad}\Lambda, \quad V \to V - \epsilon\,\partial\Lambda/\partial(ct). \quad (2\cdot85\cdot2)$$

We therefore choose

$$\mathbf{A} = \tfrac{1}{2}\mathbf{H}\times(\mathbf{r}-[\mathbf{r}]), \quad [\mathbf{r}] = \int\mathbf{r}\Psi^*\Psi\,d\tau, \quad (2\cdot85\cdot3)$$

for which the appropriate Λ is

$$\Lambda = \tfrac{1}{2}\mathbf{r}.\mathbf{H}\times[\mathbf{r}]. \quad (2\cdot85\cdot4)$$

Now

$$\frac{\partial[\mathbf{r}]}{\partial t} = \int\mathbf{r}\frac{\partial}{\partial t}(\Psi^*\Psi)\,d\tau = [\mathbf{v}]$$

by using the equation of continuity. Hence

$$\partial\Lambda/\partial t = \tfrac{1}{2}\mathbf{r}.\mathbf{H}\times[\mathbf{v}] \quad (2\cdot85\cdot5)$$

and, if $\mathbf{H} = (0, 0, H)$, the Schrödinger equation becomes

$$\left[\frac{h}{2\pi i}\frac{\partial}{\partial t} - \frac{h^2}{8\pi^2 m}\nabla^2 + V + \frac{\epsilon hH}{4\pi imc}\left\{(x-[x])\frac{\partial}{\partial y} - (y-[y])\frac{\partial}{\partial x}\right\}\right.$$
$$\left. + \frac{\epsilon H}{2c}(x[v_y] - y[v_x]) + \frac{\epsilon^2 H^2}{8mc^2}\{(x-[x])^2 + (y-[y])^2\}\right]\Psi = 0. \quad (2\cdot85\cdot6)$$

We can find the motion of a wave packet for a sufficiently short time by neglecting the terms in H^2 and by ignoring the possibility of excitation from one band to another. We therefore substitute

$$\Psi(\mathbf{r}, t) = \int\alpha(\mathbf{k}, t)\,\psi_\mathbf{k}(\mathbf{r})\,d\mathbf{k} \quad (2\cdot85\cdot7)$$

into (2·85·6), multiply by $\psi_{\mathbf{k}}(\mathbf{r})^*$, integrate and proceed as in § 2·83. We then obtain

$$\frac{\partial |\alpha|^2}{\partial t} = -\frac{\pi \epsilon H}{hc}\left\{(v_x+[v_x])\frac{\partial |\alpha|^2}{\partial k_2} - (v_y+[v_y])\frac{\partial |\alpha|^2}{\partial k_1}\right\}. \qquad (2·85·8)$$

If $|\alpha|^2$ is only large near $\mathbf{k}=\mathbf{k}_0$, $[\mathbf{v}]$ and $\mathbf{v}_{\mathbf{k}_0}$ will be the same for sufficiently small times, and the initial rate of change of $|\alpha|^2$ is given by

$$\frac{\partial |\alpha|^2}{\partial t} = -\frac{2\pi \epsilon H}{hc}\left(v_x\frac{\partial |\alpha|^2}{\partial k_2} - v_y\frac{\partial |\alpha|^2}{\partial k_1}\right) = \frac{2\pi \epsilon}{hc}(\mathbf{v}\times\mathbf{H}).\mathrm{grad}_{\mathbf{k}}|\alpha|^2.$$
$$(2·85·9)$$

If we had taken $\mathbf{A}$ as $(-H\{y-[y]\}, 0, 0)$, equation (2·85·8) would have $v_x+[v_x]$ replaced by $2v_x$ and $v_y+[v_y]$ replaced by $2[v_y]$, so that (2·85·9) would be unchanged, as it must be.

We can apply the above results to the effect of a constant electric field $\mathscr{E}$ and a constant magnetic field $\mathbf{H}$ on the distribution function $f(\mathbf{k}, t)$ determining the number of electrons per unit volume with wave vectors lying in the range $d\mathbf{k}$. Combining equations (2·83·3) and (2·85·9), we see that

$$\frac{\partial f}{\partial t} = \frac{2\pi \epsilon}{h}\left(\mathscr{E} + \frac{1}{c}\mathbf{v}\times\mathbf{H}\right).\mathrm{grad}_{\mathbf{k}}f, \qquad (2·85·10)$$

and this is the expression to be used in setting up the Boltzmann equation. It is only valid for times which are small compared with $2mc/\epsilon H$.

THE SELF-CONSISTENT FIELD

2·9. All the preceding results have been obtained on the assumption that we have a single electron moving in a fixed field which has the periodicity of the crystal lattice. In actual fact we have to consider all the valency electrons as a single assembly, though we can treat the inner (core) electrons as giving rise to a fixed field. In conduction problems it is essential to treat the valency electrons as if the motion of each were, at least to a first approximation, independent of the motion of the others. We do not, however, entirely neglect the other electrons. We replace their effect by a smeared field, which is their field averaged in such a way that correlations between the positions of the electrons are neglected.

For many purposes this assumption is insufficient and we therefore have to consider how the theory can be generalized.

2·91. Since a complete account of the self-consistent field has been given by Brillouin (1934a), we shall only introduce the main ideas and quote the results which are required. Suppose that we have N electrons with coordinates $\mathbf{r}_i$ and a number of positive ions $g, g', \ldots$ in fixed positions.

The stationary states of the electrons are given by the Schrödinger equation

$$(\mathscr{H} - E)\Psi \equiv \left[-\frac{h^2}{8\pi^2 m} \sum_{i=1}^{N} \nabla_i^2 + \tfrac{1}{2}\sum_{i,j} \mathscr{G}_{ij} + \sum_{g,i} V_{gi} + \tfrac{1}{2}\sum_{g \neq g'} V_{gg'} - E \right]\Psi = 0,$$
(2·91·1)

where $\mathscr{G}_{ij} = e^2/r_{ij}$ and V_{gi}, $V_{gg'}$ are the mutual potential energies of the ion g and the electron i and of the ions g, g' respectively. The simplest assumption to make is that the total wave function Ψ is a product of N functions, each of which contains the coordinates of one electron only, so that $\Psi = \Pi\psi(k_i, \mathbf{r}_i)$, where k_i represents the quantum numbers defining the state. This leads to a Schrödinger equation in which each electron moves in the field of the ions and of the smeared field of the other electrons—the self-consistent field V_H proposed by Hartree, where

$$V_H(\mathbf{r}_i) = \sum_{j=1}^{N} \int \frac{e^2}{r_{ij}} |\psi(k_j, \mathbf{r}_j)|^2 d\tau_j.$$
(2·91·2)

The assumption that Ψ is a simple product is inadequate, and Ψ should be written in the antisymmetrical form

$$\Psi = (N!)^{-\frac{1}{2}} \begin{vmatrix} \phi_1(\mathbf{r}_1, s_1) & \cdots & \phi_N(\mathbf{r}_1, s_1) \\ \vdots & & \vdots \\ \phi_1(\mathbf{r}_N, s_N) & \cdots & \phi_N(\mathbf{r}_N, s_N) \end{vmatrix},$$
(2·91·3)

where s_i is the spin coordinate of the ith electron. In certain simple important cases it is sufficient to assume that each ϕ is separable into a product $\alpha(s)\,\psi(\mathbf{r})$. We then obtain the so-called symmetrized Hartree field. In this case the total energy

$$E_{\text{total}} = \int \Psi^* \mathscr{H}\Psi \, d\tau$$

can be shown to be given by

$$E_{\text{total}} = -\frac{h^2}{8\pi^2 m} \sum_i \int \psi(k_i, \mathbf{r}_i)^* \nabla_i^2 \psi(k_i, \mathbf{r}_i)\, d\tau_i$$

$$+ \sum_{g,i} \int V_{gi} |\psi(k_i, \mathbf{r}_i)|^2\, d\tau_i + \tfrac{1}{2}\sum_{g \neq g'} V_{gg'}$$

$$+ \tfrac{1}{2}\sum_{i,j} (k_i k_j \mid \mathscr{G} \mid k_i k_j) - \frac{1}{2}\sum_{i,j}^{N_1} (k_i k_j \mid \mathscr{G} \mid k_j k_i) - \frac{1}{2}\sum_{i,j}^{N_2} (k_i k_j \mid \mathscr{G} \mid k_j k_i),$$
(2·91·4)

where

$$(k_i k_j \mid \mathscr{G} \mid k_i' k_j') = \iint \frac{e^2}{r_{ij}} \psi(k_i, \mathbf{r}_i)\, \psi(k_i', \mathbf{r}_i)^*\, \psi(k_j, \mathbf{r}_j)\, \psi(k_j', \mathbf{r}_j)^*\, d\tau_i d\tau_j$$
(2·91·5)

and where there are N_1 electrons with spin $\frac{1}{2}$ and N_2 with $-\frac{1}{2}$, there being N_2 states occupied by two electrons with opposite spins and the system having a total spin of $\frac{1}{2}(N_1 - N_2)$. The last two terms in (2·91·4) represent the effect of the exchange forces, which are absent in Hartree's original self-consistent field.

2.92. A still more general treatment of the exchange forces has been given by Fock and Dirac. It is assumed that, when the system has no resultant spin, $\psi(k_i, \mathbf{r}_i)$ satisfies the equation

$$\left[-\frac{h^2}{8\pi^2 m} \nabla_i^2 + \sum_g V_{gi} + V_H(\mathbf{r}_i) - \mathfrak{A} - E_{iF} \right] \psi_{iF} = 0, \qquad (2\cdot92\cdot1)$$

where $\mathfrak{A}$ is an operator defined by

$$\mathfrak{A}\psi(k, \mathbf{r}) = \sum_{k'}^{\frac{1}{2}N} \int \frac{\epsilon^2}{|\mathbf{r} - \mathbf{r}'|} \psi(k', \mathbf{r}) \, \psi(k', \mathbf{r}')^* \, \psi(k, \mathbf{r}') \, d\tau'. \qquad (2\cdot92\cdot2)$$

In this case the total energy is

$$E_{\text{total}} = \sum_{i=1}^{N} E_{iF} + \frac{1}{2} \sum_{g+g'} V_{gg'} - \frac{1}{2} \sum_{i,j}^{N} (k_i k_j \mid \mathscr{G} \mid k_i k_j)$$

$$+ \frac{1}{2} \sum_{i,j}^{N_1} (k_i k_j \mid \mathscr{G} \mid k_j k_i) + \frac{1}{2} \sum_{i,j}^{N_2} (k_i k_j \mid \mathscr{G} \mid k_j k_i), \qquad (2\cdot92\cdot3)$$

where $N_1 = N_2 = \frac{1}{2}N$.

2.93. Neither the Hartree nor the Fock methods are free from criticism, since they tend to over-emphasize the importance of polar states; that is, the equations give solutions in which the electrons approach nearer to one another than they do in reality. It is possible to use more complicated types of wave functions which are free from this defect, but such wave functions can only be used in relatively simple cases on account of the complexity of the calculations.

In the theorems of this chapter it has been assumed implicitly that the self-consistent field has the same symmetry as the lattice, although the wave functions of the individual electrons are not periodic in the unit cell. A general proof that this assumption is correct has been given by Brillouin (1934b).

REFERENCES

Bloch, F. (1928). The quantum mechanics of electrons in crystal lattices. *Z. Phys.* **52**, 555.

Brillouin, L. (1930). Free electrons in metals and the role of the Bragg reflexions. *J. Phys. Radium,* **1**, 377.

Brillouin, L. (1934a). *Les champs self-consistents de Hartree et de Fock* (Paris).

Brillouin, L. (1934b). Fock's self-consistent field for the electrons in metals. *J. Phys. Radium,* **5**, 413.

Courant, R. and Hilbert, D. (1924). *Methoden der mathematischen Physik,* **1** (Berlin).

Haupt, O. (1914). Oscillation theorems. *Math. Ann.* **76**, 67.

Haupt, O. (1919). Linear homogeneous differential equations of the second order with periodic coefficients. *Math. Ann.* **79**, 278.

Jones, H. and Zener, C. (1934). A general proof of certain fundamental equations in the theory of metallic conduction. *Proc. Roy. Soc.* A, **144**, 101.

Kronig, R. de L. and Penney, W. G. (1931). Quantum mechanics of electrons in crystal lattices. *Proc. Roy. Soc.* A, **130**, 499.

Peierls, R. (1930). The theory of the electrical and thermal conductivity of metals. *Ann. Phys., Lpz.* (5), **4**, 121.

Seitz, F. (1940). *The modern theory of solids* (New York).

Chapter III

METALLIC STRUCTURES

METALS AND INSULATORS

3·1. The results of the previous chapter show that all the electrons in a solid can be considered as free, since to each wave function there corresponds a current, but this does not mean that every solid ought to be a metal. A solid will be a conductor only if the total current due to all the electrons can be made different from zero by the application of a small electric field. In the equilibrium state there is no resultant current in any solid, since the number of electrons moving in one direction is equal to the number moving in the opposite direction, and the electrons occupy the lowest possible energy levels, there being two electrons with opposite spins in each level. If an electric field is applied to the crystal, it will tend to accelerate the electrons, and to transfer them into neighbouring energy states, but this transfer can only take place if these neighbouring states are unoccupied. Therefore the only electrons which can contribute to the resultant current are those with the greatest energies. Let us suppose first that the valency electrons are such that they do not fill an energy band. There are then unoccupied energy levels in this band with energies arbitrarily near the energies of the least tightly bound electrons, so that under the influence of an electric field these electrons will be accelerated, and will produce a resultant current, which would increase without limit were it not for the thermal motion of the atoms. A solid of this type is a metal. Let us next suppose that the valency electrons exactly fill an energy band, there being a region of disallowed energy levels between this band and the next. An electric field will not now produce a resultant current, since the only unoccupied states are those in the higher band, and an electron will be unable to acquire the large amount of energy necessary to raise it into this band unless the field strength is of the order of 10^6 V. per cm. A solid of this type is an insulator.

We saw in § 2·13 that in a linear lattice each band is separated from the rest by a region of forbidden energies, and that the number of levels in each band is exactly equal to the number of atoms in the lattice. If this were a general result the distinction between metals and insulators would be a very simple one. All elements of odd atomic number would be metals and those of even atomic number would be insulators. We have, however, seen that the band structure in a three-dimensional crystal is much more complicated, and it is difficult to give any general rules, since so much depends upon the overlapping of the bands, which is a quantitative question. Each energy level can accommodate two electrons, and it therefore requires an even number of electrons to fill a band. This gives us the only rule of universal application. Every element of odd atomic number which has a simple translation lattice must be a metal, and every non-metallic element must either be of even atomic number or have a crystal structure based upon a unit cell containing an even number of atoms.

Although all the electrons in a solid can move through the lattice without resistance, the electrons in a closed group cannot contribute to the resultant current, and it would be merely confusing to regard them as free electrons. We therefore define the free electrons as those electrons which do not belong to a closed group in the solid.

The number of free electrons in a metal is of the same order as the number of atoms, and its variation with temperature is negligible since the excitation energy of the electrons forming the closed groups is high. An insulator, on the other hand, contains no free electrons at $T = 0$, but at any other temperature there will be a few free electrons produced by the thermal agitation.

ENUMERATION OF THE METALLIC STRUCTURES

3·2. The number of possible crystal lattices is very large, but fortunately metals possess structures of a comparatively simple type, and we need only consider cubic, hexagonal, rhombohedral and tetragonal lattices. It is sometimes convenient to use rectangular and sometimes oblique cartesian coordinates to describe the lattices. The advantages of rectangular axes are obvious, while the desirability of using oblique axes is due to the fact that they often render possible the use of a smaller unit cell than when rectangular coordinates are employed. The 'simple translation lattices' or Bravais lattices have the property that any lattice point can be obtained from any other by a displacement

$$g_1 \mathbf{a}_1 + g_2 \mathbf{a}_2 + g_3 \mathbf{a}_3 \quad (g_1, g_2, g_3 \text{ integers})$$

along three non-coplanar lines $\mathbf{a}_1, \mathbf{a}_2, \mathbf{a}_3$. Not every lattice is of this type, the commonest example of one which is not being the hexagonal close-packed lattice. We have then to displace two or more atoms to obtain the whole lattice, and it is necessary to give the positions of these atoms, that is, the 'basis', as it is called, to describe the crystal. The crystal symmetries do not therefore always correspond to the axes. There are in fact fourteen Bravais lattices, and only in seven is the crystal symmetry the same as the axis symmetry. In order to have simple coordinate systems it is often convenient to regard the other seven Bravais lattices as composite lattices with a basis.

3·21. *Cubic structures.* No metal has a simple cubic structure, but many metals have either a face-centred cubic or a body-centred cubic form.

(i) *The face-centred cubic lattice.* Consider a cube of side a. In the face-centred cubic structure there are atoms at the corners of the cube and at the centres of each face, so that (referred to rectangular axes along the sides of the cube) the atom at $(0, 0, 0)$ has as its nearest neighbours in the lattice the twelve atoms at the face centres with coordinates

$$(\pm \tfrac{1}{2}a, \pm \tfrac{1}{2}a, 0), \quad (\pm \tfrac{1}{2}a, 0, \pm \tfrac{1}{2}a), \quad (0, \pm \tfrac{1}{2}a, \pm \tfrac{1}{2}a)$$

which are at a distance $d = \tfrac{1}{2}\sqrt{2}\,a$. We can therefore describe the face-centred cubic lattice as a simple cubic lattice with a basis, there being four atoms in the unit cell, namely the atom at $(0, 0, 0)$ and those of its three nearest neighbours which have positive coordinates. The basis, which is always expressed in terms of the lattice constant, is $(0, 0, 0), (0, \tfrac{1}{2}, \tfrac{1}{2}), (\tfrac{1}{2}, 0, \tfrac{1}{2}), (\tfrac{1}{2}, \tfrac{1}{2}, 0)$. The lattice is, however, a Bravais lattice, as is illustrated in fig. III 1. If the axes of the simple cubic lattice are

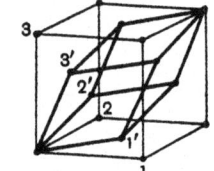

Fig. III 1. The elementary cell of a face-centred cubic lattice.

$$A_1, A_2, A_3 \quad (|A_1| = |A_2| = |A_3|)$$

and those of the Bravais lattice are a_1, a_2, a_3, then

$$a_1 = \tfrac{1}{2}(A_2 + A_3), \quad a_2 = \tfrac{1}{2}(A_3 + A_1), \quad a_3 = \tfrac{1}{2}(A_1 + A_2). \tag{3·21·1}$$

(It is easily verified that with these expressions for a_1, a_2, a_3 the vector $g_1 a_1 + g_2 a_2 + g_3 a_3$ gives all the lattice points of the face-centred cubic lattice.) Also $a_2 \times a_3 = \tfrac{1}{4}(A_2 + A_3 - A_1)a$ and $a_1 . (a_2 \times a_3) = \tfrac{1}{4}a^3$, so that the volume of the unit cell of the face-centred lattice is one-quarter the volume of the cell of the simple cubic lattice from which it is derived. (This is an immediate consequence of there being four atoms per unit cell when the face-centred lattice is considered to be a lattice with a basis.) The reciprocal vectors are

$$b_1 = (-A_1 + A_2 + A_3)/a^2, \quad b_2 = (A_1 - A_2 + A_3)/a^2, \quad b_3 = (A_1 + A_2 - A_3)/a^2. \tag{3·21·2}$$

These form a body-centred cubic lattice (see below).

(ii) *The body-centred cubic lattice.* The body-centred lattice (fig. III 2) has atoms at the centre of each cube as well as at the corners of the cube. The atom at $(0, 0, 0)$ has eight nearest neighbours at the points $(\pm \tfrac{1}{2}a, \pm \tfrac{1}{2}a, \pm \tfrac{1}{2}a)$ which are at a distance $d = \tfrac{1}{2}\sqrt{3}\,a$, while it has six second nearest neighbours at the points $(\pm a, 0, 0), (0, \pm a, 0), (0, 0, \pm a)$ which are at a distance a.

The body-centred cubic lattice can be described as a simple cubic lattice with a basis consisting of two atoms, the basis being $(0, 0, 0)$, $(\frac{1}{2}, \frac{1}{2}, \frac{1}{2})$. We can, however, describe it as a Bravais lattice by taking axes $\mathbf{a}_1$, $\mathbf{a}_2$, $\mathbf{a}_3$ defined by

$$\mathbf{a}_1 = \tfrac{1}{2}(-\mathbf{A}_1 + \mathbf{A}_2 + \mathbf{A}_3), \quad \mathbf{a}_2 = \tfrac{1}{2}(\mathbf{A}_1 - \mathbf{A}_2 + \mathbf{A}_3), \quad \mathbf{a}_3 = \tfrac{1}{2}(\mathbf{A}_1 + \mathbf{A}_2 - \mathbf{A}_3).$$
$$(3\cdot21\cdot3)$$

It is easily verified that $\mathbf{a}_1 . (\mathbf{a}_2 \times \mathbf{a}_3) = \frac{1}{2}a^3$ and that the reciprocal vectors are

$$\mathbf{b}_1 = (\mathbf{A}_2 + \mathbf{A}_3)/a^2, \quad \mathbf{b}_2 = (\mathbf{A}_3 + \mathbf{A}_1)/a^2, \quad \mathbf{b}_3 = (\mathbf{A}_1 + \mathbf{A}_2)/a^2. \quad (3\cdot21\cdot4)$$

These form a face-centred lattice, and we see that face- and body-centred lattices are reciprocal to one another.

A list of face-centred and body-centred cubic metals is given in Tables III 1 and III 2. (For these and other structures, see e.g. Dehlinger, 1935 and Hume-Rothery, 1944.)

(iii) *The diamond lattice.* The diamond lattice is the first example of a lattice with a true basis, and which therefore cannot be reduced to a Bravais lattice by any choice of the axes. It can be described most simply as a face-centred cubic lattice

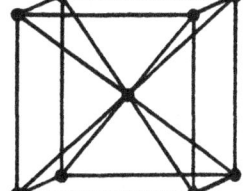

Fig. III 2. The body-centred cubic lattice.

with basis $(0, 0, 0)$ and $(\frac{1}{4}, \frac{1}{4}, \frac{1}{4})$. Treated as a simple cubic lattice the unit cell contains eight atoms, each atom having four nearest neighbours, arranged tetrahedrally, at a distance $d = \frac{1}{4}\sqrt{3}\,a$. Four elements, carbon (diamond), silicon, germanium and (grey) tin, have lattices of this type. The parameters are given in Table III 3.

Table III 1. *The parameters of face-centred cubic metals*

	Al	α-Ca	γ-Fe	β-Co	β-Ni	Cu
a in 10^{-8} cm.	4·04	5·56	3·65	3·55	3·52	3·61
d in 10^{-8} cm.	2·86	3·93	2·58	2·51	2·49	2·55
Atomic volume in 10^{-24} c.c.	16·50	43·0	12·19	11·24	10·90	11·76

	Sr	Rh	Pd	Ag	β-La	β-Ce
a in 10^{-8} cm.	6·07	3·80	3·88	4·08	5·30	5·14
d in 10^{-8} cm.	4·30	2·68	2·75	2·88	3·73	3·62
Atomic volume in 10^{-24} c.c.	56	13·6	14·7	16·97	37·0	33·6

	Ir	Pt	Au	β-Tl	Pb	Th
a in 10^{-8} cm.	3·83	3·92	4·07	4·84	4·94	5·08
d in 10^{-8} cm.	2·71	2·77	2·88	3·42	3·49	3·59
Atomic volume in 10^{-24} c.c.	14	15	16·85	28·36	30	32·8

The parameters of γ-iron refer to 885° C.

Table III 2. *The parameters of body-centred cubic metals*

	Li	Na	K	V	α-Cr	α-Fe	Rb
a in 10^{-8} cm.	3·51	4·30	5·33	3·01	2·88	2·86	5·62
d in 10^{-8} cm.	3·04	3·72	4·62	2·61	2·49	2·48	4·87
Atomic volume in 10^{-24} c.c.	21·5	39·5	76	13·5	11·92	11·7	88·7

	β-Zr	Nb	Mo	Cs	Ba	Ta	α-W
a in 10^{-8} cm.	3·61	3·30	3·14	6·05	5·02	3·30	3·15
d in 10^{-8} cm.	3·13	2·86	2·72	5·24	4·34	2·84	2·73
Atomic volume in 10^{-24} c.c.	23·7	18·0	15·48	110·7	63·0	17·66	15·75

The parameters of Rb and Cs refer to $-173°$ C. and those for β-Zr to $867°$ C.

Table III 3. *The parameters of diamond-type lattices*

	C	Si	Ge	Sn
a in 10^{-8} cm.	3·56	5·42	5·62	6·46
d in 10^{-8} cm.	1·54	2·35	2·43	2·80
Atomic volume in 10^{-24} c.c.	5·51	19·9	22·2	33·7

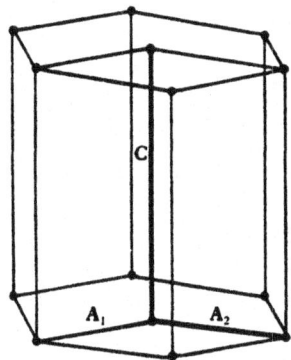

Fig. III 3. The unit cell of a simple hexagonal lattice.

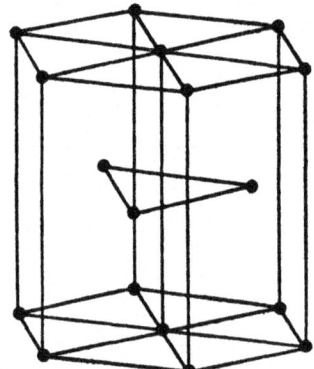

Fig. III 4. The ideal hexagonal close-packed lattice.

3·22. *Hexagonal structures.* The simple hexagonal lattice is shown in fig. III 3. The unit cell is a hexagonal prism with atoms at all the corners and at the centres of the hexagonal faces. If we take the origin at the centre of the hexagonal base, it is usual to take the axes to be A_1, A_2 and C as shown. A_1 and A_2 are equal in magnitude (usually denoted by a) and the angle between them is $120°$. C is at right angles to A_1 and A_2 and is of magnitude c. In order to display the hexagonal symmetry, a third redundant axis

$$A_3 = -A_1 - A_2$$

is sometimes employed. (Note that $A_1 . A_2 = A_2 . A_3 = A_3 . A_1 = -\tfrac{1}{2}a^2$.)

No metal has a simple hexagonal structure, but derivatives of the structure are common, none of which are Bravais lattices.

(i) *The hexagonal close-packed structure.* This type of lattice consists of two simple hexagonal lattices, the basis being $(0, 0, 0)$, $(\frac{1}{3}, \frac{2}{3}, \frac{1}{2})$. Each atom has six neighbours at a distance a which lie in the basal plane through the atom (perpendicular to the hexagonal axis). There are six others at a distance $d = \sqrt{(\frac{1}{3}a^2 + \frac{1}{4}c^2)}$, three lying above and three below the basal plane. The ideal case of close packing occurs when all twelve neighbours are at the same distance away, i.e. when the axial ratio c/a is $(8/3)^{\frac{1}{2}} = 1\cdot63$ (fig. III 4). If c/a differs very considerably from this value the number of nearest neighbours must be considered to be six only. The volume of the unit cell is $\frac{1}{2}\sqrt{3}\,a^2c$, and the volume per atom is half of this.

It is sometimes convenient to use rectangular axes $O1$, $O2$, $O3$, where $3O1$ and $2O3$ bisect the angles between $\mathbf{A}_1$ and $\mathbf{A}_2$. Referred to these axes we have

$$\mathbf{A}_1 = (\tfrac{1}{2}a, -\tfrac{1}{2}\sqrt{3}a, 0), \quad \mathbf{A}_2 = (\tfrac{1}{2}a, \tfrac{1}{2}\sqrt{3}a, 0), \quad \mathbf{C} = (0, 0, c), \quad (3\cdot22\cdot1)$$

and the basis is $(0, 0, 0)$, $(\tfrac{1}{2}a, \tfrac{1}{6}\sqrt{3}a, \tfrac{1}{2}c)$. The reciprocal vectors are

$$\mathbf{b}_1 = \left(\frac{1}{a}, -\frac{1}{\sqrt{3}a}, 0\right), \quad \mathbf{b}_2 = \left(\frac{1}{a}, \frac{1}{\sqrt{3}a}, 0\right), \quad \mathbf{b}_3 = \left(0, 0, \frac{1}{c}\right). \quad (3\cdot22\cdot2)$$

A list of metals having close-packed hexagonal structures is given in Table III 4.

Table III 4. *The parameters of hexagonal close-packed metals*

	Be	Mg	β-Ca	Ti	β-Cr	α-Co	α-Ni
a in 10^{-8} cm.	2·27	3·20	3·98	2·95	2·72	2·51	2·49
c/a	1·58	1·62	1·64	1·60	1·63	1·62	1·64
d in 10^{-8} cm.	2·22	3·19	3·99	2·91	2·71	2·50	2·49
Atomic volume in 10^{-24} c.c.	8·16	23·0	45	17·7	15·2	11·24	11

	Zn	Y	α-Zr	Ru	Cd	α-Ce	Pr
a in 10^{-8} cm.	2·66	3·66	3·22	2·69	2·97	3·65	3·66
c/a	1·86	1·59	1·59	1·59	1·89	1·63	1·62
d in 10^{-8} cm.	2·91	2·63	3·18	2·65	3·30	3·55	3·64
Atomic volume in 10^{-24} c.c.	15·2	33·7	23·2	13·4	21·4	34·3	34

	Nd	Er	Hf	Re	Os	α-Tl
a in 10^{-8} cm.	3·66	3·74	3·20	2·76	2·72	3·45
c/a	1·61	1·63	1·59	1·61	1·58	1·60
d in 10^{-8} cm.	3·62	3·73	3·14	2·74	2·68	3·39
Atomic volume in 10^{-24} c.c.	34	36·7	22·5	14·8	14·0	28·5

The parameters of β-Ca refer to 450° C.

(ii) *The graphite structure.* The graphite structure consists of four simple hexagonal lattices, the basis being $(0, 0, 0)$, $(0, 0, \frac{1}{2})$, $(\frac{1}{3}, \frac{2}{3}, u)$, $(\frac{2}{3}, \frac{1}{3}, u + \frac{1}{2})$ referred to the hexagonal axes. The quantity u is very nearly zero. The atoms are arranged in sheets, those in one sheet forming hexagons (without a central atom) with side $d = a/\sqrt{3}$, so that each atom has three neighbours at this distance. Half of the atoms in a sheet have two neighbours at a distance of $e = \frac{1}{2}c$ lying perpendicularly above and below the sheet. The parameters are shown in Table III 5.

Table III 5. *The lattice parameters of graphite*

a	c	c/a	d	e	Atomic volume
2·46 A.	6·78 A.	2·76	1·42 A.	3·4 A.	8·9 A.³

(iii) *The selenium structure.* The elements selenium and tellurium have hexagonal lattices in which the atoms are arranged in spiral chains as shown in fig. III 5. The lattices can be described as simple hexagonal lattices with three atoms as a basis at the points $(u, 0, 0)$, $(0, u, \frac{1}{3})$, $(-u, -u, \frac{2}{3})$ referred to the hexagonal axes. Each atom has two close neighbours at distance d in each chain and four more distant neighbours at distance e in adjacent chains. The parameters are given in Table III 6.

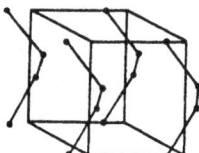

Fig. III 5. The structure of selenium.

Table III 6. *The lattice parameters of selenium and tellurium*

	a	c	c/a	u	d	e	Atomic volume
Se	4·34 A.	4·95 A.	1·14	0·217	2·32 A.	3·46 A.	26·9 A.³
Te	4·44 A.	5·91 A.	1·33	0·269	2·86 A.	3·74 A.	33·7 A.³

3·23. *Rhombohedral structures.* The rhombohedral lattice has three equal axes A_1, A_2, A_3 of length a inclined at the same angle α to one another. The volume of the unit cell is

$$a^3(1 - 3\cos^2\alpha + 2\cos^3\alpha)^{\frac{1}{2}} = a^3(1 - \cos\alpha)(1 + 2\cos\alpha)^{\frac{1}{2}}.$$

If we wish to use rectangular axes a_1, a_2, a_3 it is convenient to choose the trigonal axis† of the crystal as $O3$ and to take one of the crystal axes to lie in the plane $3\,0\,1$. Elementary coordinate geometry then gives

$$\left.\begin{aligned}
A_1 &= (\tfrac{2}{3}\sqrt{3}\sin\tfrac{1}{2}\alpha,\ 0,\ \tfrac{1}{3}\sqrt{3}(1 + 2\cos\alpha)^{\frac{1}{2}})\,a, \\
A_2 &= (-\tfrac{1}{3}\sqrt{3}\sin\tfrac{1}{2}\alpha,\ \sin\tfrac{1}{2}\alpha,\ \tfrac{1}{3}\sqrt{3}(1 + 2\cos\alpha)^{\frac{1}{2}})\,a, \\
A_3 &= (-\tfrac{1}{3}\sqrt{3}\sin\tfrac{1}{2}\alpha,\ -\sin\tfrac{1}{2}\alpha,\ \tfrac{1}{3}\sqrt{3}(1 + 2\cos\alpha)^{\frac{1}{2}})\,a,
\end{aligned}\right\} \quad (3\cdot23\cdot1)$$

† A trigonal axis is such that a rotation of $2\pi/3$ round the axis leaves the crystal invariant. In a rhombohedral crystal the angle between the trigonal axis and each of the crystal axes is $\cos^{-1}[\tfrac{1}{3}\sqrt{3}(1 + 2\cos\alpha)^{\frac{1}{2}}]$.

and the reciprocal vectors are given by

$$\left.\begin{aligned}
\mathbf{b}_1 &= \left(\frac{1}{\sqrt{3}\sin\frac{1}{2}\alpha},\; 0,\; \frac{1}{\sqrt{3}(1+2\cos\alpha)^{\frac{1}{2}}}\right)a^{-1}, \\[2mm]
\mathbf{b}_2 &= \left(-\frac{1}{2\sqrt{3}\sin\frac{1}{2}\alpha},\; \frac{1}{2\sin\frac{1}{2}\alpha},\; \frac{1}{\sqrt{3}(1+2\cos\alpha)^{\frac{1}{2}}}\right)a^{-1}, \\[2mm]
\mathbf{b}_3 &= \left(-\frac{1}{2\sqrt{3}\sin\frac{1}{2}\alpha},\; -\frac{1}{2\sin\frac{1}{2}\alpha},\; \frac{1}{\sqrt{3}(1+2\cos\alpha)^{\frac{1}{2}}}\right)a^{-1},
\end{aligned}\right\} \qquad (3\cdot23\cdot2)$$

so that $\quad |\mathbf{b}_1| = |\mathbf{b}_2| = |\mathbf{b}_3| = \left(\dfrac{1+\cos\alpha}{(1-\cos\alpha)(1+2\cos\alpha)}\right)^{\frac{1}{2}}\dfrac{1}{a}.$

There are no metals with a simple rhombohedral structure, but the semi-metals arsenic, antimony and bismuth have structures based on the rhombohedral lattice.

(i) *The bismuth structure.* The bismuth lattice can be described in two ways, as a lattice in which the rhombohedral angle is nearly 90°, the unit cell containing eight atoms, or as a rhombohedral lattice with an angle α which is nearly 60°, the unit cell containing two atoms. In the latter description the basis of the composite lattice is (u, u, u), $(-u, -u, -u)$, where u is nearly $\frac{1}{4}$. The atoms are arranged in double layers, each atom having three near neighbours at a distance d in the double layer in which it is situated, and three others at a greater distance d' in an adjacent layer. The parameters are given in Table III 7.

Table III 7. *The parameters of the bismuth-type lattices*

	a	α	u	d	d'	Atomic volume
As	4·15A.	53° 49′	0·226	2·51A.	3·15A.	21·7A.^{3}
Sb	4·50A.	57° 5′	0·233	2·90A.	3·36A.	29·7A.^{3}
Bi	4·74A.	57° 16′	0·237	3·11A.	3·47A.	35·2A.^{3}

(ii) Mercury has a rhombohedral structure which is nearly cubic. Each atom has six nearest neighbours and six others at a slightly greater distance. So far no results of theoretical interest have been obtained for this structure.

3·24. Tetragonal structures. The tetragonal lattice has its principal axes at right angles, with $|\mathbf{A}_1| = |\mathbf{A}_2| \neq |\mathbf{A}_3|$. The most important tetragonal structure is that of white tin.

(i) *The white tin structure.* This structure is most conveniently described as being a very deformed diamond lattice with $c/a = 0\cdot384$, where

$$|\mathbf{A}_1| = |\mathbf{A}_2| = a \quad \text{and} \quad |\mathbf{A}_3| = c.$$

It consists of two face-centred tetragonal lattices with the basis $(0, 0, 0)$, $(\frac{1}{4}, \frac{1}{4}, \frac{1}{4})$, the unit cell containing eight atoms. Each atom has four nearest neighbours at distance d arranged tetrahedrally, and two others along the c axis at a distance c which is not much greater than d. If c/a were $\sqrt{\frac{2}{15}}$, all six atoms would be at the same distance, while if c/a were 1 the structure would be the diamond structure. The parameters are shown in Table III 8.

Table III 8. *The parameters of white tin*

a	c	c/a	d	Atomic volume
8·25A.	3·17A.	0·384	3·02A.	27A.[3]

(ii) There are other tetragonal structures, but they are of little theoretical importance. They comprise the structures of indium and γ-manganese, which are nearly face-centred cubic, with $c/a = 1·078$ for indium and $c/a = 0·934$ for γ-manganese.

3·25. There are a number of complicated metallic structures which have so far not proved amenable to theoretical treatment. The most important is gallium which has a rhombic structure, while other types include α- and β-manganese, β-tungsten and uranium (see, for example, Hume-Rothery, 1944, Part III).

ACCURATE DETERMINATION OF THE WAVE FUNCTIONS

3·3. The methods discussed in Chapter II are only capable of giving qualitative results, and we now consider more powerful methods of determining the energy levels and wave functions of the conduction electrons. If we draw planes round a lattice point which bisect at right angles the lines joining this point to its nearest and next nearest neighbours, we obtain a polyhedron, which for a body-centred lattice is the truncated octahedron shown in fig. II 8 (iv).

Now the wave function $u_0(\mathbf{r})$ of the valency electrons which has the lowest energy has zero wave vector, and its period is therefore the unit cell. By symmetry the normal component of $\operatorname{grad} u_0(\mathbf{r})$ must vanish at the mid-points of the polyhedron described above, and this criterion allows us to calculate $u_0(\mathbf{r})$ approximately, as was first pointed out by Wigner and Seitz (1933). For a body-centred lattice the polyhedron is nearly a sphere, and the simplification introduced by Wigner and Seitz is to calculate $u_0(\mathbf{r})$ subject to the condition that $\partial u_0/\partial r = 0$ on the sphere $r = r_s$, where $\frac{4}{3}\pi r_s^3$ is the atomic volume.

The diameters of the orbits of the core electrons are smaller than the distance of the atoms apart, so that the charge in one cell is due to the ion and to the valency electrons in that cell. To a high degree of approximation we can consider the charge distribution to be spherically symmetrical, and the charge in any cell then produces zero electric force at points outside the cell. We can therefore calculate $u_0(\mathbf{r})$ as the solution of the Schrödinger equation for a valency electron moving in the field of the metal ion at the centre of the particular cell in question. The procedure for sodium, for example, is as follows. The potential due to the ion is known numerically from self-consistent field calculations. A value of the energy E of the valency electron is assumed, and a spherically symmetrical solution of the wave equation is determined for this energy starting from the origin. The numerical integration is continued outwards until a value of r is reached for which $\partial u/\partial r$ is zero. This value of r is taken to be the value of r_s corresponding to E. The process is repeated for different values of E, and so E is constructed as a function of r_s. (Note that since the wave function of the

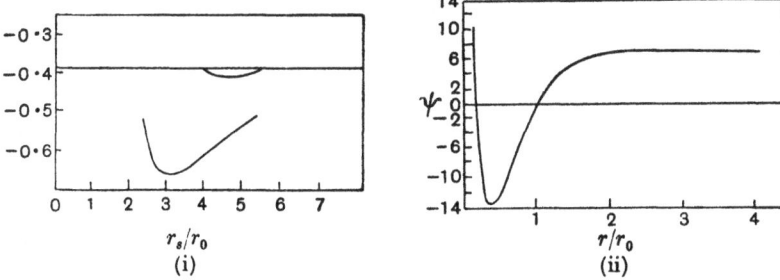

Fig. III 6. (i) The energy of the lowest $3s$-state (lower curve) and the total energy per atom (upper curve, see p. 74) in sodium as functions of the lattice constant. (ii) The wave function for the $3s$-state in metallic sodium for $r_s = 4r_0$.

valency electron in the free atom is a $3s$ function, u_0 must have two nodes and have a non-zero value for $r = 0$. Wave functions not satisfying these conditions must be excluded.)

The results obtained by Wigner and Seitz for sodium are shown in fig. III 6. The lower curve in fig. III 6 (i) shows the energy $E(r_s)$ as a function of r_s/r_0, where r_0 is the radius of the first Bohr orbit in hydrogen, while fig. III 6 (ii) shows the wave function as a function of r/r_0 for $r_s = 4r_0$. It will be noted that, over the major portion of the unit cell and particularly outside the region occupied by the core electrons, the wave function is very nearly constant, so that to treat the valency electrons as free electrons is not such a bad approximation as might have been expected.

There are various approximations in Wigner and Seitz's method which can be removed by introducing refinements and generalizations. Before discussing these in detail we shall consider a simple case which illustrates the principles without involving many complications.

3·31. *Wave functions in a linear lattice.* Consider a linear chain of three-dimensional atoms at the same distance apart. If x is the coordinate along the chain, the wave functions must be quasi-periodic in x (whereas they should decrease exponentially for large values of y and z). Writing therefore $\psi_k(\mathbf{r}) = e^{ikx} u_k(x, y, z)$, where $u_k(x, y, z) = u_k(x + a, y, z)$, consider the wave function in the 'cell' $-\tfrac{1}{2}a \leqslant x \leqslant \tfrac{1}{2}a$ belonging to the atom at $x = 0$. This wave function must join on smoothly to the wave function in the cell $\tfrac{1}{2}a \leqslant x \leqslant \tfrac{3}{2}a$ belonging to the atom at $x = a$, i.e. ψ and its derivative must be continuous at $x = \tfrac{1}{2}a$. These conditions can be written

$$\psi_{x=\frac{1}{2}a-0} = \psi_{x=\frac{1}{2}a+0}, \quad (\partial\psi/\partial x)_{x=\frac{1}{2}a-0} = (\partial\psi/\partial x)_{x=\frac{1}{2}a+0}. \tag{3·31·1}$$

Now, owing to the periodicity,

$$u_{x=\frac{1}{2}a+0} = u_{x=-\frac{1}{2}a+0} \quad \text{and} \quad (\partial u/\partial x)_{x=\frac{1}{2}a+0} = (\partial u/\partial x)_{x=-\frac{1}{2}a+0},$$

so that the conditions (3·31·1) can be written in the following form, which involves the wave function in one cell only:

$$\psi(\tfrac{1}{2}a) = e^{ika} \psi(-\tfrac{1}{2}a), \tag{3·31·2}$$

$$(\partial\psi/\partial x)_{x=\frac{1}{2}a} = e^{ika} (\partial\psi/\partial x)_{x=-\frac{1}{2}a}. \tag{3·31·3}$$

To solve these equations for a given value of the energy E we clearly require a wave function involving at least two arbitrary constants. We therefore take two solutions of the Schrödinger equation for the valency electron in the field of the metallic ion, one of them being the s-function for the energy E and the other being a p-function, and in fact the real p-function whose nodal plane is perpendicular to the chain. If we consider the assembly to be a three-dimensional one, we have not only to satisfy the conditions (3·31·2) and (3·31·3) along the chain but also to ensure that ψ vanishes as y and z tend to infinity. This latter criterion requires continuity conditions to be satisfied at points off the chain, and the proper behaviour in the y and z directions cannot be obtained with a solution involving only two arbitrary constants. We therefore consider the assembly to be a one-dimensional one, and we can then write

$$\psi(E, x) = As(E, x) + Bp(E, x), \tag{3·31·4}$$

where s is an even and p an odd function of x. The continuity conditions (3·31·2) and (3·31·3) then give the equations

$$\left. \begin{array}{l} iAs(E, \tfrac{1}{2}a) \sin\tfrac{1}{2}ka - Bp(E, \tfrac{1}{2}a) \cos\tfrac{1}{2}ka = 0, \\ As'(E, \tfrac{1}{2}a) \cos\tfrac{1}{2}ka - iBp'(E, \tfrac{1}{2}a) \sin\tfrac{1}{2}ka = 0, \end{array} \right\} \tag{3·31·5}$$

where $s' = \partial s/\partial x$ and $p' = \partial p/\partial x$. Eliminating A and B from these equations we find as the compatibility condition

$$\tan^2 \tfrac{1}{2}ka = -\frac{s'(E, \tfrac{1}{2}a)\, p(E, \tfrac{1}{2}a)}{s(E, \tfrac{1}{2}a)\, p'(E, \tfrac{1}{2}a)}. \tag{3·31·6}$$

Since k must be real for a propagated wave, the energy bands occupy the regions determined by the inequality

$$s(E, \tfrac{1}{2}a)\, s'(E, \tfrac{1}{2}a)\, p(E, \tfrac{1}{2}a)\, p'(E, \tfrac{1}{2}a) \leqslant 0, \qquad (3·31·7)$$

and the limits of the energy bands are given by $s' = 0$, $s = 0$, $p = 0$, $p' = 0$. Fig. III 7 gives the configuration of the bands for one particular potential as calculated by Hund and Mrowka (1935). It will be noted that the bands never overlap, in agreement with the general theorem that overlapping of allowed energy bands is a phenomenon which only occurs in two or three dimensions.

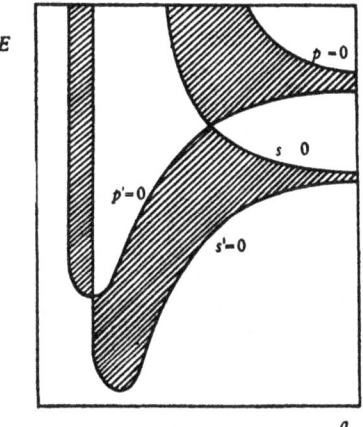

Fig. III 7. The energy bands in a simple linear chain.

If we treat the linear chain as belonging to a three-dimensional world, the above solution is only an approximate one, since the boundary conditions perpendicular to the chain are not taken into account. It is, however, possible to use the solution to obtain qualitative information about certain types of two- and three-dimensional lattices. Some interesting applications will be found in Hund and Mrowka's papers.

3·311. *A linear lattice with a basis.* As a generalization of the preceding results we can consider a lattice in which the atoms are at two different distances apart, a and b (fig. III 8). In this

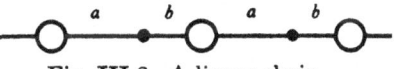

Fig. III 8. A linear chain with a basis.

case the unit cell has length $a + b$ and contains two atoms. A minimum of four constants is then required to obtain a suitable wave function in one cell, and we therefore put

$$\psi = As(E, x) + Bp(E, x)$$

in the part of the cell associated with one atom and

$$\psi = Cs(E, x) + Dp(E, x)$$

in the part of the cell associated with the neighbouring atom. We have then to impose continuity conditions upon ψ and $\partial\psi/\partial x$ midway between the atoms, and also to satisfy the periodicity conditions (3·31·2) and (3·31·3) at the ends of the cell. On writing down the four equations expressing these conditions and eliminating A, B, C, D we find that (3·31·6) has to be replaced by

$$\tan^2 \tfrac{1}{2}k(a+b) = -\frac{(s_1 p_2 + s_2 p_1)(s_1' p_2' + s_2' p_1')}{(s_1 p_2' + s_2' p_1)(s_1' p_2 + s_2 p_1')}, \qquad (3·311·1)$$

where s_1, p_1, s_1', p_1' are the values of the functions at $x = \frac{1}{2}a$, while s_2, p_2, s_2', p_2' are the values at $x = \frac{1}{2}b$.

Now, for a fixed value of a/b, the curve $s_1'p_2' + s_2'p_1' = 0$, for example, gives two values of E for each value of a, one being near the root of $s'(E, \frac{1}{2}a) = 0$ and the other near the root of $p'(E, \frac{1}{2}a) = 0$ if a/b is nearly unity. Hence each of the factors on the right of (3·311·1), considered as a function of E, has two zeros, and there are four values of E for each value of $k(a + b)$ in the range $(0, \pi)$. The major qualitative difference between the simple lattice and the lattice with a basis is, therefore, that each band of the simple lattice is split into two in the composite lattice (compare the comparable phenomenon which occurs in the vibration of a composite lattice, § 6·14). The splitting takes place along the curves given by

$$s_1 p_2' + s_2' p_1 = 0, \quad s_1' p_2 + s_2 p_1' = 0.$$

These curves are coincident for a simple chain, but they are different for a composite chain, and the energy region between them is one in which a wave cannot be propagated. The form of the energy bands for a constant ratio of a/b is shown in fig. III 9.

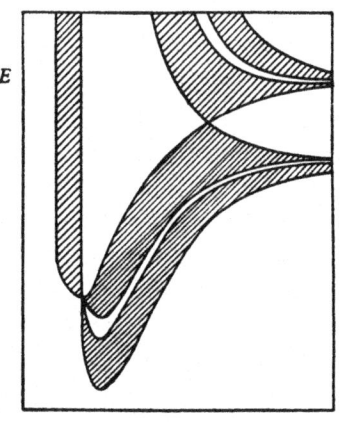

Fig. III 9. The energy bands in a linear lattice with a basis.

A formula similar to but more complicated than (3·311·1) will apply if we have two different atoms in the unit cell. In this case there will be twice as many functions $s_1, s_2, \ldots$ to consider, since not only do they refer to two different values of x but also because they are wave functions determined from two different wave equations.

3·32. *The wave functions of a three-dimensional crystal.* The approximate method given in the preceding section has been generalized by Slater (1934) to apply to real crystals. For a given value of the energy E, the most general solution of the Schrödinger equation in one cell of the lattice consists of an infinite series of spherical harmonics multiplied by functions of r, subject to the sole boundary condition that the solution is finite at the origin. If we then choose the coefficients of the various spherical harmonics to satisfy the continuity conditions for ψ and grad ψ at all points of the boundary of the cell, we obtain the complete solution to the problem.

In practice it is necessary to use only a finite series of spherical harmonics and to satisfy the continuity conditions at a limited number of points. In discussing body-centred cubic lattices, Slater satisfied the continuity conditions at the eight points half-way between the central atom and its eight nearest neighbours. He then required eight harmonics; he chose the zero harmonic, the three harmonics x/r, y/r, z/r of order 1, the three harmonics xy/r^2, yz/r^2, zx/r^2 of order 2 and the harmonic xyz/r^3 of order 3. The results obtained for sodium are shown in fig. III 10. When r_s is infinite the energy levels are discrete, being those of the normal sodium atom, while for any

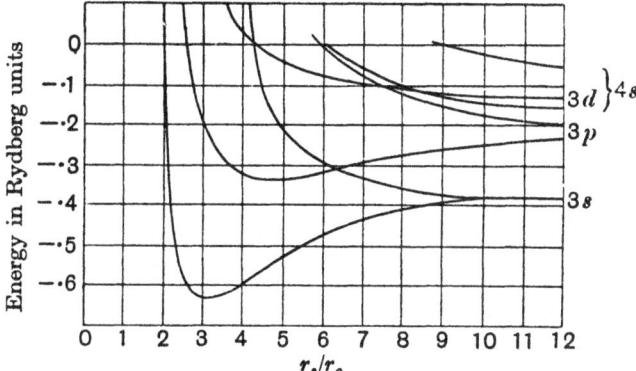

Fig. III 10. The energy bands in sodium.

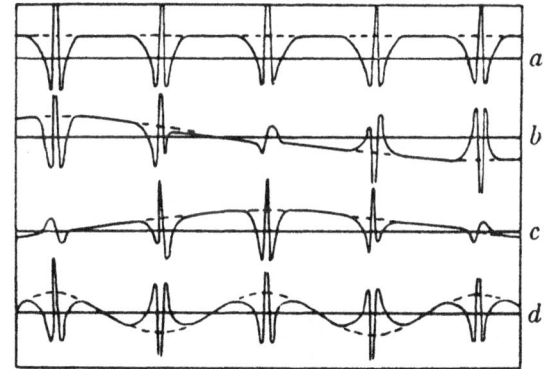

Fig. III 11. Wave functions for the $3s$-electrons in sodium as functions of the distance in the $(1, 1, 1)$ direction. a, wave function at the bottom of the band; b, c, real and imaginary parts of the function whose wave-length is eight times the lattice constant; d, wave function at the top of the band.

finite value of r_s the typical band structure appears and, for sufficiently small r_s, the bands overlap. (In the diagram only the extreme energy levels corresponding to the top and bottom of each band are shown.) The fact that, for the value of r_s which actually occurs, the first and second bands (the $3s$- and $3p$-bands) overlap is not of much importance since sodium is monovalent and the valency electrons only half fill the first band.

Some of the wave functions calculated by Slater are shown in fig. III 11;

the thick lines represent the actual wave functions, while the dotted lines represent the wave functions of perfectly free electrons, i.e. $e^{i\mathbf{k}\cdot\mathbf{r}}$. It will be seen that over most of the crystal the wave functions are well represented by those for free electrons, particularly for small values of k, the deviations occurring only inside the atomic cores. It is for this reason that many of the properties of metals can be explained so successfully by assuming that the valency electrons behave as if they were perfectly free. A further important point is illustrated by the fact that the curves b and c have s-like properties at some points and p-like properties at others.† This means that, when the energy bands overlap, the states can no longer be regarded as being derived from single atomic states; the states have become mixed and there is no sharp distinction between the symmetry properties of the wave functions belonging to different bands.

Numerous elaborations of Slater's work have been published, the most recent being by Howarth and Jones (1952) and by Kohn (1952).

The Cohesion of Metals

3·4. The cohesive forces in solids can be classified into four main types, the van der Waals, the covalent, the electrostatic and the metallic. Solid argon is a typical example of a structure held together by van der Waals forces, while in diamond the carbon atoms each have four neighbours arranged tetrahedrally so that the structure can be considered as being built up according to the classical valency concepts of organic chemistry. In salts such as sodium chloride, the metal atoms are positively and the halogen atoms negatively ionized, and the cohesive forces are the long-range Coulomb forces. In metals, however, none of the above concepts apply, and in particular the coordination number (the number of nearest neighbours) bears no relation to the valency attributable to an isolated atom. The typical metallic structures have coordination number 8 (body-centred cubic) or 12 (face-centred cubic and hexagonal close packed), but the semimetals have structures bordering on the covalent.

The semi-metals selenium and tellurium of group 6 have a chain structure (fig. III 5). They have a primary coordination number of 2 but have a secondary coordination number of 6 (the four second nearest neighbours are not much farther away than the nearest neighbours, especially in tellurium). It is therefore reasonable to describe the chains as being formed by covalent forces while the various chains are held together by van der

† A p-wave function has a node at the centre of an atom while an s-wave function has not.

Waals forces. However, the metallic character is by no means lacking, and the whole of the cohesive force cannot be attributable to covalent and van der Waals forces.

Arsenic, antimony and bismuth of group 5 all have a primary coordination number of 3 with a secondary coordination number of 6. Here again it is reasonable to describe the binding as largely covalent, but the metallic character is much more marked than in selenium and tellurium, and, particularly in bismuth where there is only 10 % difference in the distance of nearest and second neighbours, the binding must be partly covalent and partly metallic.

The elements of group 4, carbon, silicon, germanium and grey tin, form pure covalent lattices with coordination number 4. The metallic character increases with increasing atomic number, being entirely absent in diamond but by no means negligible in grey tin where the energy gap between the filled and conduction bands is of the order of 0·1 eV., while the fifth element of the group, lead, has a typically metallic face-centred cubic structure.

The metals of groups 1, 2 and 3 all have coordination numbers 8 or 12, and the cohesive forces cannot be considered to be localized in bonds but must be attributed to the interaction of the conduction electrons with the metallic ions. In the extreme limit the conduction electrons can be considered to be shared equally by all the atoms, the metallic ions being immersed in a sea of conduction electrons, and the forces responsible for the cohesion of metals can, in this sense, be thought of as being of infinitely long range.

The way in which the binding energy originates has been considered in § 3·3. Owing to its being surrounded by a large number of neighbouring atoms, the valency electron of a metal atom is pushed back into the core, and its energy is thereby decreased as shown by the lower curve in fig. III 6 (i). This curve, however, only applies to the conduction electron with zero wave number and greatly overestimates the binding energy. The calculation of the total cohesion is extremely complicated and has not yet been put on an entirely satisfactory basis, so that only the most important principles will be outlined here.

3·41. *Calculation of the binding energy.* The wave functions

$$\psi_{\mathbf{k}}(\mathbf{r}) = e^{i\mathbf{k}\cdot\mathbf{r}} u_{\mathbf{k}}(\mathbf{r})$$

are single-electron wave functions, and to obtain the total energy we must use the correct antisymmetrical wave function for the whole assembly of conduction electrons. This leads to the expression given in (2·91·4), which in the present notation is

$$U_{\text{total}} = -\frac{h^2}{8\pi^2 m} \sum_{k,s} \int \psi_{k,s}(\mathbf{r})^* \nabla^2 \psi_{k,s}(\mathbf{r}) \, d\tau + \sum_{k,s} \int |\psi_{k,s}(\mathbf{r})|^2 V(\mathbf{r}) \, d\tau$$

$$+ \tfrac{1}{2} \sum_{g \neq g'} V_{g,g'} + \frac{1}{2} \int\int \sum_{k,s} |\psi_{k,s}(\mathbf{r})|^2 \frac{\epsilon^2}{|\mathbf{r} - \mathbf{r}'|} \sum_{k',s'} |\psi_{k',s'}(\mathbf{r}')|^2 \, d\tau \, d\tau'$$

$$- \frac{1}{2} \int\int \sum_{k} \sum_{k'} \sum_{s} \psi_{k,s}(\mathbf{r}) \psi_{k,s}(\mathbf{r}')^* \frac{\epsilon^2}{|\mathbf{r} - \mathbf{r}'|} \psi_{k',s}(\mathbf{r})^* \psi_{k',s}(\mathbf{r}') \, d\tau \, d\tau',$$

$$(3 \cdot 41 \cdot 1)$$

where $V(\mathbf{r})$ is the potential energy of an electron due to the ions and $V_{g,g'}$ is the mutual energy of the ions g and g'. The wave functions are normalized to unity in the whole crystal, and the summation over s denotes a summation over the spin states.

In the cellular method of determining the wave functions, ψ_k is found as the solution of the equation

$$\frac{h^2}{8\pi^2 m} \nabla^2 \psi_k + \{E_k - V_g(\mathbf{r})\} \psi_k = 0 \qquad (3 \cdot 41 \cdot 2)$$

in the cell g, where $V_g(\mathbf{r})$ is the potential energy of an electron due to the ion at the centre of the cell g, i.e. $V(\mathbf{r}) = \Sigma V_g(\mathbf{r})$. With the approximations used by Wigner and Seitz in their calculation of the energies and wave functions, considerable simplification of these expressions is possible. We consider the various terms separately. The first integral can be split into integrals over the N elementary cells of the lattice, and each cell must contribute equally to the integrals since the integrands are periodic with the elementary cell as period. Now ψ_k is normalized to $1/N$ in the unit cell so that this term gives

$$\sum_{k,s} E_k - N \sum_{k,s} \int |\psi_{k,s}(\mathbf{r})|^2 V_g(\mathbf{r}) \, d\tau_g, \qquad (3 \cdot 41 \cdot 3)$$

where $d\tau_g$ denotes an integration over the cell g. Also, if $g \neq g'$,

$$V_{g,g'} + \sum_{k,s} \int |\psi_{k,s}(\mathbf{r})|^2 V_{g'}(\mathbf{r}) \, d\tau_g \qquad (3 \cdot 41 \cdot 4)$$

is the potential energy of the ion at the point g' due to the ion at the point g together with the charge of the conduction electrons inside the cell g. Now the charge density and the cell are considered to be spherically symmetrical and each cell is electrically neutral, so that $(3 \cdot 41 \cdot 4)$ must be zero, since any spherical distribution with no resultant charge produces zero potential at external points. Similarly,

$$V_{g'}(\mathbf{r}) + \epsilon^2 \sum_{k',s} \int \frac{|\psi_{k',s}(\mathbf{r}')|^2}{|\mathbf{r} - \mathbf{r}'|} \, d\tau_{g'} \qquad (3 \cdot 41 \cdot 5)$$

is the potential energy of an electron at the point $\mathbf{r}$ due to the ion at the point g' together with the charge of the conduction electrons inside the cell g', and this again must be zero if $\mathbf{r}$ is not in the cell g'. Therefore, if $g \neq g'$,

$$\sum_{\mathbf{k},s} \int |\psi_{\mathbf{k},s}(\mathbf{r})|^2 V_{g'}(\mathbf{r})\, d\tau_g + \epsilon^2 \sum_{\mathbf{k},s} \sum_{\mathbf{k}',s'} \int\int \frac{|\psi_{\mathbf{k},s}(\mathbf{r})|^2\, |\psi_{\mathbf{k}',s'}(\mathbf{r}')|^2}{|\mathbf{r}-\mathbf{r}'|}\, d\tau_g\, d\tau_{g'} = 0,$$
(3·41·6)

and combining (3·41·3), (3·41·4) and (3·41·6) we see that the only non-zero term coming from the second, third and fourth terms of (3·41·1) is that part of the fourth term for which $g = g'$. Hence (3·41·1) reduces to

$$U_{\text{total}} = \sum_{\mathbf{k},s} E_{\mathbf{k}} + \tfrac{1}{2} N \int\int \sum_{\mathbf{k},s} |\psi_{\mathbf{k},s}(\mathbf{r})|^2 \frac{\epsilon^2}{|\mathbf{r}-\mathbf{r}'|} \sum_{\mathbf{k}',s'} |\psi_{\mathbf{k}',s'}(\mathbf{r}')|^2\, d\tau_g\, d\tau_g'$$

$$-\frac{1}{2}\int\int \sum_{\mathbf{k}}\sum_{\mathbf{k}'}\sum_{s} \psi_{\mathbf{k},s}(\mathbf{r})\,\psi_{\mathbf{k},s}(\mathbf{r}')^* \frac{\epsilon^2}{|\mathbf{r}-\mathbf{r}'|}\, \psi_{\mathbf{k}',s}(\mathbf{r})^*\, \psi_{\mathbf{k}',s}(\mathbf{r}')\, d\tau\, d\tau'.$$
(3·41·7)

The second term is N times the mutual energy of the spherically symmetrical conduction electron charge inside an elementary cell, which can be calculated as follows. If the charge inside the sphere of radius r is $M(r)$, then if we suppose the charge to be built up by laying down successive spherical shells, starting from the centre, the potential at the surface, when the radius is r, is $M(r)/r$. The work done in increasing the radius to $r + dr$ is $M(r)\, dM(r)/r$, and so the total potential energy is

$$\int_0^{r_s} \frac{M(r)}{r}\, dM(r),$$
(3·41·8)

where r_s is the radius of the completed sphere. Now $M(r) = \tfrac{4}{3}\pi r^3 \rho$ where ρ is the density, and so (3·41·8) is

$$\tfrac{16}{3}\pi^2 \rho^2 \int_0^{r_s} r^4\, dr = \tfrac{16}{15}\pi^2 \rho^2 r_s^5 = \frac{3}{5}\frac{\epsilon^2}{r_s},$$
(3·41·9)

$-\epsilon$ being the total charge. Also $\tfrac{4}{3}\pi r_s^3$ is the atomic volume and is therefore equal to $1/n$, where n is the number of electrons per unit volume.

The first term of (3·41·7) is the sum of the energies of the one-electron wave functions associated with the conduction electrons. Wigner and Seitz assumed that $E_{\mathbf{k}} = E_0 + h^2 |\mathbf{k}|^2/(8\pi^2 m)$, where E_0 is the energy calculated by them for the state with $\mathbf{k} = 0$, i.e. they assumed that $u_{\mathbf{k}}$ is independent of $\mathbf{k}$. Now as usual the sum over $\mathbf{k}$ is to be replaced by $(L^3/8\pi^3)\int d\mathbf{k}$, where L^3 is the total volume of the crystal, and the total number of states $\tfrac{1}{2}N$ with one spin for a monatomic metal is equal to $(L^3/8\pi^3)\tfrac{4}{3}\pi k_0^3$, where k_0 is the wave number of the highest occupied state.

Also n, the number of electrons per unit volume, is equal to N/L^3, so that

$$k_0^3 = 3\pi^2 N/L^3 = 3\pi^2 n. \qquad (3·41·10)$$

Hence with the above expression for E_k we have

$$\sum_{k,s} E_k = \frac{L^3}{8\pi^3} \int_0^{k_s} 2E_k\, 4\pi k^2\, dk = NE_0 + \frac{L^3 h^2 k_0^5}{40\pi^4 m} = N\left\{E_0 + \frac{3h^2}{40m}\left(\frac{3n}{\pi}\right)^{\frac{2}{3}}\right\}. \qquad (3·41·11)$$

The last term in (3·41·1) giving the effect of the exchange forces is more difficult to evaluate. With certain assumptions (see below, § 3·44), a value

$$-\frac{3Ne^2}{4}\left(\frac{3n}{\pi}\right)^{\frac{1}{3}} \qquad (3·41·12)$$

can be ascribed to it. We therefore finally arrive at the expression

$$\frac{1}{N} U_{\text{total}} = E_0 + \frac{3h^2}{40m}\left(\frac{3n}{\pi}\right)^{\frac{2}{3}} + \tfrac{3}{5}e^2\left(\frac{4\pi n}{3}\right)^{\frac{1}{3}} - \tfrac{3}{4}e^2\left(\frac{3n}{\pi}\right)^{\frac{1}{3}}. \qquad (3·41·13)$$

For body-centred monovalent elements such as the alkalis, $n = 2/a^3$, where a is the side of the unit cube, and

$$\frac{1}{N} U_{\text{total}} = E_0 + 4·556 \frac{e^2 r_0}{a^2} + 0·287 \frac{e^2}{a}, \qquad (3·41·14)$$

where $r_0 = h^2/(4\pi^2 m e^2)$.

3.42. *Numerical results.* Numerical values of (3·41·14) as a function of the lattice constant a have been calculated for sodium by Wigner and Seitz. The energy has a very flat minimum near $a = 9r_0$ and yields a dissociation energy which is only just positive. This lack of agreement with the observed facts is not surprising in view of the crudeness of some of the approximations used in arriving at (3·41·14), and Wigner and Seitz (1934) therefore introduced a number of refinements without, however, making fundamental changes in the basis of the calculation. Their results are shown in fig. III 6 (i), where the upper curve gives U_{total}/N as a function of r_s. The minimum occurs when $r_s = 4·76 \times 10^{-8}$ cm., the corresponding value of E_0 is $-7·44$ eV., and the total energy per atom is $-5·50$ eV. With the same energy zero, the energy of a neutral sodium atom is the negative of the ionization energy, i.e. $-5·12$ eV., so that the calculated value of the dissociation energy of metallic sodium into neutral atoms is $0·38$ eV., the observed value being $1·13$ eV.

Wigner (1934) has tried to improve the fundamental basis of the calculation by including the so-called 'correlation forces', which are discussed in § 3·441, and which lead to a lessening of the repulsive forces between the electrons. When these are included and other corrections are made, the numerical results are as shown in Table III 9.

Table III 9. *Observed and calculated values of the lattice constants and cohesive energies of the alkalis*

	Lattice constant in 10^{-8} cm.		Cohesive energy per atom in eV.	
	Observed	Calculated	Observed	Calculated
Li	3·46	3·50	1·7	1·57
Na	4·25	4·51	1·13	1·06
K	5·20	5·82	1·0	0·715

The agreement between the observed and calculated values given in Table III 9 is only obtained after an immense amount of calculation and after many corrections have been made, some of which are not too firmly based. It is difficult to have any great confidence in the exact numerical values on account of the intrinsic difficulties in treating the many-body problem, and it is best to regard the achievements of the theory as showing that the metallic type of binding gives rise to cohesive forces of the right order of magnitude. In the present state of the theory it is not possible to predict the crystal structure from the properties of the metallic atoms or to discuss the allotropic transformation of those metals which have more than one crystal structure. In Wigner and Seitz's method, the assumption that the elementary cell can be replaced by a sphere makes the energy a function of the atomic volume only and not of the type of crystal structure. Therefore to calculate, for example, the difference in the energy between body-centred and face-centred sodium would require still further refinements and an accuracy which is not at present attainable.

3·43. *The elastic constants.* In order to determine the elastic constants it is necessary to calculate the change in the energy when small deformations of the lattice are made. Detailed investigations have been made by Fuchs (1935, 1936 a, b), to whose papers reference should be made since the calculations are too complicated and of insufficient general interest to be given here.

Table III 10. *Observed and calculated values of the elastic constants in units of 10^{11} dynes/cm.²*

	Observed					Calculated				
	$1/\kappa$	c_{11}	c_{12}	$c_{11}-c_{12}$	c_{44}	$1/\kappa$	c_{11}	c_{12}	$c_{11}-c_{12}$	c_{44}
Li	—	—	—	—	—	1·30	1·53	1·19	0·34	1·35
Na	0·85	0·95	0·80	0·15	0·59	0·88	0·97	0·83	0·14	0·59
K	0·40	0·46	0·37	0·09	0·26	0·41	0·45	0·39	0·06	0·26
Cu	13·9	18·6	13·5	5·1	8·2	14·1	17·5	12·4	5·1	8·9

The values for Na and K refer to $-183°$ C. and have been determined by Bender (1939). The observed values for Cu have been extrapolated to $T=0$ by Fuchs.

In cubic crystals there are three independent elastic constants which, in Voigt's notation (1928), are c_{11}, c_{12} and c_{44}. The compressibility κ is related to the first two by the relation $\kappa = 3/(c_{11} + 2c_{12})$. The results are shown in Table III 10.

The agreement is much better than for the binding energy, which is to be expected, since it is easier to calculate the change in energy due to a deformation than the absolute value of the energy itself. For the alkalis the curvature of the E_0 curve is small near the minimum of the total-energy curve, and the compressibility is determined largely by the variation with lattice constant of the kinetic energy of the electrons, so that in this case the compressibility can be calculated by elementary methods. This does not apply to the other elastic constants, nor does it hold for copper where the relatively large size of the copper ion reduces the compressibility and is the dominating factor.

It will be seen from the table that the monovalent metals, though cubic in structure, do not behave isotropically, for which the condition is $c_{11} - c_{12} = 2c_{44}$. Neither do they satisfy the Cauchy-Poisson relation $c_{12} = c_{44}$, which holds if the atoms behave like points whose interactions can be represented by central forces.

3·44. *The exchange and correlation forces.* Unlike the other contributions to the energy the exchange forces cannot be reduced to integrals referring to one cell only, and their calculation is one of great difficulty. It is, however, possible to obtain an explicit expression for the exchange force in the limiting case of perfectly free electrons, and this could be taken to apply to a metal such as sodium in which the function $u_{\mathbf{k}}(\mathbf{r})$ is constant over a large part of the elementary cell.

To calculate the exchange force for free electrons, consider a large region of volume L^3, in which the wave functions are normalized. Then

$$(\mathbf{k}\mathbf{k}' \mid \epsilon^2/r_{12} \mid \mathbf{k}'\mathbf{k}) = \frac{1}{L^6} \int\int e^{i(\mathbf{k}-\mathbf{k}')\cdot\mathbf{r}_1} \frac{\epsilon^2}{|\mathbf{r}_1 - \mathbf{r}_2|} e^{i(\mathbf{k}'-\mathbf{k})\cdot\mathbf{r}_2} d\tau_1 d\tau_2. \quad (3\cdot44\cdot1)$$

Now

$$V(\mathbf{r}) = \int \frac{\epsilon^2}{|\mathbf{r} - \mathbf{r}_2|} e^{i(\mathbf{k}'-\mathbf{k})\cdot\mathbf{r}_2} d\tau_2$$

is the potential due to a charge density $\epsilon^2 e^{i(\mathbf{k}'-\mathbf{k})\cdot\mathbf{r}}$, and it therefore satisfies Poisson's equation $\nabla^2 V + 4\pi\epsilon^2 e^{i(\mathbf{k}'-\mathbf{k})\cdot\mathbf{r}} = 0$, the solution of which is

$$V(\mathbf{r}) = 4\pi\epsilon^2 \frac{e^{i(\mathbf{k}'-\mathbf{k})\cdot\mathbf{r}}}{|\mathbf{k} - \mathbf{k}'|^2}.$$

Hence (3·44·1) becomes

$$(\mathbf{k}\mathbf{k}' \mid \epsilon^2/r_{12} \mid \mathbf{k}'\mathbf{k}) = \frac{1}{L^6} \int \frac{4\pi\epsilon^2 d\tau_1}{|\mathbf{k} - \mathbf{k}'|^2} = \frac{4\pi\epsilon^2}{L^3 |\mathbf{k} - \mathbf{k}'|^2}. \quad (3\cdot44\cdot2)$$

Now a sum with respect to $\mathbf{k}$ is equivalent to the integral $L^3/(8\pi^3)\int d\mathbf{k}$. Hence

$$\sum_{\mathbf{k}'} (\mathbf{k}\mathbf{k}' \mid \epsilon^2/r_{12} \mid \mathbf{k}'\mathbf{k}) = \frac{\epsilon^2}{2\pi^2} \iiint \frac{dk_1' dk_2' dk_3'}{\mid \mathbf{k} - \mathbf{k}' \mid^2}.$$

The integral is easily evaluated in terms of polar coordinates k', θ, ϕ in the $\mathbf{k}'$ space, and is

$$\int_0^{2\pi} d\phi \int_0^{\pi} \sin\theta \, d\theta \int_0^{k_0} \frac{k'^2 \, dk'}{k^2 + k'^2 - 2kk' \cos\theta} = \pi \left(2k_0 + \frac{k^2 - k_0^2}{k} \log \frac{\mid k_0 - k \mid}{k_0 + k} \right),$$

where the electrons occupy the energy levels $0 \leqslant k$, $k' \leqslant k_0$. Hence

$$\sum_{\mathbf{k}'} (\mathbf{k}\mathbf{k}' \mid \epsilon^2/r_{12} \mid \mathbf{k}'\mathbf{k}) = \frac{\epsilon^2 k_0}{\pi} \left(1 + \frac{k^2 - k_0^2}{2kk_0} \log \frac{\mid k_0 - k \mid}{k_0 + k} \right). \qquad (3\cdot44\cdot3)$$

The total energy of the exchange forces is obtained by summing the last expression over $\mathbf{k}$. An elementary integration gives

$$\tfrac{1}{2} \sum_{\mathbf{k}} \sum_{\mathbf{k}'} (\mathbf{k}\mathbf{k}' \mid \epsilon^2/r_{12} \mid \mathbf{k}'\mathbf{k}) = \frac{L^3 \epsilon^2 k_0^4}{8\pi^3} = \frac{L^3 \epsilon^2}{8\pi^3} (3\pi^2 n)^{\frac{4}{3}} = \frac{3N\epsilon^2}{8} \left(\frac{3n}{\pi} \right)^{\frac{1}{3}}, \qquad (3\cdot44\cdot4)$$

by $(3\cdot41\cdot10)$. This is the contribution due to electrons with one spin, and since there are two electrons per state with opposite spins, the total contribution to the energy by the exchange forces is double that given by $(3\cdot44\cdot4)$.

3·441. The expressions for the exchange forces given in the preceding section could be improved by taking into account the deviations of the wave functions from plane waves. More drastic alterations in procedure, however, are essential if we are to have a completely consistent theory. The simple Hartree theory of the self-consistent field over-emphasizes the importance of polar states in which more than one electron is simultaneously attached to one metal ion, and the exchange forces are introduced to correct this. The exchange forces arise through the use of an antisymmetrical wave function which ensures that electrons with parallel spins do not occupy the same orbits, and therefore tends to diminish the importance of polar states in which the electrons concerned have parallel spins. Unfortunately, neither the symmetrized Hartree equations nor the Fock equations sufficiently reduce the importance of polar states.

Wigner (1934) has attempted to improve the theory in the following way. The normal self-consistent field methods assume that the elements of the determinantal wave function are separable into a spin function and a coordinate function. This assumption is known to be incorrect; it would imply, for example, that the coordinate wave function for the helium atom in its normal state is a product of two wave functions each containing the coordinates of one electron only. Actually all that can be said is that the wave function is symmetrical in the coordinates of the two electrons, and to obtain good numerical results it is necessary to introduce explicitly into the wave function the distance between the two electrons and in this way take account of the correlations between the positions of the electrons. The self-consistent field method smoothes out these correlations. Wigner therefore tried to introduce these correlations into the metallic wave functions by taking a type of function in which the mutual distances of the electrons are present. The theory is complicated and semi-empirical, and it is difficult to assess its accuracy.

The tentative result is that if n_1 and n_2 are the number of electrons per unit volume with positive and negative spins respectively, the correlation energy per unit volume is

$$U_{\text{corr.}} = n_1 g(n_2) + n_2 g(n_1), \qquad (3\cdot441\cdot1)$$

where

$$g(n) = -\frac{0\cdot58\epsilon^2}{10\cdot3r_0 + n^{-\frac{1}{3}}}$$

$(r_0 = h^2/4\pi^2 m\epsilon^2)$, provided that n_1 and n_2 are nearly equal. The total energy per unit volume is then given by

$$U = (n_1 + n_2)\,E_0 + \frac{h^2}{80\pi^4 m}\{(6\pi^2 n_1)^{\frac{5}{3}} + (6\pi^2 n_2)^{\frac{5}{3}}\}$$

$$+ \frac{3}{5}\left(\frac{4\pi}{3}\right)^{\frac{1}{3}} \epsilon^2 (n_1 + n_2)^{\frac{4}{3}} - \frac{\epsilon^2}{8\pi^3}\{(6\pi^2 n_1)^{\frac{4}{3}} + (6\pi^2 n_2)^{\frac{4}{3}}\} + n_1 g(n_2) + n_2 g(n_1),$$

$$(3\cdot441\cdot2)$$

which is the generalization of $(3\cdot41\cdot13)$. (To verify this, note that it reduces to $(3\cdot41\cdot13)$ if $n_1 = n_2 = \frac{1}{2}n$ and if the correlation energy is omitted; also that the electrostatic energy is determined by the total density of the electrons, i.e. by $n_1 + n_2$, whereas the other contributions to the energy are given by adding the separate contributions due to the electrons with a given direction of spin.) This is the expression used by Wigner, with $n_1 = n_2 = \frac{1}{2}n$, to obtain the cohesive energies quoted in § 3·42.

3·442. As an illustration of the effect of the exchange forces if they are interpreted literally, it is instructive to consider Fock's equation (§ 2·92) for free electrons. In this case the exchange term $\mathfrak{A}\psi(\mathbf{k}, \mathbf{r})$ is

$$\mathfrak{A}\psi_{\mathbf{k}}(\mathbf{r}) = \frac{\epsilon^2}{L^{\frac{3}{2}}} \sum_{\mathbf{k}'} e^{i\mathbf{k}'\cdot\mathbf{r}} \int \frac{e^{i(\mathbf{k}-\mathbf{k}')\cdot\mathbf{r}'}}{|\mathbf{r}-\mathbf{r}'|}\,d\tau'.$$

The integral has already been calculated in § 3·44, and the above expression is

$$\mathfrak{A}\psi_{\mathbf{k}}(\mathbf{r}) = \frac{4\pi\epsilon^2}{L^{\frac{3}{2}}} \sum_{\mathbf{k}'} \frac{e^{i\mathbf{k}\cdot\mathbf{r}}}{|\mathbf{k}-\mathbf{k}'|^2} = \frac{4\pi\epsilon^2}{L^3} \sum_{\mathbf{k}'} \frac{1}{|\mathbf{k}-\mathbf{k}'|^2}\psi_{\mathbf{k}}(\mathbf{r}).$$

The summation over $\mathbf{k}'$ is carried out exactly as in §3·44, and we find

$$\mathfrak{A}\psi_{\mathbf{k}}(\mathbf{r}) = C_{\mathbf{k}}\psi_{\mathbf{k}}(\mathbf{r}),$$

where $C_{\mathbf{k}}$ is given by (3·44·3), i.e.

$$C_{\mathbf{k}} = \frac{\epsilon^2}{\pi} k_0 \left(1 + \frac{k^2 - k_0^2}{2kk_0} \log \frac{|k_0 - k|}{k_0 + k}\right). \tag{3·442·1}$$

The Hartree field must be considered to be zero in order to maintain electrical neutrality, and the Fock equation is therefore

$$\left(-\frac{h^2\nabla^2}{8\pi^2 m} - C_{\mathbf{k}} - E_{\mathbf{k}}\right)\psi_{\mathbf{k}}(\mathbf{r}) = 0, \tag{3·442·2}$$

the exchange forces appearing directly in the energy levels

$$E_{\mathbf{k}} = \frac{h^2 |\mathbf{k}|^2}{8\pi^2 m} - C_{\mathbf{k}}. \tag{3·442·3}$$

This is exactly what one would expect, but it has an important effect upon the density of states, which is defined by the relation (2·71·2)

$$\mathfrak{n}(E)\,dE = k^2\,dk/2\pi^2,$$

when E is a function of $k = |\mathbf{k}|$ only. Therefore

$$\mathfrak{n}(E) = \frac{k^2}{2\pi^2\,dE/dk} = \frac{2mk}{h^2}\bigg/\left(1 - \frac{4\pi^2 m}{h^2}\frac{1}{k}\frac{\partial C_{\mathbf{k}}}{\partial k}\right). \tag{3·442·4}$$

Now

$$-\frac{\partial C_{\mathbf{k}}}{\partial k} = \frac{\epsilon^2}{2\pi}\left(1 + \frac{k_0^2}{k^2}\right)\left(\log\frac{k_0 + k}{k_0 - k} - \frac{2kk_0}{k^2 + k_0^2}\right),$$

and it is readily shown by differentiation that the expression in the last bracket increases steadily from zero as k increases. The effect of the exchange forces is therefore to diminish $\mathfrak{n}(E)$, and, moreover, since $-\partial C_{\mathbf{k}}/\partial k$ is infinite for $k = k_0$, to make $\mathfrak{n}(E) = 0$ there. Since most of the properties of metals depend upon the behaviour of $\mathfrak{n}(E)$ near $k = k_0$, this has a very great influence, and if taken literally would upset most of the theories of the metallic state. The part played by the exchange forces is therefore extremely obscure, and it is not possible at the moment to take them into account properly. The general tendency has been to neglect them entirely, but when this cannot be done because the phenomena are dominated by them, it is necessary to proceed with great caution and to try if possible to include the correlation forces as well (Bardeen, 1936; Wohlfarth, 1950), which, it has been suggested, will cancel the effect of the exchange forces on $\mathfrak{n}(E)$ and restore the results of the simple theory.

The Energy Zones of Metals

3·5. The methods outlined in § 3·3 suffice to determine the energy levels and the wave functions of any metallic structure, but the calculations are very laborious and the results are not obtained in a simple and easily understandable form. A certain amount of information, mainly of a qualitative nature, can be obtained by considering the Brillouin zones, the disposition of which is determined entirely by the crystal structure. The exact arrangement of the energy levels in the zones, however, and the energy discontinuities at the edges of the zones can only be obtained by numerical integration of the Schrödinger equations, but there are two principles that enable some progress to be made without undertaking the complete numerical integration. The first principle is that, provided **k** is not near a zone boundary, the energy levels will bear some resemblance to those of free electrons. The second is that the energy discontinuities are correlated with the X-ray reflexions, since both are determined by the relevant Fourier coefficients of the potential of the lattice. In particular, if an X-ray reflexion is entirely absent, the corresponding Fourier coefficient must be zero and the energy discontinuity must also be zero. We now proceed to survey the zone structure of metals in the light of these principles and to give further details of the energy levels of particular metals. The main results required are the following:

(i) The zone boundaries are given by

$$\mathfrak{g}_b \cdot \mathbf{k} + \pi \,|\, \mathfrak{g}_b \,|^2 = 0 \quad (V_{\mathfrak{g}} \neq 0), \tag{3·5·1}$$

where $V_{\mathfrak{g}}$ is the $\mathfrak{g}$ Fourier coefficient of the potential energy of a conduction electron due to the lattice field.

(ii) If the lattice is a composite one with basis $0, \mathbf{s}^{(1)}, \mathbf{s}^{(2)}, ..., \mathbf{s}^{(n)}$, and if the potential energy of an electron due to the simple lattice is

$$\sum_{\mathfrak{g}} v_{\mathfrak{g}} e^{2\pi i \mathfrak{g}_b \cdot \mathbf{r}},$$

then the potential energy due to the composite lattice is

$$V = \sum_{\mathfrak{g}} v_{\mathfrak{g}} e^{2\pi i \mathfrak{g}_b \cdot \mathbf{r}} \left[1 + \sum_{t=1}^{n} \exp \left\{ 2\pi i \mathfrak{g}_b \cdot (s_1^{(t)} \mathbf{a}_1 + s_2^{(t)} \mathbf{a}_2 + s_3^{(t)} \mathbf{a}_3) \right\} \right].$$

For the position vector of the tth atom in the basis, relative to the origin, is, by definition of the basis, $s_1^{(t)} \mathbf{a}_1 + s_2^{(t)} \mathbf{a}_2 + s_3^{(t)} \mathbf{a}_3$, and the potential energy due to the n lattices is given by the above expression. A more convenient form is obtained by writing

$$V \equiv \sum_{\mathfrak{g}} V_{\mathfrak{g}} e^{2\pi i \mathfrak{g}_b \cdot \mathbf{r}} = \sum S_{\mathfrak{g}} v_{\mathfrak{g}} e^{2\pi i \mathfrak{g}_b \cdot \mathbf{r}}, \tag{3·5·2}$$

where $$S_g = 1 + \sum_{t=1}^{n} e^{2\pi i g \cdot s^{(t)}}. \qquad (3 \cdot 5 \cdot 3)$$

S_g is called the structure factor.

The discontinuity in the energy at a zone boundary is determined by V_g (and, if the approximation of nearly free electrons holds, it is equal to $2 |V_g|$). If $S_g = 0$, the corresponding zone boundary disappears.

(iii) If we consider a cyclical lattice consisting of $(2G+1)^3$ unit cells, the periodicity conditions which the wave function $e^{i \mathbf{k} \cdot \mathbf{r}} u_{\mathbf{k}}(\mathbf{r})$ must satisfy are

$$e^{i(2G+1)\mathbf{k} \cdot \mathbf{a}_1} = e^{i(2G+1)\mathbf{k} \cdot \mathbf{a}_2} = e^{i(2G+1)\mathbf{k} \cdot \mathbf{a}_3} = 1, \qquad (3 \cdot 5 \cdot 4)$$

and the permissible values of $\mathbf{k} . \mathbf{a}_1$, $\mathbf{k} . \mathbf{a}_2$, $\mathbf{k} . \mathbf{a}_3$ are given by

$$2\pi(0, \pm 1, \pm 2, ..., \pm G)/(2G+1),$$

i.e. there are $(2G+1)^3$ in all. This is equivalent to the statement that the region of $\mathbf{k}$ space defined by $-\pi \leqslant \mathbf{k} . \mathbf{a}_1$, $\mathbf{k} . \mathbf{a}_2$, $\mathbf{k} . \mathbf{a}_3 \leqslant \pi$ contains, in the limit of infinite G, as many electronic states (for one direction of the spin of the electron) as there are unit cells, i.e. one electronic state per unit cell. Now the corners of the parallelepiped whose faces are $\mathbf{k} . \mathbf{a}_1 = \pm \pi$, $\mathbf{k} . \mathbf{a}_2 = \pm \pi$, $\mathbf{k} . \mathbf{a}_3 = \pm \pi$ have the position vectors $\pi(\pm \mathbf{b}_1 \pm \mathbf{b}_2 \pm \mathbf{b}_3)$, and its volume is therefore $8\pi^3$ times the volume of the unit cell of the reciprocal lattice, i.e. $8\pi^3 \mathbf{b}_1 . (\mathbf{b}_2 \times \mathbf{b}_3) = 8\pi^3/\Delta$, where $\Delta = \mathbf{a}_1 . (\mathbf{a}_2 \times \mathbf{a}_3)$ is the volume of the unit cell of the metal. Hence the number of electronic states per unit cell (for one direction of the spin of the electron) associated with any region of $\mathbf{k}$ space is equal to the volume of the region divided by $8\pi^3/\Delta$.

CUBIC STRUCTURES

3·6. We shall consider the cubic structures to be simple cubic lattices with a basis, since this obviates the use of oblique axes. If the side of the cube is a, the reciprocal vectors of the simple cubic lattice are $\mathbf{b}_1 = (1, 0, 0)/a$, $\mathbf{b}_2 = (0, 1, 0)/a$, $\mathbf{b}_3 = (0, 0, 1)/a$, and the boundaries of the Brillouin zones are given by

$$a(g_1 k_1 + g_2 k_2 + g_3 k_3) + \pi(g_1^2 + g_2^2 + g_3^2) = 0. \qquad (3 \cdot 6 \cdot 1)$$

The first zone is the region $-\pi \leqslant a(k_1, k_2, k_3) \leqslant \pi$, and it contains one energy level per unit cell, the levels having a two-fold degeneracy on account of the spin of the electron.

3·61. *Face-centred cubic structures.* The face-centred cubic lattice has the basis $(0, 0, 0)$, $(\frac{1}{2}, \frac{1}{2}, 0)$, $(\frac{1}{2}, 0, \frac{1}{2})$, $(0, \frac{1}{2}, \frac{1}{2})$, and hence

$$S_g = 1 + e^{\pi i(g_1 + g_2)} + e^{\pi i(g_2 + g_3)} + e^{\pi i(g_3 + g_1)}, \qquad (3 \cdot 61 \cdot 1)$$

which is zero for the $(1, 0, 0)$, $(1, 1, 0)$ and similar planes, the first non-vanishing S_g's arising from the $(\pm 1, \pm 1, \pm 1)$ and the $(\pm 2, 0, 0)$, $(0, \pm 2, 0)$,

(0, 0, ± 2) planes. The first Brillouin zone is the truncated octahedron shown in fig. II 8 (iv), p. 38, and its boundaries are given by

$$a(k_1 \pm k_2 \pm k_3) \pm 3\pi = 0, \quad ak_1 = \pm 2\pi, \quad ak_2 = \pm 2\pi, \quad ak_3 = \pm 2\pi. \quad (3·61·2)$$

The volume of the truncated octahedron is $32\pi^3/a^3$, and the first zone can therefore accommodate eight electrons per unit cell. Since there are four atoms per unit cell, the valency electrons of a monatomic metal just fill half the first zone. Now the hexagonal faces of the zone boundary are at a distance $\sqrt{3}\,\pi/a$ from the origin while the square faces are at a distance $2\pi/a$. It is therefore reasonable to suppose that the occupied levels in monatomic metals are approximately spherical, since the radius of the sphere which would contain them is $k_0 = (12/\pi)^{\frac{1}{3}}\,\pi/a = 1·56\pi/a$, and the sphere does not cut the zone boundaries. On the other hand, the sphere with radius k_0 approaches fairly near to the hexagonal faces, and we must therefore expect some distortion of the energy surfaces in those neighbourhoods. Thus although the k_0 energy level is probably spherical over most of its surface there are likely to be distortions, particularly near the (1, 1, 1) and similar directions, and one or both of the principal curvatures may even be negative over small portions of the surface (compare figs. II 10 and II 11).

3·611. *The energy levels of the noble metals.* Krutter (1935) has carried out detailed investigations of the energy levels of copper by the cellular method discussed in § 3·32. The main difference between copper and the alkalis is that atoms of the latter are strictly monovalent whereas copper atoms have a variable valency. This is due to the fact that the $3d$ electrons in copper are not very much more tightly bound than the $4s$ electron, and the electronic configuration $(3d)^9\,(4s)^2$ of the free atom has an energy only about 1·5 eV. higher than that of the configuration $(3d)^{10}\,4s$. Since the excitation energy of a $3d$ electron is of the same order as the metallic binding energy we cannot treat the $3d$ electron shell as acting merely as a force field and we must consider the possibility of a d electron being excited. In the alkalis on the other hand, the core electrons are so tightly bound that they can always be treated as being equivalent to a fixed force field.

In computing the wave functions for metallic copper it is necessary to treat them as linear combinations of $4s$ and $3d$ functions (and, if great accuracy is required, of $4p$ and higher functions as well). For large lattice constants the resulting $4s$ and $3d$ bands are narrow and do not overlap, but for lattice constants of the order of that actually occurring the $4s$ and $3d$ bands overlap considerably as shown in fig. III 12. (The wave functions cannot then be described as pure $4s$ or pure $3d$ wave functions, but it is convenient to label the bands to correspond with the atomic levels to which they degenerate when the lattice constant becomes infinite.) With over-lapping bands of this type it is inconvenient to use the free-electron wave vector to describe the states, and the reduced wave vector is much more suitable. The Brillouin zones which then have to be considered are six truncated octahedra defined by (3·61·2), five relating to the 3d bands and

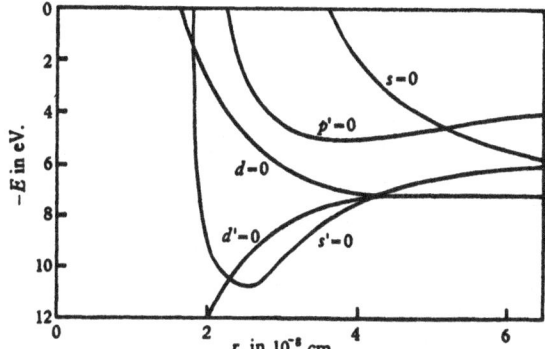

Fig. III 12. The energy bands in copper.

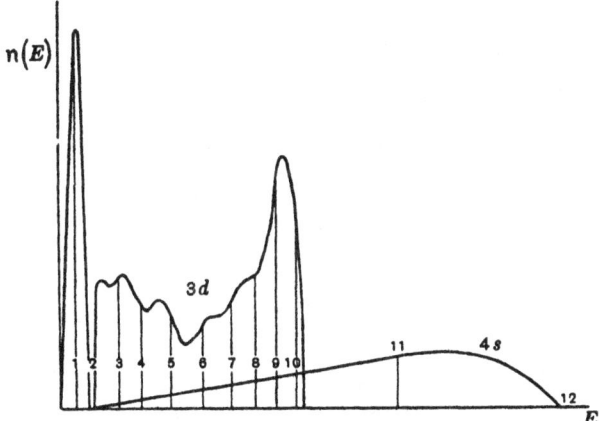

Fig. III 13. The density of levels in the 3d- and 4s-bands.

one relating to the 4s band. The density of states in these bands as a function of the energy is shown in fig. III 13 for the lattice constant which actually occurs, the curve marked 3d referring to the total density of states in all the five 3d bands. The numbers 1 to 12 denote the limits to which the bands would be filled if there were 1 to 12 electrons per atom available to fill the 3d and 4s bands and if each level were occupied by two electrons. (It should be noted that in computing the 4s levels, the 4p wave functions have been neglected so that the 4p band does not appear. This would overlap the 4s band just above the 11-electron line, so that the energy levels shown cannot be used when there are more than 11 electrons present.) The most important feature that the diagram shows is that the 4s band is a broad one and that it is overlapped by a relatively narrow 3d band in which the density of levels is very high. Near the top of the 3d band it is possible to fit the $n(E)$ curve by a parabola. We can then write $n(E) = \text{constant}\,(A - E)^{\frac{1}{2}}$, so that for such energies the 3d band is an inverted band of standard form (see § 2·7). We may also define an effective mass m^* by equating the constant in the above relation to $2\pi\varpi_d(2m^*)^{\frac{3}{2}}/h^3$ (see equation (2·711·1)), where ϖ_d is the weight factor of the 3d band. (In Krutter's calculation the degeneracy of the 3d states is not removed, so that $\varpi_d = 5$.)

In copper there are 11 electrons to be accommodated in the 3d and 4s bands, so that the 3d states are entirely filled and for most purposes need not be considered explicitly. The red colour of the metal is, however, due to the existence of a visible absorption frequency in which a 3d electron is excited into the upper band.

3·612. *The energy levels of the transition metals.* The transition metals, which occur just before the noble metals in the periodic table, are all face-centred cubic and their system of energy levels must be similar to those of the noble metals, though detailed calculations are lacking. The level system of copper, as illustrated in figs. III 12 and III 13, can therefore be used for a qualitative, and even semi-quantitative, discussion of the behaviour of nickel (Slater, 1936). It will be seen from fig. III 13 that there is a striking difference between nickel and copper, in that in nickel the ten 3d electrons do not completely fill the 3d band, but, on account of the overlapping of the bands, are distributed between the 3d and 4s bands. The situation is complicated by the fact that nickel is ferromagnetic and therefore that some states only accommodate one electron and not two, so that the position of the highest occupied level cannot be deduced immediately from the figure. A detailed discussion must be left until the theories of the specific heat, the paramagnetism and the ferromagnetism of metals have been given (Chapters VI and VII), but the result of all these considerations is that the 4s band in nickel contains about 0·6 electron per atom, and that there are a corresponding number of unfilled levels in the 3d band.

According to Krutter's results, the width of the 3d band is about 5·5 eV. Calculations by Fletcher (1952), based upon the approximation of tightly bound electrons, have, however, given a figure of 2·7 eV. for the width of the 3d band in nickel. It would therefore seem that the numerical results should not be considered to have a high degree of accuracy.

Similar considerations apply to palladium and platinum. The number of electrons in the 5s band in palladium seems to be about 0·6 electron per atom, while the number in the 6s band of platinum is less certain but is probably about 0·3 electron per atom.

3·613. *Divalent metals.* The energy levels of metallic calcium have been investigated by Manning and Krutter (1937). The two valency electrons per atom are just sufficient to fill the first Brillouin zone, and the alkaline earths would be insulators if the first and second zones did not overlap. The detailed calculations show that overlap occurs but the degree of overlap is not great; it does not occur in the principal crystallographic directions but in the $(0, 2, 1)$ and similar directions. The density of the occupied energy levels in the two zones is shown in fig. III 14, from which it would appear that the metallic character of calcium is unlikely to be very pronounced.

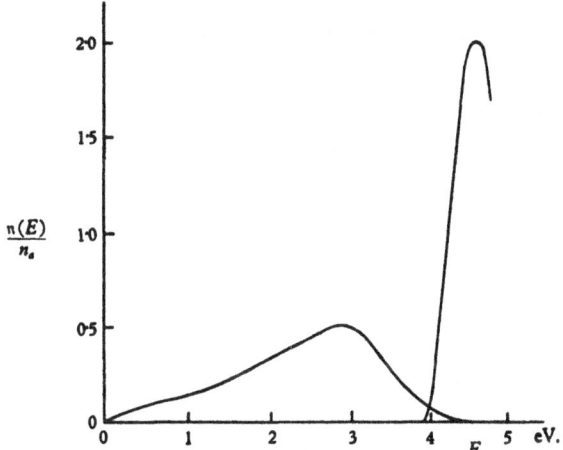

Fig. III 14. The density of levels per atom in calcium.

3·62. *Body-centred cubic structures.* The body-centred cubic lattice has the basis $(0, 0, 0)$, $(\tfrac{1}{2}, \tfrac{1}{2}, \tfrac{1}{2})$ and hence

$$S_{g} = 1 + e^{\pi i(g_1 + g_2 + g_3)}, \qquad (3\cdot62\cdot1)$$

which is zero when the sum $g_1 + g_2 + g_3$ is odd. There are therefore no reflexions from the $(1, 0, 0)$ and similar planes, as is well known, and the first Brillouin zone is bounded by the planes $(\pm 1, \pm 1, 0)$, $(0, \pm 1, \pm 1)$, $(\pm 1, 0, \pm 1)$, which in **k** space have the equations

$$a(k_1 \pm k_2) \pm 2\pi = 0, \quad a(k_2 \pm k_3) \pm 2\pi = 0, \quad a(k_3 \pm k_1) \pm 2\pi = 0. \quad (3\cdot62\cdot2)$$

The zone is the dodecahedron shown in fig. II 8 (ii), p. 38. By mensuration the volume of the dodecahedron is found to be $16\pi^3/a^3$, and the first zone can therefore accommodate four electrons per unit cell of the crystal. But there are two atoms per unit cell, and hence the valency electrons of a monatomic metal just fill half the first zone.

3·621. *The energy levels of the alkalis.* The most important body-centred metals are the alkalis, and for them it is reasonable to assume that the occupied energy levels are approximately spherical. The faces of the dodecahedron forming the boundary of the Brillouin zone are at a distance $\sqrt{2}\,\pi/a$ from the origin, whereas the sphere of volume $8\pi^3/a^3$, which can accommodate two electrons per unit cell, has radius $k_0 = (6/\pi)^{\frac{1}{3}}\,\pi/a = 1\cdot24\pi/a$, and k_0 is considerably less than $\sqrt{2}\,\pi/a$. The occupied energy levels in the alkalis do not approach the zone boundaries as closely as do the levels in the face-centred noble metals, and the treatment of the valency electrons as perfectly free is a better approximation for the alkalis than for any other metals. The energy levels of sodium, calculated by Slater's method, have already been discussed in detail, and the results for the other alkalis are so similar as to require no further comment. The original papers can be consulted for the details of the calculations (e.g. Millman, 1935; Seitz, 1935).

3·622. *Multivalent metals.* Of the multivalent body-centred cubic metals the one for which the most accurate calculations have been carried out is tungsten (Manning and Chodorow, 1939), which is more properly classed as a transition metal in which the electrons to be considered are four electrons in $5d$ states and two in $6s$ states. If we use the reduced wave-vector to enumerate the states, the Brillouin zones are six of the polyhedra shown in fig. II 8 (ii), five being associated with the d levels and one with the s levels, although the energy levels overlap so considerably that such a labelling is almost meaningless. The density of states in the various zones and the sum of the densities is shown in fig. III 15. Since the $5d$ shell is not complete in tungsten many of the $5d$ levels and most of the $6s$ levels are unfilled, there being only about 0·1 electron per atom in the s levels. As a consequence of this the electronic specific heat should be unusually large (see § 6·3).

Most of the other metals having a body-centred structure are transition elements with electronic configurations similar to that of tungsten. We may therefore expect the energy levels in these metals to be very similar to those given in fig. III 15. In particular, the energy-level system for tantalum, which has one less electron than tungsten, must be almost identical with that of tungsten, the only difference being that less of the energy levels are filled, as shown in the figure. The energy levels of body-centred barium, on the other hand, are likely to be similar to those of face-centred calcium, the energy discontinuities at the boundary of the first Brillouin zone being such that a small number of electrons overlap into the second zone.

The most important remaining body-centred metal is α-iron, the variety which is stable at ordinary temperatures. A detailed investigation of its band structure has been carried out by Manning (1943), the density of states as found by him being shown in fig. III 16. For comparison, the density of states calculated by Greene and Manning (1943) for γ-iron, the face-centred variety which is stable between 906 and 1400° C., is shown in fig. III 17.

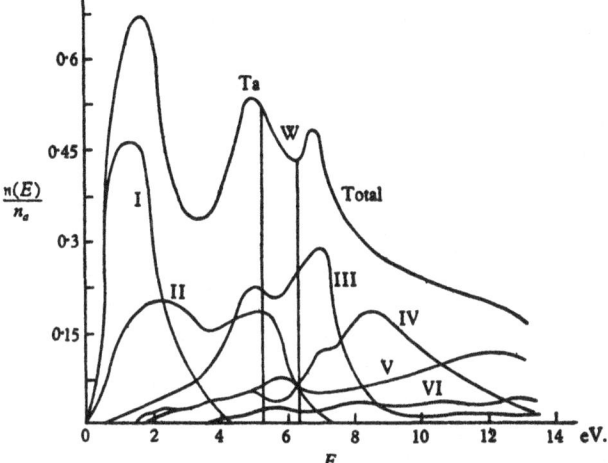

Fig. III 15. The density of levels per atom in tantalum and tungsten.

3·63. The diamond lattice consists of two face-centred cubic lattices with basis $(0, 0, 0)$, $(\frac{1}{4}, \frac{1}{4}, \frac{1}{4})$, the unit cell containing eight atoms. Therefore S_g is equal to $(3\cdot61\cdot1)$ multiplied by $1 + e^{\frac{1}{2}\pi i(g_1 + g_2 + g_3)}$. In this case the important zone is the dodecahedron bounded by planes such as

$$a(k_1 \pm k_2) \pm 4\pi = 0,$$

and its volume is sixteen times that of the fundamental cube, so that it can accommodate 32 electrons per cell, which is the number of valency electrons present in the elements with this structure. In diamond, this zone must be separated from the next by large regions of forbidden energies since diamond is an insulator. This is borne out by the calculations of Kimball (1935) and of Hund and Mrowka (1935), which are much more informative but much more complex than the qualitative considerations based upon the structure of the Brillouin zones. It should be borne in mind that the arrangement of the Brillouin zones is not unique (see p. 36), and they are only of value when it is known that the electrons behave as if they were nearly free or when information concerning the detailed structure of the energy bands is completely lacking.

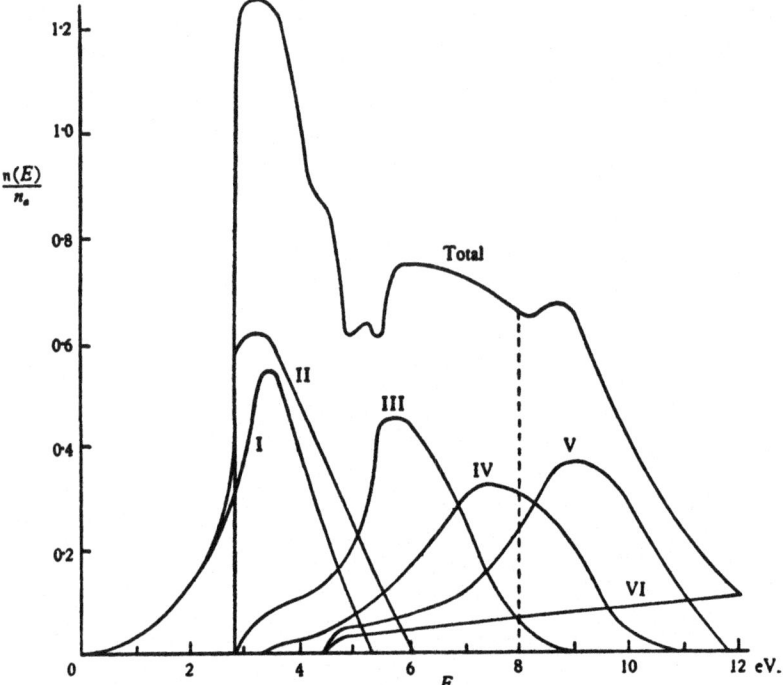

Fig. III 16. The density of levels per atom in α-iron.

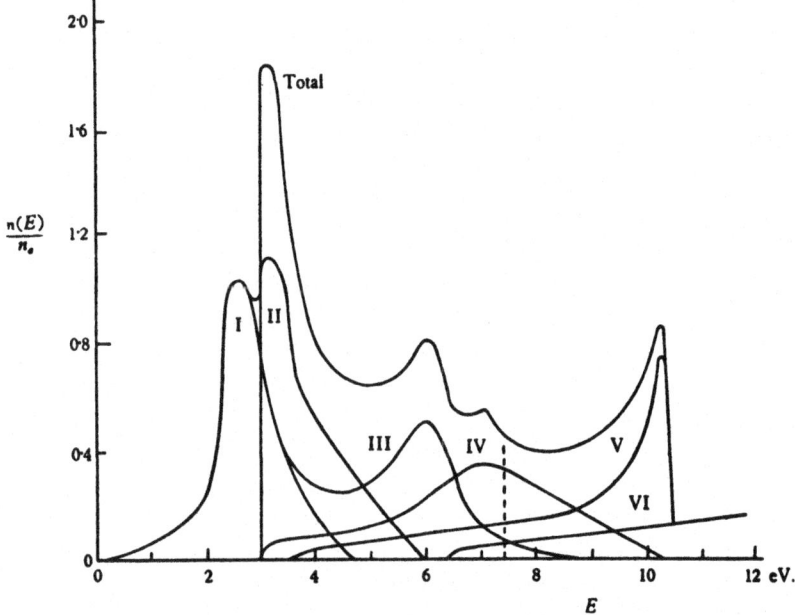

Fig. III 17. The density of levels per atom in γ-iron.

HEXAGONAL STRUCTURES

3·7. The expressions given in §3·22, equations (3·22·2), show that for a hexagonal lattice

$$\mathbf{g}_b = \left(\frac{g_1 + g_2}{a}, \frac{-g_1 + g_2}{\sqrt{3}\,a}, \frac{g_3}{c} \right) \tag{3·7·1}$$

when rectangular coordinates are used, so that the equations

$$\mathbf{g}_b \cdot \mathbf{k} + \pi \,|\, \mathbf{g}_b \,|^2 = 0$$

of the zone boundaries are

$$\frac{(g_1 + g_2)\,k_1}{a} + \frac{(-g_1 + g_2)\,k_2}{\sqrt{3}a} + \frac{g_3 k_3}{c} + \pi \left\{ \frac{4(g_1^2 + g_1 g_2 + g_2^2)}{3a^2} + \frac{g_3^2}{c^2} \right\} = 0. \tag{3·7·2}$$

The volume of the unit cell of the crystal is $\frac{1}{2}\sqrt{3}\,a^2 c$, and, by §3·5(iii), the number of (doubly degenerate) states per unit cell is equal to the volume of $\mathbf{k}$ space divided by $16\pi^3/(\sqrt{3}\,a^2 c)$.

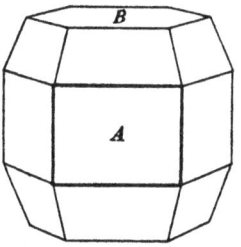

Fig. III 18. The first Brillouin zone of a hexagonal close-packed lattice.

3·71. Since the close-packed hexagonal lattice has the basis $(0, 0, 0)$, $(\frac{1}{3}, \frac{2}{3}, \frac{1}{2})$, the structure factor S_g is given by

$$S_g = 1 + e^{\pi i (\frac{2}{3} g_1 + \frac{4}{3} g_2 + g_3)}. \tag{3·71·1}$$

The first energy zone is therefore bounded by the planes

$$(\pm 1, 0, 0), \ (0, \pm 1, 0), \ (1, -1, 0), \ (-1, 1, 0), \ (\pm 1, 0, \pm 1),$$

$$(0, \pm 1, \pm 1), \ (0, 0, \pm 2), \ (1, -1, \pm 1), \ (-1, 1, \pm 1),$$

the equations of which are

$$\left. \begin{array}{ll} \pm\dfrac{k_1}{a} \pm \dfrac{k_2}{\sqrt{3}\,a} = \dfrac{4\pi}{3a^2}, & \dfrac{k_2}{a} = \pm\dfrac{2\pi}{\sqrt{3}a^2}, \\[3mm] \pm\dfrac{k_1}{a} \pm \dfrac{k_2}{\sqrt{3}a} \pm \dfrac{k_3}{c} = \left(\dfrac{4}{3a^2} + \dfrac{1}{c^2}\right)\pi, \quad \dfrac{k_3}{c} = \pm\dfrac{2\pi}{c^2}, \quad \pm\dfrac{2k_2}{\sqrt{3}a} \pm \dfrac{k_3}{c} = \left(\dfrac{4}{3a^2} + \dfrac{1}{c^2}\right)\pi. \end{array} \right\}$$

$$\tag{3·71·2}$$

(The zone is shown in fig. III 18.) The half of the zone above the $k_1 k_2$ plane consists of a hexagonal prism surmounted by a truncated hexagonal pyramid. The base of the hexagonal prism has side $\frac{4}{3}\pi/a$, and therefore its area is $8\pi^2/(\sqrt{3}\,a^2)$, while its height is π/c. The pyramid has total height $\frac{4}{3}\pi c/a$, and it is truncated at a distance $\dfrac{4\,\pi c}{3\,a^2}\left(1 - \dfrac{3a^2}{4c^2}\right)$ from the vertex.

The volume of the truncated pyramid is therefore

$$\frac{1}{3}\frac{8\pi^2}{\sqrt{3}\,a^2}\frac{4\pi c}{3a^2}\left\{1-\left(1-\frac{3a^2}{4c^2}\right)^3\right\}=\frac{8\pi^3}{\sqrt{3}\,a^2c}\left(1-\frac{3}{4}\frac{a^2}{c^2}+\frac{3}{16}\frac{a^4}{c^4}\right),$$

and the volume of the whole zone is

$$\frac{16\pi^3}{\sqrt{3}\,a^2c}\left(2-\frac{3}{4}\frac{a^2}{c^2}+\frac{3}{16}\frac{a^4}{c^4}\right).$$

Now there are two atoms per unit cell, and so the zone can accommodate

$$2-\frac{3}{4}\frac{a^2}{c^2}+\frac{3}{16}\frac{a^4}{c^4} \qquad (3\cdot71\cdot3)$$

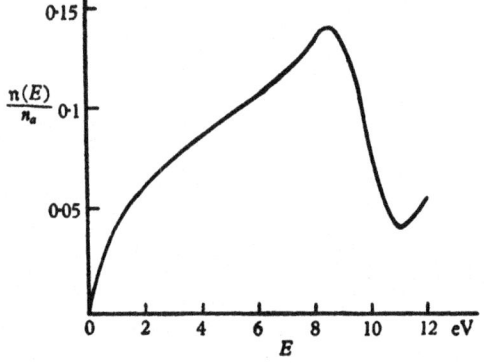

Fig. III 19. The density of levels per atom in beryllium.

electrons per atom. There are, however, other zones whose boundaries lie near those of the first. For example, the hexagonal prism formed by the planes $(\pm1,0,0)$, $(0,\pm1,0)$, $(1,-1,0)$, $(-1,1,0)$, $(0,0,\pm2)$ can accommodate exactly two electrons per atom.

The axial ratios and the numbers of levels per atom in the first Brillouin zone of the divalent hexagonal metals are shown in Table III 11. It will be seen that in all cases the valency electrons must overlap into the second Brillouin zone.

Table III 11. *The number of energy levels in the first zone of hexagonal metals*

	Be	Mg	Zn	Cd
Axial ratio c/a	1·58	1·62	1·86	1·89
Number of levels per atom in the first zone	1·73	1·74	1·80	1·81

The only one of the above metals for which detailed calculations exist is beryllium, the energy levels having been computed by Herring and Hill (1940). Their results for the density of states are shown in fig. III 19. It will be noticed that $n(E)$ reaches a maximum at about 8 eV. and then decreases rapidly, which indicates that the energy discontinuities at the zone boundaries are large and that the overlapping of the zones is not very pronounced. If we fit the lower part of the $n(E)$ curve by a parabola and determine the effective mass m^* of the electrons, it is found that $m^* = 1·6$ approximately.

3·72. *The graphite structure.* The graphite lattice has the basis $(0, 0, 0)$, $(0, 0, \frac{1}{2})$, $(\frac{1}{3}, \frac{2}{3}, u)$, $(\frac{2}{3}, \frac{1}{3}, u + \frac{1}{2})$, and the structure factor is therefore

$$S_g = 1 + e^{\pi i \sigma_3} + e^{2\pi i u \sigma_3} (e^{\pi i (\frac{2}{3}\sigma_1 + \frac{4}{3}\sigma_2)} + e^{\pi i (\frac{4}{3}\sigma_1 + \frac{2}{3}\sigma_2 + \sigma_3)}). \tag{3·72·1}$$

The energy zone bounded by the planes $(\pm 2, 0, 0)$, $(0, \pm 2, 0)$, $(2, -2, 0)$, $(-2, 2, 0)$, $(0, 0, \pm 2)$ is a hexagonal prism of cross-section $32\pi^2/(\sqrt{3}\,a^2)$ and height $4\pi/c$. It can therefore just accommodate all the sixteen valency electrons belonging to the atoms in the unit cell. The zone is very compressed along the hexagonal axis, the ratio of the height to the greatest cross-sectional diameter being $\frac{3}{2}a/c = 0·54$. As a consequence the electrons behave like very tightly bound electrons for directions of $\mathbf{k}$ which are not perpendicular to the hexagonal axis, or nearly so, and it is only the forces between atoms in the same sheet that can be considered to be of a metallic nature, while the forces between atoms in different sheets must be of the van der Waals type.

3·73. *The selenium structure.* Since the selenium structure has the basis $(u, 0, 0)$, $(0, u, \frac{1}{3})$, $(-u, -u, \frac{2}{3})$, the structure factor is

$$S_g = e^{2\pi i u \sigma_1} + e^{2\pi i (u \sigma_2 + \frac{1}{3}\sigma_3)} + e^{2\pi i (-u\sigma_1 - u\sigma_2 + \frac{2}{3}\sigma_3)}. \tag{3·73·1}$$

Since selenium and tellurium are poor conductors we would expect there to be a zone capable of holding all the eighteen valency electrons in the unit cell. This zone is a prism bounded by the planes $(2, -1, 0)$, $(-2, 1, 0)$, $(-1, 2, 0)$, $(1, -2, 0)$, $(1, -1, 0)$, $(-1, 1, 0)$, $(0, 0, \pm 3)$. There are, however, other zones containing approximately six electrons per atom which intersect the above zone, and without calculating the energy discontinuities it is impossible to predict which are the important zones. We should therefore expect the energy discontinuities not to be very large, and this is borne out by the fact that selenium and tellurium seem to be intrinsic semiconductors.

Rhombohedral Structures

3·8. If we use the rectangular axes defined by equations (3·23·1) and (3·23·2) the equations of the zone boundaries are

$$(2g_1 - g_2 - g_3)\, k_1 + \sqrt{3}\,(g_2 - g_3)\, k_2 + (g_1 + g_2 + g_3)\,\sqrt{2}\left(\frac{1-\cos\alpha}{1+2\cos\alpha}\right)^{\frac{1}{2}} k_3$$

$$+\frac{\pi\sqrt{6}\,(1+\cos\alpha)\,(g_1^2 + g_2^2 + g_3^2) - 2\cos\alpha(g_1 g_2 + g_2 g_3 + g_3 g_1)}{a\quad(1+2\cos\alpha)\,(1-\cos\alpha)^{\frac{1}{2}}} = 0. \quad (3·8·1)$$

Also, the number of (doubly degenerate) states per unit cell is equal to the volume of **k** space divided by $8\pi^3/\{a^3(1-\cos\alpha)\,(1+2\cos\alpha)^{\frac{1}{2}}\}$.

The main interest in the zones of rhombohedral structures lies in their effect on the properties of the bismuth group of semi-metals. To account for their abnormal properties it is necessary that there should be a zone capable of accommodating all or nearly all the five valency electrons. Such a zone exists and it can accommodate exactly five electrons per atom. This is most easily shown by using the rhombohedral axes whose mutual inclination is nearly 60°. The bismuth structure then consists of two simple rhombohedral lattices with the basis $(-u, -u, -u)$ and (u, u, u) where u is nearly $\frac{1}{4}$, the structure factor being

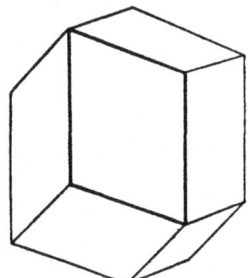

Fig. III 20. The Brillouin zone for the bismuth structure.

$$S_\xi = e^{2\pi i u(g_1 + g_2 + g_3)} + e^{-2\pi i u(g_1 + g_2 + g_3)}. \quad (3·8·2)$$

With these axes the requisite Brillouin zone is bounded by the six planes such as $(1, -1, 0)$ and the six planes such as $(2, 2, 1)$, $(-2, -2, -1)$. It consists of a hexagonal prism capped at both ends by a partial pyramid (see fig. III 20). The volume can be found by mensuration in many different ways, one method being the following. Consider the half-zone for which $k_3 > 0$ and consider it to be split up into six equal triangular prisms. The base of one such prism has coordinates $(0, 0, 0)$, $(\frac{1}{3}p, 0, 0)$, $(\frac{1}{2}p, \frac{1}{6}\sqrt{3}\,p, 0)$, where $p = 2\sqrt{6}\,\pi/\{(1-\cos\alpha)^{\frac{1}{2}}a\}$, and its opposite face has coordinates

$$\{0, 0, (27 - 21\cos\alpha)\,r\},\quad \{\tfrac{1}{3}p, 0, (25 - 25\cos\alpha)\,r\},\quad \{\tfrac{1}{2}p, \tfrac{1}{6}\sqrt{3}\,p, (23 - 29\cos\alpha)\,r\},$$

where $\quad r = \pi/\{5\sqrt{3}\,(1-\cos\alpha)\,(1+2\cos\alpha)^{\frac{1}{2}}a\}.$

Elementary coordinate geometry gives the volume of the prism, after some simplification, as $10\pi^3/\{3a^3(1-\cos\alpha)\,(1+2\cos\alpha)^{\frac{1}{2}}\}$, so that the whole zone can accommodate exactly ten electrons per unit cell.

The energy discontinuities round the zone must be sufficiently large to make the number of electrons in the next zone, and the number of vacant levels in the primary zone, very small, of the order of 10^{-3} per atom (see § 8·56), but quantitative calculations of the energy discontinuities are lacking. It is, however, worth noting that, whereas $S_{1,-1,0} = 2$ whatever the value of u, $|S_{221}| = 2|\cos 10\pi u|$ and S_{221} is zero if $u = \frac{1}{4}$. The values of u for As, Sb and Bi are 0·226, 0·233 and 0·237 respectively, the corresponding values of $|S_{221}|$ being 1·369, 1·018 and 0·794. The energy discontinuities in the $(2, 2, 1)$ and similar directions therefore decrease with increasing atomic number, and the metallic properties correspondingly increase.

THE X-RAY SPECTRA OF METALS

3·9. Most of the physical properties of metals are determined either by the energy-level system as a whole (for example, the cohesion of metals) or by the highest occupied level. The X-ray spectra, however, are capable of giving direct evidence about the individual energy levels in the conduction and higher bands.

If one of the core electrons is removed from a metal by, for example, electron impact in an X-ray tube, one of the conduction electrons can make a transition into the vacant level, emitting an X-ray, and since the conduction electrons have a considerable energy spread (of the order of 3–10 eV.), there will be a continuous X-ray emission spectrum (Houston, 1931). This is normally in the soft X-ray region of 100 Å. or more. In the X-ray absorption spectra, on the other hand, a core electron must be excited into one of the vacant levels in the conduction or higher bands. We therefore see that, associated with any given inner $K, L, \ldots$ level, there is an X-ray emission spectrum whose width corresponds to the energy spread of the conduction electrons, and that, if thermal effects are neglected, there is a sharp edge to the emission band on the short-wave side, the position of the edge corresponding to the highest occupied energy level. The X-ray absorption spectrum, on the other hand, has a sharp edge on the long-wave side, coinciding with the short-wave edge of the emission band, and the spectrum extends indefinitely into the short-wave region.

The breadths of the emission bands for a number of metals are given in Table III 12, certain corrections being made to the observed breadths which are mentioned below. If we assume that all the valency electrons are free with effective mass m^*, we can determine m^* by equating the band width to the Fermi energy ζ, that is to $(h^2/8m^*)(3n/\pi)^{\frac{2}{3}}$. The values of m^* so derived are also shown in the table, and it will be seen that in all cases m^*/m is very near 1. When m^* is derived from other physical properties, this is not usually true except for the alkalis, but in most phenomena m^* refers to the effective mass at the energy ζ, whereas here it refers to the average effective mass over all the valency electrons and we would therefore not expect it to be widely different from m.

Table III 12. *Widths of X-ray emission bands*

	Reduced observed width in eV.	Number of valency electrons per c.c. × 10⁻²²	m^*/m
Li	3·7	4·8	1·2
Na	2·5	2·6	1·2
Cu	7·0	8·5	1
K	1·9	1·3	1·1
Be	13·8	24	1
Mg	6·2	8·7	1·1
Ca	3·0	4·6	1·4
Zn	11	13·1	0·85
Al	11·8	18·5	0·97

The mean breadths of the bands of the ferromagnetic metals, and of those which immediately precede them in the periodic table, have been measured by Skinner (quoted by Kingston, 1951). They are as follows: Ti, 5·6 eV.; Va, 5·6 eV.; Cr, 6·3 eV.; Mn, 6·0 eV.; Fe, 4·4 eV.; Co, 5·8 eV.; Ni, 4·7 eV. In all these cases the electrons concerned are the $3d$ electrons, and the calculated densities of states are of the general type given in fig. III 13. Now the width of the occupied levels in nickel, based upon Krutter's calculations, is about 5·5 eV., and for the preceding metals it must be less, so that the observed widths are usually greater, and in some cases considerably greater, than those calculated.

3·91. Measurements of the emission and absorption intensities would give direct information about the density of states were it not for the fact that the transition probability varies considerably with the energy and depends upon the structure of the energy band (Jones, Mott and Skinner, 1934). It is therefore necessary to calculate the dipole moment determining the transition, averaged over all directions and polarizations, or rather the quantity

$$\frac{1}{4\pi^3}\int \sum_i |(0\,|\,\partial/\partial x_i\,|\,\mathbf{k})|^2\,d\mathbf{k} = \frac{1}{4\pi^3}\int \sum_i \left|\int \psi_0^* \frac{\partial\psi_\mathbf{k}}{\partial x_i}\,d\mathbf{r}\right|^2 d\mathbf{k}, \qquad (3\cdot91\cdot1)$$

where ψ_0 is the wave function of the bound state (an atomic $s, p, d, \ldots$ function) and $\psi_\mathbf{k}$ is the wave function of the conduction state (an occupied state for emission, and a vacant state for absorption). The integration with respect to $\mathbf{k}$ must be taken over those states which satisfy the conservation of energy, i.e. over the states with energies lying between E and $E + dE$, where the frequency of the X-ray is given by $h\nu = E_0 + E$, E_0 being the depth of the bound state below the conduction band. The transition probability is therefore proportional to

$$\text{Av} \sum_i |(0 \,|\, \partial/\partial x_i \,|\, \mathbf{k})|^2 \mathfrak{n}(E), \qquad (3\text{·}91\text{·}2)$$

where the average has to be taken over all directions of $\mathbf{k}$; and absorption and emission experiments do not determine $\mathfrak{n}(E)$ directly but only the quantity (3·91·2). Since our knowledge of the actual wave functions in the conduction and higher bands is meagre, the separation of $\mathfrak{n}(E)$ from (3·91·2) presents considerable difficulties in practice, and the details will not be discussed here. We shall only outline the most obvious points, and refer the reader to the review article by Skinner (1939) for further information and for a list of references, especially those relating to the experimental data.

In the first place, the optical selection rules make the forms of the K and L emission spectra very different near the threshold. There are three L spectra, L_1, L_2 and L_3, in the first of which the inner state is $2s$, while in the other two it is $2p$, L_2 and L_3 only differing in that they are associated with different directions of the spin of the electron (L_2 is associated with the state $^2P_{\frac{1}{2}}$ and L_3 with $^2P_{\frac{3}{2}}$). The L_1 spectra are not observed experimentally in the soft X-ray region, presumably because of a transition between the core electrons, and the L_2, L_3 spectra are nearly coincident and can be treated as identical for our present purposes. The matrix element associated with the K spectrum is determined by an integral of the type

$$\int e^{-\lambda r} \frac{\partial \psi_{\mathbf{k}}}{\partial x} d\mathbf{r}, \qquad (3\text{·}91\text{·}3)$$

whereas the integral determining the L spectrum is of the type

$$\int x \, e^{-\lambda r} \frac{\partial \psi_{\mathbf{k}}}{\partial x} d\mathbf{r}, \qquad (3\text{·}91\text{·}4)$$

the wave function of the inner level being an atomic s or p function uninfluenced by the crystal forces. It is sufficient to assume that $\psi_{\mathbf{k}} = e^{i\mathbf{k}\cdot\mathbf{r}}$, and the integrals are then easily evaluated by using polar coordinates. For small values of k, the matrix element (3·91·3) is proportional to k, while the matrix element (3·91·4) is independent of k. We therefore have

$$\text{Av} \sum_i |(K \,|\, \partial/\partial x_i \,|\, \mathbf{k})|^2 \mathfrak{n}(E) \propto E^{\frac{3}{2}}, \quad \text{Av} \sum_i |(L \,|\, \partial/\partial x_i \,|\, \mathbf{k})|^2 \mathfrak{n}(E) \propto E^{\frac{1}{2}},$$

$$(3\text{·}91\text{·}5)$$

for small values of E, so that the energy dependence of the emission intensity is determined by the nature of the X-ray level (O'Bryan and Skinner, 1934).

In the second place the optical selection rules enable us in theory to split up $\mathfrak{n}(E)$ into two parts. In the method used by Slater, and outlined in

§ 3·32, for calculating the energy levels of the conduction band, the wave function of a conduction electron is a linear combination of the form

$$\psi = a_s\psi_s + a_p\psi_p + a_d\psi_d + \cdots,$$

where only the a's depend upon $\mathbf{k}$. Now

$$(K \mid \partial/\partial x_i \mid \mathbf{k}) = a_p(\mathbf{k})\,(\psi_K \mid \partial/\partial x_i \mid \psi_p),$$

$$(L \mid \partial/\partial x_i \mid \mathbf{k}) = a_s(\mathbf{k})\,(\psi_L \mid \partial/\partial x_i \mid \psi_s) + a_d(\mathbf{k})\,(\psi_L \mid \partial/\partial x_i \mid \psi_d),$$

and $\quad (\psi_K \mid \partial/\partial x_i \mid \psi_p), \quad (\psi_L \mid \partial/\partial x_i \mid \psi_s), \quad (\psi_L \mid \partial/\partial x_i \mid \psi_d),$

being matrix elements between atomic levels, are all of the same order of magnitude. Therefore any major differences in the transition probabilities in the K and L spectra are determined by $a_s(\mathbf{k})$, $a_p(\mathbf{k})$ and $a_d(\mathbf{k})$.

Hence a_p determines the transition probability to an atomic s level, and a_s and a_d determine the transition probability to an atomic p level. We can write formally

$$\mathfrak{n}(E) = \mathfrak{n}_s(E) + \mathfrak{n}_p(E) + \mathfrak{n}_d(E) + \cdots,$$

where the quantities $\mathfrak{n}_s(E)$, $\mathfrak{n}_p(E)$, $\mathfrak{n}_d(E)$, ... are the partial densities of states for the energy levels of the conduction electrons which have $s, p, d, \ldots$ characteristics inside the atoms; they, like the transition probabilities, are determined by $\mid a_s(\mathbf{k}) \mid^2$, $\mid a_p(\mathbf{k}) \mid^2$, $\mid a_d(\mathbf{k}) \mid^2$, ... We can therefore obtain $\mathfrak{n}_p(E)$ from the K spectra, and $\mathfrak{n}_s(E) + \mathfrak{n}_d(E)$ from the L spectra.

In practice it is not easy to determine the shape of the emission bands near the long-wave limits, and the observed curves show decided tails. It is thought that this is caused by the broadening of the valency levels due to an Auger or similar effect, and it is necessary to correct the observed band widths and shapes by methods which are discussed by Skinner (1940). The widths obtained in this way are those given in Table III 12 under the heading 'reduced widths'.

A few typical emission spectra for metals are shown in fig. III 21, and it will be seen that the variation (3·91·5) is roughly obeyed, and that the shapes of the K and L spectra are very different. The influence of the filling of the Brillouin zones can also be seen, especially in the L spectra of magnesium and aluminium. It will be noted that there are indications in the sodium spectrum that the electrons overlap into the $3p$ band. If this is so, the energy levels of sodium must differ from those for free electrons more radically than is suggested by the other properties of the metal. Similar conclusions hold for the absorption spectra, and when the results of both spectra are amalgamated we obtain the $\mathfrak{n}(E)$ curves shown in fig. III 22.

The emission spectra for graphite, diamond and silicon are shown in fig. III 23, and it will be seen that they are quite different from those for metals. The general shape of the curves is exactly that to be expected from completely filled bands, and is convincing evidence that the energy levels of insulators and metals are very similar and that the essential difference between the two is that in insulators the electrons form a closed group whereas in metals they do not.

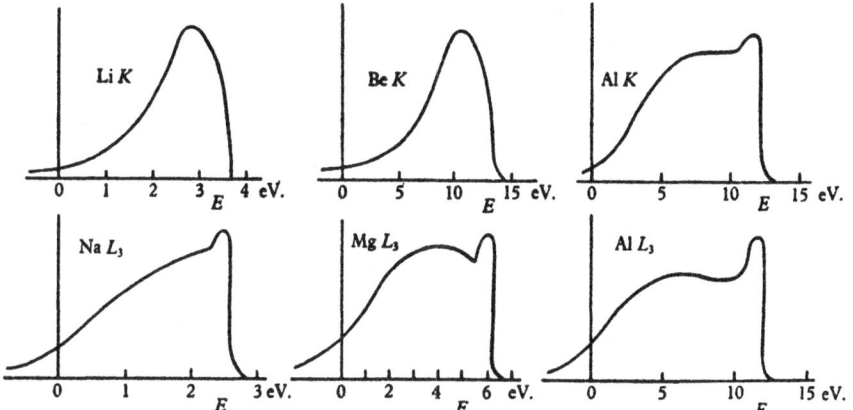

Fig. III 21. The K and L emission spectra of some typical metals. The origin is placed at what is considered to be the true long-wave limit.

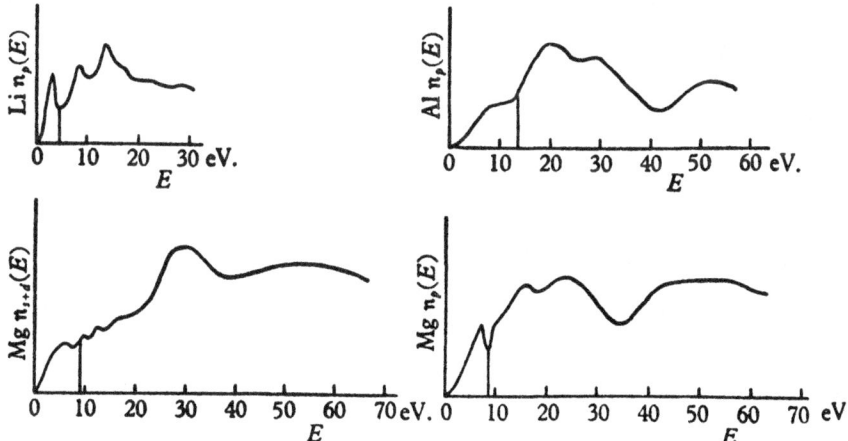

Fig. III 22. Experimental curves for the partial densities of levels.

3·92. For sufficiently hard X-rays (with wave-lengths of the order of 1 to 2 Å.) the core electrons must be excited into fairly high conducting states. It is found (Coster and Veldkamp, 1931) that there are fluctuations in the absorption coefficient even for frequencies well above the threshold (fig. III 24). It was first pointed out by Kronig (1932) that this is a direct confirmation of the fact that the positions of the discontinuities in the energy spectrum of a periodic lattice have no upper limit. It is possible to calculate the frequencies at which the discontinuities occur and their shape, if we assume that the electrons are nearly free, and this should be a good approxi-

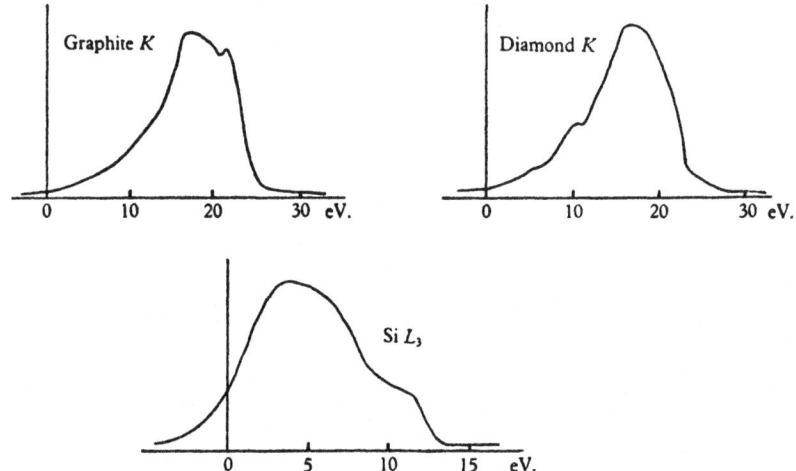

Fig. III 23. The K and L emission spectrum of some typical non-metals. The origin is placed at what is considered to be the true long-wave limit.

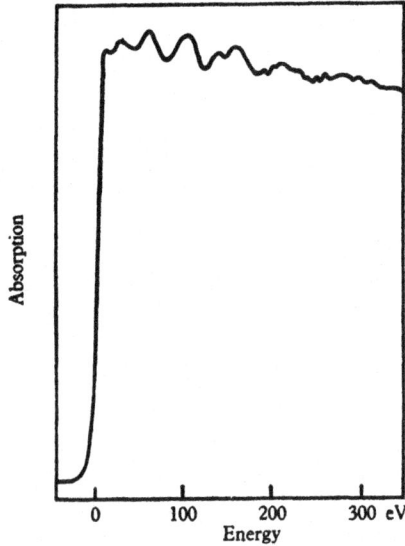

Fig. III 24. The fine structure of the K absorption spectrum of copper.

mation for sufficiently high energies. The boundaries of the Brillouin zones are, however, so numerous in the high-energy region that the result is not very illuminating, and it is difficult to obtain results of decisive theoretical importance.

REFERENCES

Bardeen, J. (1936). Electron exchange in the theory of metals. *Phys. Rev.* 50, 1098.

Bender, O. (1939). Elasticity measurements on single crystals of alkali metals at low temperatures. *Ann. Phys., Lpz.* (5), 34, 359.

Coster, D. and Veldkamp, J. (1931). The absorption coefficient of X-rays in the neighbourhood of the K absorption edge of copper and zinc. *Z. Phys.* 70, 306.

Dehlinger, U. (1935). Gitteraufbau metallischer Systeme. *Handbuch der Metallphysik*, 1 (Leipzig).

Fletcher, G. C. (1952). Density of states curve for the $3d$ electrons in nickel. *Proc. Phys. Soc.* A, 65, 192.

Fuchs, K. (1935). The cohesive forces of metallic copper. *Proc. Roy. Soc.* A, 151, 585.

Fuchs, K. (1936a). The elastic constants of monovalent metals. *Proc. Roy. Soc.* A, 153, 622.

Fuchs, K. (1936b). The elastic constants and specific heats of the alkali metals. *Proc. Roy. Soc.* A, 157, 444.

Greene, J. B. and Manning, M. F. (1943). Electronic energy bands in face-centred iron. *Phys. Rev.* 63, 203.

Herring, C. and Hill, A. G. (1940). The constitution of metallic beryllium. *Phys. Rev.* 58, 132.

Houston, W. V. (1931). The structure of soft X-ray lines. *Phys. Rev.* 38, 1797.

Howarth, D. J. and Jones, H. (1952). The cellular method of determining electronic wave functions and eigenvalues in crystals, with applications to sodium. *Proc. Phys. Soc.* A, 65, 355.

Hume-Rothery, W. (1944). *The structure of metals and alloys*. Part III (London).

Hund, F. and Mrowka, B. (1935). The electronic states in a crystal lattice. *Ber. sächs. Akad. Wiss.* 87, 185, 325.

Jones, H., Mott, N. F. and Skinner, H. W. B. (1934). The theory of the form of the X-ray emission bands of metals. *Phys. Rev.* 45, 379.

Kimball, G. E. (1935). The electronic structure of diamond. *J. Chem. Phys.* 3, 560.

Kingston, R. H. (1951). Spectroscopy of the solid state: potassium and calcium. *Phys. Rev.* 84, 944.

Kohn, W. (1952). Variational methods for periodic lattices. *Phys. Rev.* 87, 472.

Kronig, R. de L. (1932). The fine structure of X-ray absorption spectra. *Z. Phys.* 75, 191.

Krutter, H. M. (1935). Energy bands in copper. *Phys. Rev.* 48, 664.

Manning, M. F. (1943). Electronic energy bands in body-centred iron. *Phys. Rev.* 63, 190.

Manning, M. F. and Chodorow, M. I. (1939). Electronic energy bands in metallic tungsten. *Phys. Rev.* 56, 787.

Manning, M. F. and Krutter, H. M. (1937). Electronic energy bands in metallic calcium. *Phys. Rev.* 51, 761.

Millman, J. (1935). Energy bands in metallic lithium. *Phys. Rev.* 47, 286.

O'Bryan, H. M. and Skinner, H. W. B. (1934). Characteristic X-rays from metals in the extreme ultra-violet. *Phys. Rev.* 45, 370.

Seitz, F. (1935). The constitution of metallic lithium. *Phys. Rev.* 47, 400.

Skinner, H. W. B. (1939). The soft X-ray spectroscopy of the solid state. *Rep. Progr. Phys.* 5, 257.

Skinner, H. W. B. (1940). The soft X-ray spectroscopy of solids. *Philos. Trans.* A, 239, 95.

Slater, J. C. (1934). Electronic energy bands in metals. *Phys. Rev.* **45**, 794.

Slater, J. C. (1936). The ferromagnetism of nickel. *Phys. Rev.* **49**, 537.

Voigt, W. (1928). *Lehrbuch der Kristallphysik* (Leipzig).

Wigner, E. (1934). The interaction of electrons in metals. *Phys. Rev.* **46**, 1002.

Wigner, E. and Seitz, F. (1933). The constitution of metallic sodium. *Phys. Rev.* **43**, 804.

Wigner, E. and Seitz, F. (1934). The constitution of metallic sodium. II. *Phys. Rev.* **46**, 507.

Wohlfarth, E. P. (1950). The influence of exchange and correlation forces on the specific heat of free electrons in metals. *Phil. Mag.* **41**, 534.

Chapter IV

THE STRUCTURE OF ALLOYS

ALLOYS OF A NON-METALLIC NATURE

4·1. When two metals are melted together they may or may not mix, and, if they do, it does not necessarily follow that, on cooling, a homogeneous solid solution will be obtained. The equilibrium diagrams giving the possible phases for all temperatures and compositions of binary mixtures of metals are of such varied types that only a brief account of the simpler facts can be given here.

Many alloys of the true metals with the semi-metals of groups 4, 5 and 6 can be classed as compounds, since they only occur for one definite composition, and the crystal structure is either of the ionic or of the homopolar type. The ionic lattices are generally of the sodium chloride type for compositions such as AB and of the calcium fluoride type for compositions such as AB_2. In the rock-salt lattice the ions form a simple cubic lattice, the positive and negative ions each forming a face-centred lattice. Alloys of this type are usually alloys of a divalent metal with either selenium or tellurium. The alloys with the compositions AB_2 are anti-isomorphic with fluorspar, i.e. the metal atoms B occupy the positions of the fluorine atoms in fluorspar and the semi-metal atoms A occupy the positions of the calcium atoms. (There are four molecules AB_2 in the cubic unit cell, the A atoms being at $(0, 0, 0)$, $(0, \frac{1}{2}, \frac{1}{2})$, $(\frac{1}{2}, 0, \frac{1}{2})$, $(\frac{1}{2}, \frac{1}{2}, 0)$, while the B atoms are at $\pm(\frac{1}{4}, \frac{1}{4}, \frac{1}{4})$, $\pm(\frac{1}{4}, \frac{3}{4}, \frac{3}{4})$, $\pm(\frac{3}{4}, \frac{1}{4}, \frac{3}{4})$, $\pm(\frac{3}{4}, \frac{3}{4}, \frac{1}{4})$.) Examples of this type of alloy are Mg_2Si, Mg_2Sn, Mg_2Pb, Mg_2Ge, Cu_2Se. The complete phase diagram of the Mg-Sn system is shown in fig. IV 1, from which it can be seen that the homogeneous alloys, which form the α-phase, have a very restricted range. Over most of the equilibrium diagram the alloys are heterogeneous mixtures.

Covalent structures can be formed when the average number of valency electrons per atom is four, the possible structures being then of the zinc-blende and wurtzite types. The zinc-blende lattice is a diamond-type lattice in which the lattice points are divided into two sets, namely, $(0, 0, 0)$, $(\frac{1}{2}, \frac{1}{2}, 0)$, $(\frac{1}{2}, 0, \frac{1}{2})$, $(0, \frac{1}{2}, \frac{1}{2})$, and $(\frac{1}{4}, \frac{1}{4}, \frac{1}{4})$, $(\frac{3}{4}, \frac{3}{4}, \frac{1}{4})$, $(\frac{3}{4}, \frac{1}{4}, \frac{3}{4})$, $(\frac{1}{4}, \frac{3}{4}, \frac{3}{4})$, the Zn atoms occupying the first set of sites and the S atoms the second, the co-ordination number being four for each atom. Alloys of this type are BeSe, ZnSe, CdSe, HgSe, BeTe, ZnTe, CdTe, HgTe, AlAs, GaAs, AlSb, GaSb, InSb. The wurtzite lattice is a hexagonal lattice with coordination number four, examples of this type of alloy being MgTe and CdSe.

The nickel-arsenide structure occurs frequently in alloys of the transition metals with the semi-metals Se, Te, As, Sb and Bi. There are four atoms in a hexagonal cell, the atoms of the transition metal being at $(0, 0, 0)$, $(0, 0, \frac{1}{2})$, while the atoms of the semi-metal are at $(\frac{1}{3}, \frac{2}{3}, u)$, $(\frac{2}{3}, \frac{1}{3}, u + \frac{1}{2})$, where u is nearly $\frac{1}{4}$. These alloys cannot be considered to be ionic or covalent, and they can best be considered to be semi-metallic compounds akin to the secondary alloys described in § 4·21.

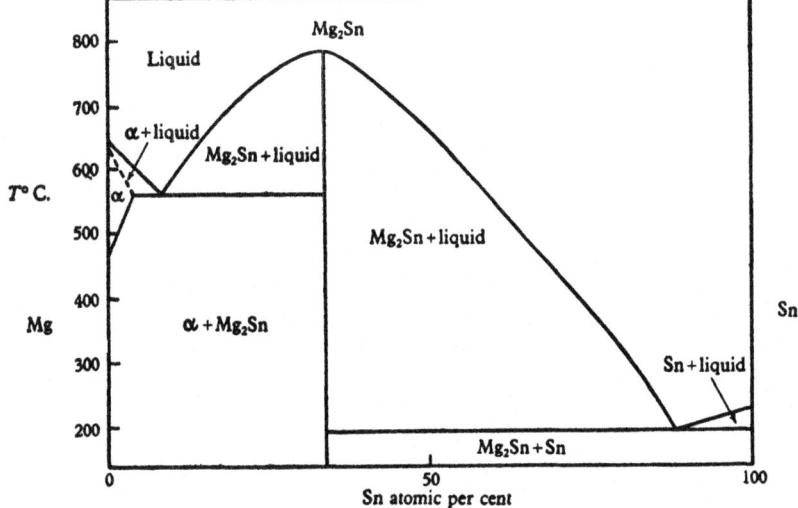

Fig. IV 1. The phase diagram of the magnesium-tin alloys.

PRIMARY AND SECONDARY ALLOYS

4·2. *Primary alloys.* When a small amount of a metal B dissolves in a metal A, the resulting alloy has the same crystal structure as A, and is formed by B atoms replacing A atoms in the lattice. Such a substitutional alloy is called a primary alloy. (There are also interstitial alloys which can only be formed when one atom is very much smaller than the other. The small atom, which is usually hydrogen, boron, carbon or nitrogen, does not displace a metal atom from its lattice but fits into the spaces which exist in the original structure.)

For some pairs of metals, which must of course have the same crystal structure, the primary alloy may exist for all concentrations, examples being AuAg, CuPt and NiMn (see fig. IV 2). This only occurs when the metals have very similar properties, and even when metals have the same crystal structure, the primary solid solution has usually only a limited solubility range. For example, copper dissolves in silver up to 6 % and silver in copper up to 2 %. These two solid solutions are primary phases, known as the α- and β-phases respectively. For concentrations of copper between 6 and 98 %, the alloys at low temperatures consist of a mixture of the α- and β-phases; the complete phase diagram is shown in fig. IV 3.

4·21. *Secondary alloys.* When the metals have different crystal structures it is impossible for a primary solid solution to extend over the whole of the

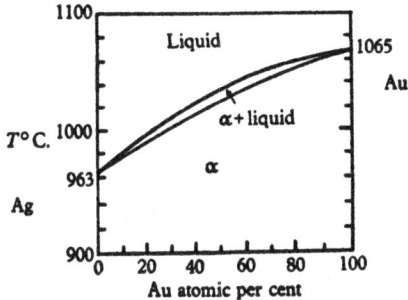

Fig. IV 2. The phase diagram of the gold-silver alloys.

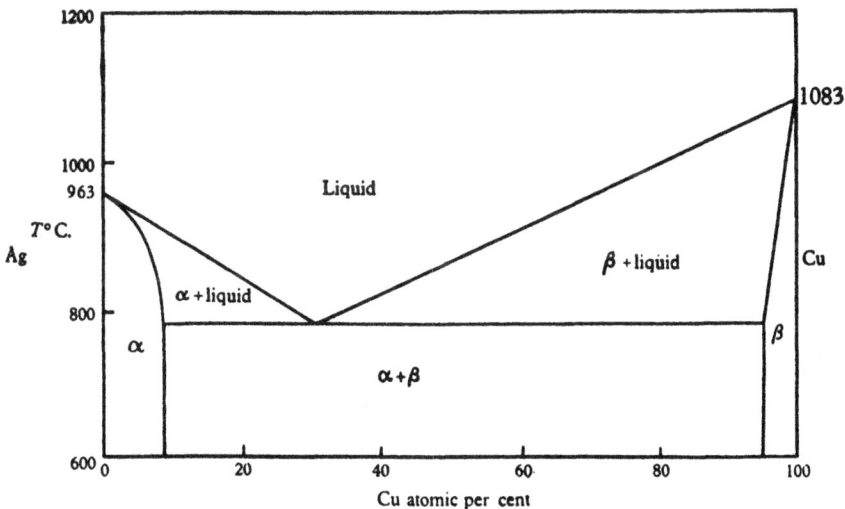

Fig. IV 3. The phase diagram of the copper-silver alloys.

concentration range. For some concentrations either the alloy must consist of a mixture of two primary solid solutions, or else an alloy must be formed whose crystal structure differs from those of the parent metals. Such alloys are called secondary or intermediate solid solutions. They were at one time thought to be intermetallic compounds, since they exist only for more or less definite compositions, are hard and brittle and have low electrical conductivities. On the other hand, their homogeneity range, though restricted, is not zero, and the compositions at which they occur are completely at variance with the classical valency rules. We must therefore conclude that the cohesive forces in secondary alloys are largely metallic in character and that the feebly metallic properties of these alloys arise in the same way as do those of bismuth, namely, by the conduction electrons forming a nearly closed group.

4·22. There are many factors which determine the structure of alloys, including the atomic radii of the atoms and their excitation energies, but one factor which is often the determining one is the number of valency electrons present. When a small quantity of a multivalent metal is added to a monovalent metal the number of electrons increases and, for sufficiently dilute alloys, the cohesive forces are increased. The available evidence suggests that, as long as the density of states increases with increasing electron concentration, the alloys are stable, but as soon as a concentration is reached for which the density of states diminishes with increasing concentration, the alloy becomes unstable and passes over to a different (homogeneous or heterogeneous) form. This is to be expected, since a high density of states means that an extra electron can be accommodated with very little increase in energy. When the density of states is small the adding of extra electrons results in a large increase in energy, and it is then likely that there are other structures which involve smaller energy changes. Since the density of states reaches a maximum when the energy levels touch the zone boundaries, we can expect the properties of alloys to be influenced strongly by the relation of the highest occupied energy level to the zone boundaries.

THE COPPER-ZINC ALLOYS

4·3. The principles enunciated above can be illustrated by means of the copper-zinc alloys, the phase diagram of which is shown in fig. IV 4. When zinc is added to copper, the first or α-phase is face-centred cubic and is a primary solid solution of zinc in copper; it persists up to 35 % of zinc. Between the limits of 35 and 46 % of zinc, a two-phase region, corresponding to the $\alpha\beta$-brasses, occurs, while for concentrations between 46 and 49 % the alloys are body-centred cubic and are known as the β-brasses. There are further two-phase regions for the concentration ranges 49–61 %, 67–79 % and 87–97 % of zinc, the intervening homogeneous phases being a cubic γ-phase, an hexagonal ϵ-phase with an axial ratio in the neighbourhood of 1·56 and an hexagonal η-phase with an axial ratio near 1·8. The η-phase, like the α-phase, is a primary solid solution, but of copper in zinc, while the β-, γ- and ϵ-phases are secondary solid solutions. (The figures given above depend slightly on the temperature and are therefore only approximate.)

4·31. *The α-phase.* The number of valency electrons in the copper-zinc alloys increases with the concentration of zinc, and so the contour of the highest occupied energy level is pushed closer to the boundary of the first Brillouin zone as the concentration of zinc increases. Now when there is one electron per atom in a face-centred cubic structure the electrons occupy a space equal in volume to a sphere of radius $4·91/a$ in **k** space (§ 3·61), while the sphere inscribed in the Brillouin zone has radius $5·44/a$. Hence if the

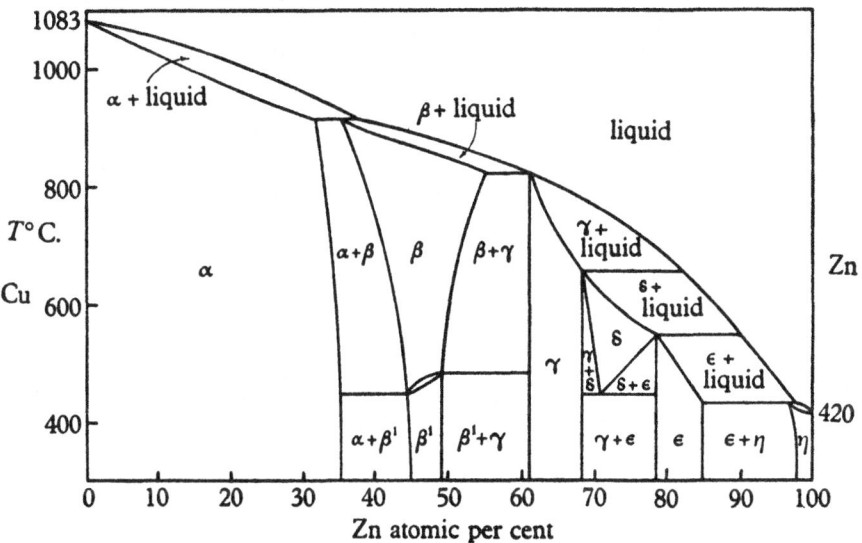

Fig. IV 4. The phase diagram of the copper-zinc alloys.

energy contours can be treated as spherical, the energy contour of the highest occupied level just touches the zone boundary when the number of electrons per atom is $(5·44)^3/(4·91)^3 = 1·36$, i.e. when 36 % of zinc has been added. At about this concentration the density of the levels available for accommodating further electrons starts diminishing, and we should therefore expect the α-phase to become unstable. The actual limit of stability of the β-phase is at 35 % zinc, and although the close agreement between the two figures is fortuitous we can make the rough generalization that the limit of stability of a phase is reached when the energy contour of the highest occupied level just touches the zone boundary.

4·32. *The β-phase.* A body-centred cubic lattice can accommodate more electrons per atom than a face-centred lattice before the density of states starts diminishing, and so the copper-zinc alloys pass over into a body-centred structure when too much energy is required to absorb more zinc into the face-centred lattice. (The upper limit of the α-phase and the lower limit of the β-phase are actually determined by their stability as compared with a heterogeneous mixture of the two phases. Thus although the α-phase has lower free energy than the β-phase at say 40 % zinc, the free-energy curves are such that the common tangent lies below both curves for concentrations between 36 and 46 %, so that the heterogeneous mixture

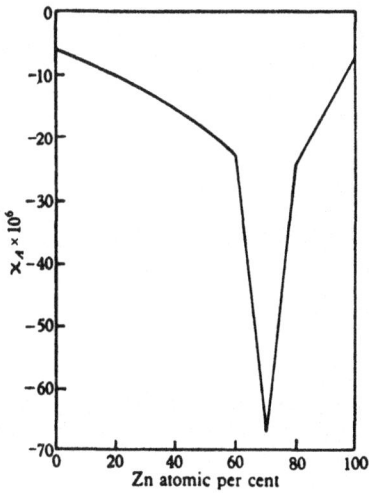

Fig. IV 5. The diamagnetic suscepti-bility per gram atom of the copper-zinc alloys.

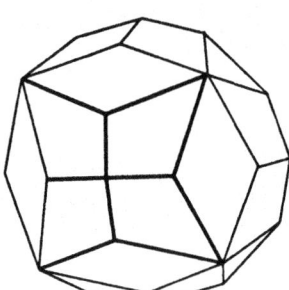

Fig. IV 6. The Brillouin zone for the γ-phase.

is the most stable configuration.) Now the sphere inscribed in the first Brillouin zone in a body-centred lattice has radius $4\cdot44/a$, while the sphere containing one electron per atom has radius $3\cdot898/a$. The energy contour of the highest occupied energy level therefore touches the zone boundary when the number of valency electrons per atom is $(4\cdot44)^3/(3\cdot898)^3 = 1\cdot48$. The actual concentration at which the β-phase becomes unstable is 49 % zinc.

4·33. *The γ-phase.* The homogeneous alloys which succeed the β-brasses have a complicated cubic structure with fifty-two atoms in the unit cell. Unlike the β-brasses, the γ-alloys behave like semi-metals in that they are hard, brittle, poor conductors of heat and electricity and are highly dia-magnetic. (See fig. IV 5. The diamagnetic susceptibility per unit volume reaches a maximum value of $-8\cdot3 \times 10^{-6}$, which is comparable with that of bismuth, namely, -14×10^{-6}.) We should therefore expect there to be a Brillouin zone only just capable of accommodating the valency electrons, so that the energy contour of the highest occupied level is everywhere near the zone boundary. Now the only planes with small indices which give strong X-ray reflexions are of the type (4, 1, 1) and (3, 3, 0). We therefore assume that the first Brillouin zone is bounded by the planes such as

$$(4k_1 \pm k_2 \pm k_3)a \pm 18\pi = 0, \quad (k_1 \pm k_2)a \pm 6\pi = 0. \quad (4\cdot33\cdot1)$$

This polyhedron, shown in fig. IV 6, has thirty-six sides at a distance $3\sqrt{2}\,\pi/a$ from the origin, and is nearly a sphere. Its volume can readily be calculated by finding the coordinates of the vertices of two of the faces, one of each of the two types, and the result is that the polyhedron can contain exactly ninety electrons per unit cell, while the inscribed sphere can contain eighty (Jones 1934a).

In the γ-phase the alloys have the approximate composition $Cu_{20}Zn_{32}$, there being eighty-four electrons per unit cell. The surface of the Fermi distribution must therefore be very close to the zone boundary, and the alloys behave like semi-metals. It is not known whether all the electrons are contained in the first zone or whether some overlap into the second zone, but the magnitudes of the Fourier coefficients of the lattice potential obtained from the X-ray data suggest that the energy gaps are not very large. (If we assume that $\Delta E_g = 2\,|V_g|$, the relation which holds for nearly free electrons, the X-ray data (Bradley and Thewlis, 1926) give $\Delta E_{411} = 2\cdot9\,\text{eV.}$, $\Delta E_{330} = 4\cdot6\,\text{eV.}$) It is therefore probable that overlap occurs into the second zone in the neighbourhood of the (4, 1, 1) direction.

4·34. *The ϵ- and η-phases.* The two hexagonal phases only differ in the axial ratio; the parameters are given in Table IV 1. The number of free electrons per atom that can be accommodated in the first Brillouin zone is given by (3·71·3) as

$$2 - \frac{3}{4}\frac{a^2}{c^2} + \frac{3}{16}\frac{a^4}{c^4}.$$

Table IV 1. *The lattice parameters of the hexagonal* CuZn *phases in* 10^{-8} cm.

	Percentage of zinc	a	c	c/a
ϵ-phase: Beginning	78·6	2·730	4·286	1·570
End	86·8	2·760	4·289	1·554
η-phase: Beginning	97·3	2·674	4·823	1·804
End	100	2·659	4·935	1·856

Therefore in the ϵ-phase the number of free electrons per atom varies from 1·786 to 1·868, while the number of electrons which can be accommodated in the first zone decreases slightly from 1·727 to 1·722. In the η-phase the number of free electrons per atom varies from 1·97 to 2, of which 1·787 to

1·792 can be accommodated in the first zone. There is therefore an overlap into the second zone throughout both phases. If we assume that the electrons are nearly free, the overlap is most likely to take place at the points on the zone boundary where $|\mathbf{k}|$ is smallest, that is, at the points such as A and B in fig. III 18, the values of $|\mathbf{k}|$ being

$$|\mathbf{k}|_A = 2\pi/(\sqrt{3}\,a), \quad |\mathbf{k}|_B = 2\pi/c.$$

Throughout the ϵ-phase, $|\mathbf{k}|_A$ is appreciably less than $|\mathbf{k}|_B$, so that the overlap probably takes place at points near A. In the η-phase, $|\mathbf{k}|_A/|\mathbf{k}|_B$ is nearly unity, and overlap will probably take place near both A and B, especially as the number of electrons to be accommodated is rather large (at least 0·18 per atom).

Jones (1934 b) has shown that it is possible to obtain an estimate of the change in the lattice parameters with composition, provided that the elastic constants are known. The agreement is fair, and the results show that the slightly different behaviour of the ϵ- and η-phases can be attributed to the difference in the way that the overlap takes place in the two phases.

THE ELECTRON CONCENTRATION IN ALLOYS

4·4. The discussion of the behaviour of the copper-zinc alloys shows that an important factor in determining the stability of a phase is the proximity of the energy contours of the highest occupied level to the zone boundaries, and that these approach one another for compositions which could not be predicted by the classical rules of valency. If the electron concentration is the dominant factor, we should expect the β-, γ- and ϵ-structures to occur whenever the electron concentration reaches the values found in the copper-zinc alloys, no matter what the particular metals are. Now in the copper-zinc alloys the β-phase has the approximate composition CuZn with an electron concentration of $\frac{3}{2}$, the γ-phase has the approximate composition Cu_5Zn_8 with an electron concentration of $\frac{21}{13}$, and the ϵ-phase has the approximate composition $CuZn_3$ with an electron concentration of $\frac{7}{4}$. The significance of these peculiar ratios of valency electrons to atoms in determining the type of structure has been known for a long time (see, for example, Hume-Rothery, 1944, Part V), and many examples of alloys satisfying the above rules have been found, the most important of which are given in Table IV 2 (cf. Dehlinger, 1935). In order to fit the transition metals into this scheme it is necessary to ascribe zero valency electrons to all of them in the metallic state.

The assignment of zero valency to all the transition metals is, however, far from satisfactory in view of their supposed electronic structure. We should expect Ni, Pd and Pt to behave as if they were zero-valent, while Co should have a valency -1, Fe -2 and so on. There are in fact a number of alloys such as Co_2Al_5 and Co_2Al_9 in which cobalt seems to act as an acceptor of electrons (Pratt and Raynor, 1951), and no satisfactory explanation has yet been given of the varied behaviour of the transition metals.

Table IV 2. *The electron concentration in alloys*

Electron concentration $\frac{3}{2}$ Body-centred cubic		Electron concentration $\frac{21}{13}$ γ-brass structure		Electron concentration $\frac{7}{4}$ Close-packed hexagonal	
Alloy	Electron concentration range	Alloy	Electron concentration range	Alloy	Electron concentration range
CuBe	1·47–1·49	Cu_5Zn_8	1·61–1·67	$CuZn_3$	1·78–1·85
CuZn	1·46–1·49	Cu_5Cd_8	?	$CuCd_3$	?
AgMg	?	Ag_5Zn_8	1·60–1·64	Cu_3Si	1·75
AgZn	1·49–1·54	Ag_5Hg_8	1·61	Cu_3Ge	?
AgCd	1·49–1·51	Au_5Zn_8	1·64–1·68	Cu_3Sn	1·75
AuZn	1·38–1·55	Cu_9Al_4	1·62–1·71	$AgZn_3$	1·7–1·86
AuCd	?	Cu_9Ga_4	?	$AgCd_3$	1·69–1·83
Cu_3Al	1·44–1·60	Cu_9In_4	?	Ag_5Al_3	1·54–1·86
Cu_3Ga	?	$Cu_{31}Si_8$	?	Ag_3Sn	?
Cu_5Sn	1·39–1·51	$Cu_{31}Sn_8$	1·61	$AuZn_3$	1·84–1·89
FeAl	?	Fe_5Zn_{21}	(1·42–1·54)	$AuCd_3$	?
CoAl	1·36–1·58	Co_5Zn_{21}	1·56–1·70	Au_5Al_3	?
NiAl	?	Ni_5Zn_{21}	?	Au_3Sn	?
		Rh_5Zn_{21}	?	$MnZn_7$	?
		Pd_5Zn_{21}	?	$FeZn_7$	1·74–1·84
		Pt_5Zn_{21}	?		
		Ni_5Cd_{21}	?		
		$Na_{31}Pb_8$	?		

Although the structures have been written as CuZn, Cu_5Zn_8, $CuZn_3$, etc., it is not implied that any chemical compound is formed. These compositions are in fact not fixed, and, as shown in Table IV 2, there is a small concentration range over which each type of structure is stable.

The above classification does not exhaust all the types of alloys. In the first place, there are alloys similar to the CuZn alloy, but which instead of being body-centred cubic have the complex β-manganese form with 20 atoms in the unit cell. Examples are Ag_3Al, Au_3Al, Cu_5Si, $CoZn_3$, all of which have an electron concentration of $\frac{3}{2}$. Then there are alloys in which the electron concentration seems to be only a minor factor and whose structure is determined by other influences which are not yet fully understood. The most important of these alloys have body-centred cubic structures which are either similar to that of caesium chloride or to that of NaTl, which is a complicated structure in which each Na and Tl atom is surrounded by four Na and four Tl atoms at the same distance away. Examples are LiAg, LiCd, LiZn, LiHg, LiAl, LiTl, NaTl, NaBi.

SUPERLATTICES

4·5. In general the different atoms in a substitutional alloy are distributed at random over the lattice points, but at low temperatures and at well-defined simple compositions the alloys are often in an ordered state, the different atoms being arranged in a definite way just as in an ionic compound like sodium chloride. The hypothesis that such ordered structures occur was first put forward by Tamman in 1919, but it was not until 1925 that Johansson and Linde were able to prove directly, by means of an X-ray analysis, that the CuAu alloy can exist in an ordered state.

It is well known that, on account of the interference of the waves scattered by the atoms at the cube corners and centres, there are no X-ray reflexions of odd orders for a body-centred cubic metal. The same is true for a disordered body-centred alloy, since there must be on the average no difference between the cube corners and the cube centres. If, however, the alloy is in an ordered state with one type of atom, called A, occupying the cube corners and the other type of atom, called B, occupying the cube centres, the odd-order X-ray reflexions must occur, since they cannot now be destroyed by the interference of the waves scattered by the atoms at the corners with those scattered by the atoms at the centres, the scattering powers of the two types of atom being different. Thus the appearance of the odd-order X-ray reflexions in a body-centred alloy is evidence that an ordered state or superlattice exists. Similarly, in a face-centred cubic alloy the appearance of the $(1, 1, 0)$ reflexions is the criterion for the existence of a superlattice.

Body-centred cubic superlattices can only occur when there are equal numbers of the two types of atom present. Examples are fairly common in the β-phases (then called β'-phases) and include CuZn, AgZn, AuCd, AgMg and FeAl. Face-centred cubic superlattices can only occur at the composition AB_3, since there are three times as many face-centres as face-corners, the two sites not being equivalent. Examples are $AuCu_3$, $PdCu_3$, $PtCu_3$, $NaPb_3$, $CaPb_3$, $CePb_3$, $CaTl_3$, $CaSn_3$, $CeSn_3$. (The superlattice which is formed in the AuCu system at the equi-atomic ratio cannot be cubic for symmetry reasons and is in fact tetragonal with $c/a = 0·93$.)

The theory of superlattices is essentially a part of the equilibrium theory of the statistical mechanics of cooperative systems, and since it does not involve any ideas specific to the electron theory of metals it will not be considered further here.

REFERENCES

Bradley, A. J. and Thewlis, J. (1926). The structure of γ-brass. *Proc. Roy. Soc. A*, **112**, 678.

Dehlinger, U. (1935). Gitteraufbau metallischer Systeme. *Handbuch der Metallphysik*, **1** (Leipzig).

Hume-Rothery, W. (1944). *The structure of metals and alloys* (London).

Johansson, C. H. and Linde, J. O. (1925). The X-ray determination of the atomic arrangement in the mixed-crystal series gold-copper and palladium-copper. *Ann. Phys., Lpz.* (4), **78**, 439.

Jones, H. (1934a). The theory of alloys in the γ-phase. *Proc. Roy. Soc. A*, **144**. 225.

Jones, H. (1934b). Applications of the Bloch theory to the study of alloys and the properties of bismuth. *Proc. Roy. Soc. A*, **147**, 396.

Pratt, J. N. and Raynor, G. V. (1951). Intermetallic compounds in ternary aluminium rich alloys containing transitional metals. *Proc. Roy. Soc. A*, **205**, 103.

Tamman, G. (1919). The chemical and galvanic properties of mixed crystals and their atomic structure. *Z. anorg. Chem.* **107**, 1.

Chapter V

SEMI-CONDUCTORS

GENERAL PRINCIPLES

5·1. Insulators are substances in which the electrons just fill an energy band at the absolute zero. They will, however, possess an electronic conductivity which varies with temperature like $e^{-b/T}$ owing to the excitation of electrons into the next, unoccupied, band, but normally b is so large that the electronic conductivity is negligible. The normal conductivity of insulators is electrolytic in character, as is shown by the polarization effects set up by the passage of an electric current, and the deduction from this is that in general it is easier to displace an ion than to excite an electron. There are, however, substances which have an appreciable electrical conductivity at room temperature, which is definitely known to be of electronic origin, but which decreases exponentially as the temperature is lowered. These substances, which are of great technical importance, are known as semi-conductors and owe their properties to the presence of minute amounts of impurities, a fact first clearly recognized by Gudden (1930, 1934).

It is possible, in principle, to distinguish two types of semi-conductors, intrinsic and extrinsic semi-conductors. Intrinsic semi-conductors are those which show electronic conductivity (at a sufficiently high temperature) in the pure state and are therefore essentially insulators in which the electronic excitation energy is reasonably low. (For practical purposes the line of demarcation between insulators and semi-conductors can be placed at a conductivity of 10^{-10} ohm^{-1} cm.$^{-1}$.) For a long time it was not definitely known whether any intrinsic semi-conductors existed, but there is now conclusive evidence that silicon and germanium, at ordinary temperatures, and numerous compounds of various types at elevated temperatures, are intrinsic semi-conductors. The more important semi-conductors are the extrinsic ones, which are elements or compounds owing their conductivity to traces of impurities, or, more usually, to the presence of one of the constituents in excess of the stoichiometric amount.

Extrinsic semi-conductors can be classified into two groups according as the impurities responsible for the conduction act as donors or acceptors of electrons. There are other equivalent classifications which are detailed below.

THE NUMBER OF FREE ELECTRONS

5·2. In an intrinsic semi-conductor there are two energy bands to be considered, the first being completely empty at $T = 0$ and the second being completely filled (Wilson, 1931 a). The distribution function for $T \neq 0$ is determined by equating the number of electrons in the two bands to the number in the lower band 2 at $T = 0$, i.e. by the condition

$$2 \int n_2(E) \, dE = 2 \int \frac{n_1(E) \, dE}{e^{(E-\zeta)/kT} + 1} + 2 \int \frac{n_2(E) \, dE}{e^{(E-\zeta)/kT} + 1}, \qquad (5·2·1)$$

where $n_1(E)$, $n_2(E)$ are the densities of states in the two bands for one direction of the electron spin. Since the number of electrons excited is always small, the denominator of the second integral on the right is nearly unity, whereas that of the first integral is very large and approximately $e^{(E-\zeta)/kT}$. Hence (5·2·1) can be written approximately as

$$\int e^{-(E-\zeta)/kT} n_1(E) \, dE = \int e^{(E-\zeta)/kT} n_2(E) \, dE. \qquad (5·2·2)$$

The only electrons that contribute effectively to the integrals are those near the bottom of the first band and near the top of the second band, and we can therefore treat the energy bands as being of standard form, band 1 being normal with an effective mass m_1 and band 2 being inverted with an effective mass m_2. Then if ΔE is the energy separation of the two bands and if the energy zero is taken at the bottom of band 1, we have

$$n_1(E) = 2\pi(2m_1)^{\frac{3}{2}} h^{-3} E^{\frac{1}{2}}, \quad n_2(E) = 2\pi(2m_2)^{\frac{3}{2}} h^{-3}(-\Delta E - E)^{\frac{1}{2}}, \quad (5·2·3)$$

and (5·2·2) becomes

$$m_1^{\frac{3}{2}} \int_0^\infty E^{\frac{1}{2}} e^{-(E-\zeta)/kT} \, dE = m_2^{\frac{3}{2}} \int_{-\infty}^{-\Delta E} (-\Delta E - E)^{\frac{1}{2}} e^{(E-\zeta)/kT} \, dE, \quad (5·2·4)$$

that is,

$$m_1^{\frac{3}{2}} e^{\zeta/kT} = m_2^{\frac{3}{2}} e^{-(\Delta E + \zeta)/kT},$$

so that

$$\zeta = -\tfrac{1}{2}\Delta E - \tfrac{3}{4}kT \log(m_1/m_2). \qquad (5·2·5)$$

(At $T = 0$ the Fermi energy ζ lies just half-way between the energy bands.) The distribution function of the electrons in band 1 is therefore the Maxwell function

$$8\sqrt{2} \, \pi h^{-3} m_1^{\frac{3}{2}} f_1(E) \, E^{\frac{1}{2}} \, dE = 8\sqrt{2} \, \pi h^{-3}(m_1 m_2)^{\frac{3}{4}} e^{-\frac{1}{2}\Delta E/kT} E^{\frac{1}{2}} e^{-E/kT} \, dE, \qquad (5·2·6)$$

and the total number of electrons excited per unit volume is

$$n = 2h^{-3}(m_1 m_2)^{\frac{3}{4}} (2\pi kT)^{\frac{3}{2}} e^{-\frac{1}{2}\Delta E/kT}. \qquad (5·2·7)$$

There is the same number of vacant levels in band 2, and the vacant places or 'holes' also have a Maxwellian distribution given by

$$8\sqrt{2}\,\pi h^{-3} m_2^{\frac{3}{2}} f_2(E')\,E'^{\frac{1}{2}}\,dE' = 8\sqrt{2}\,\pi h^{-3}(m_1 m_2)^{\frac{3}{4}}\,e^{-\frac{1}{2}\Delta E/kT}\,E'^{\frac{1}{2}}\,e^{-E'/kT}\,dE',$$

$$(5\cdot2\cdot8)$$

where $E' = -\Delta E - E$.

5·21. The ways in which impurities can produce free electrons can readily be understood by referring to fig. V 1, in which two energy bands of a crystal are shown. The band F is completely filled with electrons when the temperature is zero, and the band E is completely empty. Consider the effect of an impurity atom D which has one valency electron, the energy level of this electron lying between the two bands. This electron cannot take part directly in conduction, but it can do so if it is first thermally excited into the empty band E, when it is free to move through the lattice. In this case the conductivity still varies as $e^{-b/T}$, but b is now connected with the energy difference between the level D and the band E and not with the energy difference between the two bands. Thus by introducing suitable impurities we may be able to make b so small that the substance has an appreciable electronic conductivity at room temperatures (Wilson, 1931 b). We distinguish four important cases.

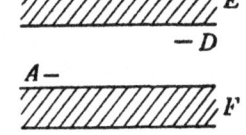

Fig. V 1. The energy levels in a semi-conductor.

(1) If the impurity atoms are electropositive and the energy levels of their valency electrons lie between the two bands, the atoms act as donors of electrons to band E, and the number of electrons in band E is of the form $e^{-b/T}$.

(2) If the impurities are electropositive and the energy levels lie in the band E, the atoms are permanently ionized and the number of electrons in band E is constant, in so far as the electrons are derived from the impurities.

(3) If the impurity atoms are electronegative they may have vacant energy levels, such as A in fig. V 1, lying between the two bands. In this case the impurity atoms can act as acceptors of electrons from the band F; holes are created in band F and their number is proportional to $e^{-b/T}$, where b is connected with the energy difference between the band F and the energy level A.

(4) If the impurity atoms are electronegative and are such that their vacant levels lie in the band F, holes are created in the band F and their number is constant.

To calculate how the number of free electrons varies with temperature in case (1) we proceed as follows. Let n_0 be the number of foreign atoms per unit volume each possessing one easily excitable electron in an energy level ΔE_1 below the bottom of the conduction band. Then, if account is taken of the effect of the electron spin in accordance with equation (A1.10) of Appendix 1, the condition for the constancy of the number of electrons can be written as

$$n_0 = \frac{n_0}{\frac{1}{2}e^{-(\Delta E_1 + \zeta)/kT} + 1} + 2\int \frac{n_1(E)\,dE}{e^{(E-\zeta)/kT} + 1}, \qquad (5·21·1)$$

where the energy zero is taken to be at the bottom of the conduction band. If we make the same assumptions as in the preceding section, we find that

$$\zeta = -\tfrac{1}{2}\Delta E_1 - \tfrac{1}{2}kT \log\{4(2\pi m_1 kT)^{\frac{3}{2}}/(n_0 h^3)\}, \qquad (5·21·2)$$

and that the distribution function for the free electrons is

$$8\sqrt{2}\pi h^{-3} m_1^{\frac{3}{2}} f(E)\,E^{\frac{1}{2}}\,dE = \frac{2}{\sqrt{\pi}}\left(\frac{2\pi m_1}{h^2 kT}\right)^{\frac{3}{2}} n_0^{\frac{1}{2}} e^{-\frac{1}{2}\Delta E_1/kT}\,E^{\frac{1}{2}}\,e^{-E/kT}\,dE. \qquad (5·21·3)$$

The total number of free electrons per unit volume is then

$$n = (2\pi m_1 kT/h^2)^{\frac{3}{4}} n_0^{\frac{1}{2}} e^{-\frac{1}{2}\Delta E_1/kT}. \qquad (5·21·4)$$

5·211. In some semi-conductors it is possible for the free electrons to become attached to atoms other than the impurity atoms from which they are originally derived. The temperature dependence of n is then somewhat different. Let N be the total number of atoms per unit volume to which the electrons can become attached. Then instead of (5·21·1) we have the condition

$$n_0 = \frac{N}{\frac{1}{2}e^{-(\Delta E_1 + \zeta)/kT} + 1} + 2\int \frac{n_1(E)\,dE}{e^{(E-\zeta)/kT} + 1}. \qquad (5·211·1)$$

We can no longer make the approximation that the denominator of the first term on the right is nearly unity, and ζ is determined by the equation

$$\tfrac{1}{2}e^{-(\Delta E_1 + \zeta)/kT} = \frac{N}{n_0 - 2(2\pi m_1 kT)^{\frac{3}{2}} h^{-3} e^{\zeta/kT}} - 1. \qquad (5·211·2)$$

If $N = n_0$ we regain (5·21·2), but, if $N \gg n_0$ and if the number of free electrons is small compared with n_0, ζ is given by

$$\zeta = -\Delta E_1 - kT \log\{2(N - n_0)/n_0\} \qquad (5·211·3)$$

approximately. In this case the distribution function for the free electrons is

$$8\sqrt{2}\pi h^{-3} m_1^{\frac{3}{2}} f(E)\,E^{\frac{1}{2}}\,dE = 2\pi(2m_1)^{\frac{3}{2}} h^{-3}\{n_0/(N - n_0)\} e^{-\Delta E_1/kT}\,E^{\frac{1}{2}}\,e^{-E/kT}\,dE, \qquad (5·211·4)$$

and the total number of free electrons is

$$n = (2\pi m_1 kT)^{\frac{3}{2}} h^{-3}\{n_0/(N - n_0)\} e^{-\Delta E_1/kT}. \qquad (5·211·5)$$

5·22. It is easy to extend the preceding calculations to more complicated cases.

If we wish to consider an intrinsic semi-conductor containing donor atoms, we take the energy zero at the bottom of band 1 and calculate ζ by combining the results of §§ 5·2 and 5·21. The condition for the constancy of the number of electrons can then be written as

$$n_0 + 2\int n_2(E)\,dE = \frac{n_0}{\frac{1}{2}e^{-(\Delta E_1 + \zeta)/kT} + 1} + 2\int \frac{n_1(E)\,dE}{e^{(E-\zeta)/kT} + 1} + 2\int \frac{n_2(E)\,dE}{e^{(E-\zeta)/kT} + 1},$$

$$(5\cdot22\cdot1)$$

which gives

$$\frac{n_0}{1 + 2e^{(\Delta E_1 + \zeta)/kT}} = 2\frac{(2\pi kT)^{\frac{3}{2}}}{h^3}\,(m_1^{\frac{3}{2}}\,e^{\zeta/kT} - m_2^{\frac{3}{2}}\,e^{-(\zeta + \Delta E)/kT}),\qquad(5\cdot22\cdot2)$$

equations (5·2·5) and (5·21·2) being special cases of this.

If there are n_0 acceptor atoms per unit volume with energy levels ΔE_2 above the top of band 2, the number of electrons 'accepted' must be calculated from equation (A 1·13). Hence, taking the energy zero at the top of band 2, we have

$$2\int n_2(E)\,dE = \frac{n_0}{2e^{(\Delta E_2 - \zeta)/kT} + 1} + 2\int \frac{n_1(E)\,dE}{e^{(E-\zeta)/kT} + 1} + 2\int \frac{n_2(E)\,dE}{e^{(E-\zeta)/kT} + 1},$$

$$(5\cdot22\cdot3)$$

which gives

$$2\frac{(2\pi m_2 kT)^{\frac{3}{2}}}{h^3}\,e^{-\zeta/kT} = \frac{n_0}{2e^{(\Delta E_2 - \zeta)/kT} + 1} + 2\frac{(2\pi m_1 kT)^{\frac{3}{2}}}{h^3}\,e^{(\zeta - \Delta E)/kT}.\quad(5\cdot22\cdot4)$$

If we neglect the intrinsic conductivity, that is, neglect the electrons in band 1, and assume that there are only a few electrons on the acceptor atoms, we have $e^{(\Delta E_2 - \zeta)/kT} \gg 1$, and

$$\zeta = \tfrac{1}{2}\Delta E_2 + \tfrac{1}{2}kT \log\{4(2\pi m_2 kT)^{\frac{3}{2}}/(n_0 h^3)\}.\qquad(5\cdot22\cdot5)$$

The free holes have a Maxwellian distribution and their total number is equal to the number of electrons on the acceptor atoms, i.e.

$$n = \tfrac{1}{2}n_0\,e^{(\zeta - \Delta E_2)/kT} = (2\pi m_2 kT/h^2)^{\frac{3}{4}}n_0^{\frac{1}{2}}\,e^{-\frac{1}{2}\Delta E_2/kT},\qquad(5\cdot22\cdot6)$$

In all cases, the total numbers of electrons and holes are given by

$$n_1 = 2(2\pi m_1 kT/h^2)^{\frac{3}{2}}\,e^{\zeta/kT},\quad n_2 = 2(2\pi m_2 kT/h^2)^{\frac{3}{2}}\,e^{-(\zeta + \Delta E)/kT},\quad(5\cdot22\cdot7)$$

if the energy zero is taken at the bottom of band 1. In general the intrinsic conductivity is unimportant; either the electrons or the holes predominate, and it is then unnecessary to write explicitly the suffixes 1 and 2.

If the free electrons are derived from impurity atoms having paired electrons such that only one electron per atom can be removed, the calculations must be based upon equation (A 1·12) of Appendix 1 instead of on equation (A 1·10).

CRITERIA FOR ESTABLISHING THE NATURE OF THE CONDUCTIVITY

5·3. *The Hall effect.* The calculation of the number of free electrons in the preceding section shows that in all the cases considered n varies as $e^{-b/T}$, and the conductivity σ will vary likewise. But such a variation with temperature of σ is insufficient to show that a substance is an electronic semi-conductor, since conductivity due to the motion of positive ions has a similar temperature variation. Electrolytic conductivity, however, is always accompanied by the transport of ions and these give rise to polarization effects and departures from Ohm's law, and the absence of such effects is conclusive proof of the existence of electronic conductivity. But the most direct proof is provided by the existence or absence of a Hall effect. An apparent transport of positive ions, obeying Faraday's law of electrolysis, has sometimes been observed which is spurious. Due to an unfortunate choice of experimental conditions, α-Ag_2S was for many years wrongly classed as an electrolytic semi-conductor on the evidence of a transport of silver ions. These did not have their origin in the Ag_2S but in the adjacent materials.† A Hall effect, on the other hand, has never been observed in electrolytic conduction and the nature of an electronic semi-conductor can be established most easily by observation of its Hall coefficient.

It is shown in § 8·62 that if the conduction in a semi-conductor is due to a few electrons in a nearly empty band, the Hall coefficient is given by $R = -3\pi/(8n\epsilon c)$, while, if it is due to a nearly full band in which there are n vacant levels or 'holes' per unit volume, the Hall coefficient is $R = 3\pi/(8n\epsilon c)$. Measurement of the Hall coefficient therefore not only shows whether the conduction is due to electrons or to 'positive holes', but also gives n directly. It has been found by such measurements that α-Ag_2S, MoS_2, ZnO, Al_2O_3, Ta_2O_5, V_2O_5 and WO_3 are 'excess conductors' in which the current is carried by electrons, and that Cu_2O, CuI, UO_2 and NiO are 'deficit conductors' in which the current is due to positive holes. The quantity kb is usually of the order of a few tenths of an electron volt.

5·31. *Oxidation and reduction semi-conductors.* In certain semi-conductors the conductivity is due to the presence of foreign atoms, for example copper and manganese atoms in the technically important zinc sulphide phosphors, but in compounds it is usually due to a stoichiometric excess of one of the constituent elements. In zinc oxide the 'impurity atoms' are probably excess zinc atoms in interstitial positions, while in

---

† For an explanation of the phenomena leading to the wrong interpretation see Wagner (1933).

cuprous oxide it is probable that certain of the copper atoms are missing and that near the vacant spots there are sufficient O^- ions (or O atoms) instead of O^{--} ions to maintain electrical neutrality. These semi-conductors can be classified according to the way in which they behave when the oxygen content is varied.

Substances in which the conductivity decreases as the oxygen content increases are called reduction semi-conductors, and they owe their conductivity to the presence of excess metal. We should expect them to be 'excess conductors' in which the metal atoms act as donors of electrons. Substances in which the conductivity increases with increasing oxygen content are called oxidation semi-conductors. We should expect them to be 'deficit conductors', the excess oxygen atoms acting as acceptors of electrons.

The rule has been formulated by several authors (see the review article by Gudden, 1934) that compounds in which the metal exerts its smallest valency are oxidation conductors, while saturated compounds in which the metal exerts its highest valency are reduction conductors. This rule has been verified for the deficit conductors Cu_2O, CuI, NiO and UO_2 and for the excess conductors α-Ag_2S, ZnO, Al_2O_3 and Ta_2O_5. It does not, however, appear to be true in all cases, and in cupric oxide, for example, the conductivity is independent of the oxygen content. Cupric oxide, therefore, is possibly an intrinsic semi-conductor.

There are semi-conductors such as CoO which do not fit in any way into the classification given here (de Boer and Verwey, 1937), and no satisfactory explanation has been given of their properties.

5·311. It is not easy to test the predicted dependence of n upon the amount of impurity present since the quantities involved are too small to be directly measurable except in special cases, and it is necessary to rely upon the variation of the conductivity of oxide semi-conductors with the oxygen pressure.

v. Baumbach and Wagner (1933) measured the conductivity of ZnO in an oxygen atmosphere between 400 and 700° C. and found that at constant temperature

$$\sigma \propto (p_{O_2})^{-1/4\cdot3}. \tag{5·311·1}$$

They suggest that the effect of the oxygen can be represented by the chemical reaction

$$O_2(gas) + 2Zn^{++}(interstitial) + 4 \text{ free electrons} \rightleftarrows 2ZnO.$$

If $[Zn^{++}(i)]$ denotes the concentration of the interstitial zinc ions and n that of the free electrons, then the law of mass action gives

$$p_{O_2}[Zn^{++}(i)]^2 n^4 = K(T),$$

and if the interstitial zinc ions and free electrons are present in equivalent numbers we have $n = 2[Zn^{++}(i)]$, and so

$$n \propto (p_{O_2})^{-1/6}. \qquad (5 \cdot 311 \cdot 2)$$

According to this equation, n varies in the right direction with p_{O_2}, although the index is wrong. If we assume that there are many more interstitial zinc atoms than are capable of giving rise to free electrons, then $[Zn^{++}(i)]$ would be practically constant and the index in the power law would be nearer to the observed value.

Similar measurements have been made on cuprous oxide by Dünwald and Wagner (1933). In this case they found that between 800 and 1000° C. the conductivity increased with the oxygen pressure according to the law

$$\sigma \propto (p_{O_2})^{1/7}. \qquad (5 \cdot 311 \cdot 3)$$

The mechanism suggested by which the oxygen creates defects in the lattice, and hence positive holes, is as follows. A gaseous oxygen molecule abstracts four electrons from the cuprous oxide to form two O^{--} ions, and at the same time four cuprous ions Cu^+ diffuse out of the interior of the cuprous oxide and, on reaching the surface, combine with the oxygen ions to form normal cuprous oxide. For each oxygen molecule absorbed at the surface there are therefore four cuprous ions and four electrons missing from the interior, a state of affairs which can be described formally by the chemical equation

$$O_2(gas) + 4 \text{ electrons} + 4Cu^+$$
$$\rightleftarrows 2Cu_2O + 4 \text{ vacant electron levels} + 4 \text{ missing } Cu^+.$$

The law of mass action then shows that the fourth power of the product of the concentration of vacant electron levels and the concentration of missing Cu^+ ions is proportional to p_{O_2}, it being assumed that the concentrations of Cu_2O, normal Cu^+ ions and electrons in the nearly full band are effectively constant. If we further assume that vacant electron levels and missing Cu^+ ions are produced in equivalent numbers, we then have

$$n^+ \propto (p_{O_2})^{1/8}, \qquad (5 \cdot 311 \cdot 4)$$

where n^+ is the concentration of vacant electron levels, i.e. the concentration of positive holes. This is in reasonable agreement with the observed law (5·311·3).

5.32. *The thermoelectric power.* It is shown in § 8·61 that the absolute thermoelectric force per degree of a semi-conductor is given by

$$\frac{d\Theta}{dT} = -\frac{1}{\epsilon}\left(2k - \frac{\zeta}{T}\right) \qquad (5 \cdot 32 \cdot 1)$$

if the current is carried exclusively by electrons, whereas the sign must be changed if the current is carried by positive holes. Since the number of free electrons (or free holes) is small, ζ/kT must necessarily be negative and reasonably large for the electron gas to be non-degenerate. Electronic semi-conductors have therefore large thermoelectric powers of the order of a millivolt per degree, excess conductors having negative and deficit conductors positive thermoelectric powers. The thermoelectric powers of metals, on the other hand, are of the order of a microvolt per degree, while electrolytic conductors show no thermoelectric effects.

Survey of some Typical Polar Semi-conductors

5·4. The detailed behaviour of semi-conductors, depending as it does upon impurities about whose state little is known, is of great complexity and cannot be adequately described except in a treatise specially devoted to the subject. We shall therefore conclude this chapter by a brief description of some of the more important types of semi-conductors, restricting the discussion to those general features which are of theoretical interest.† All the properties of semi-conductors which depend upon surface phenomena, such as the technically important properties of rectifiers, amplifiers and barrier layer photocells, are, however, beyond the scope of this book.

5·41. *Silver sulphide.* Silver sulphide exists in two modifications, a rhombic β form and a cubic α form. The rhombic form, which is stable below 179° C., behaves like a typical electronic semi-conductor in which some of the current is carried by positive ions. The α form, on the other hand, has a pure electronic conductivity about 300 times larger than that of the β modification just below the transition point and has a Hall coefficient of $-7\cdot2 \times 10^{-21}$ in Gaussian units, which is practically independent of the temperature (Klaiber, 1929). The number of free electrons in the specimen measured by Klaiber is therefore about 10^{19} per c.c., and it seems probable that α-Ag_2S is an example of a semi-conductor of the second type enumerated in § 5·21 in which the impurity atoms (in this case, silver atoms) are permanently ionized.

The conductivity as measured by Klaiber was about 400 ohm^{-1} cm.$^{-1}$, i.e. $3\cdot6 \times 10^{14}$ e.s.u., and $R\sigma$ was $2\cdot6 \times 10^{-6}$ Gauss^{-1}. Now, according to

---

† To supplement the general account given here, the reader will find detailed descriptions of individual semi-conductors in, for example, *Semi-conducting materials,* edited by H. K. Henisch (1951). In particular, evidence is given there that many compounds, including the lead compounds and alloys used as infra-red detectors, are intrinsic semi-conductors at high temperatures.

equations (8·61·2) and (8·62·6), of pp. 232 and 234,

$$Ro = -\frac{\epsilon}{2c}\left(\frac{\pi}{2mkT}\right)^{\frac{1}{2}} l, \qquad (5·41·1)$$

and this gives l of the order of 10^{-6} cm., which is unexpectedly high, since it is greater than the free path in metals.

5·42. *Zinc oxide.* Zinc oxide is a typical semi-conductor in which the current is carried by electrons which originate from interstitial excess zinc atoms. As in all semi-conductors the conductivity is very sensitive to the previous history, especially the heat treatment, of the particular specimen being measured (v. Baumbach and Wagner, 1933; Fritsch, 1935). Not only does the number of impurity atoms vary but the activation energy is far from constant, decreasing considerably as the amount of impurity is increased. An 'average' value for kb is 0·4 eV.

The Hall coefficient is negative, showing that the current is carried by electrons. A typical example is $R = -4 \times 10^{-18}$ Gaussian units at room temperature, corresponding to $n = 2 \times 10^{16}$ electrons per c.c. approximately. The specimen concerned had $\sigma = 8 \times 10^{-2}$ ohm^{-1} cm.$^{-1}$, i.e. $7·2 \times 10^{10}$ e.s.u., so that Ro was -3×10^{-7} Gauss^{-1}, giving l about 10^{-7} cm.

Fritsch found that the above specimen in a copper/zinc oxide thermo-couple had a thermoelectric force of about $-0·2$ mV. per degree, which is of the right order of magnitude. On the other hand, he found that $d\Theta/dT$ was independent of the temperature from -20 to $60°$ C., which is not in agreement with equation (5·32·1), according to which $d\Theta/dT$ is given by

$$d\Theta/dT = \text{constant} \mp kb/\epsilon T \qquad (5·42·1)$$

(minus for excess and plus for deficit conductors).

v. Baumbach and Wagner measured the thermoelectric power of two different specimens in the thermocouple Ag/ZnO/Ag with the junctions at 550 and 650° C., one specimen being in pure oxygen and the other in air. For specimen 1 in pure oxygen, the thermoelectric force was -86×10^{-5} V./degree, while for specimen 2, in a partial pressure of 0·21 atmosphere of oxygen, the thermoelectric force was -84×10^{-5} V./degree. For the differential thermocouple formed by the two different specimens of zinc oxide, the thermoelectric force would therefore be $d\Theta_{12}/dT = -2 \times 10^{-5}$ V./degree. Now from equation (5·32·1) and the general relation

$$e^{\zeta/kT} = \tfrac{1}{2}nh^3(2\pi mkT)^{-\frac{3}{2}},$$

we have, for two specimens of the same excess semi-conductor,

$$\frac{d\Theta_{12}}{dT} = \frac{k}{\epsilon}\log\frac{n_1}{n_2}. \qquad (5·42·2)$$

The ratio of the conductivities was $\sigma_1/\sigma_2 = 1/1·44$, and if we assume that n_1/n_2 has the same value, equation (5·42·2) gives $d\Theta_{12}/dT = -3 \times 10^{-5}$ V./degree, which is in satisfactory agreement with the observed value, taking into account the fact that it is the difference of two large numbers.

5·43. *Cuprous oxide.* Cuprous oxide is of great technical importance in connexion with rectifiers and barrier layer photoelectric cells (Grondahl, 1933; Henisch, 1949; Torrey and Whitmer, 1948), and it was the first semi-conductor to receive detailed attention. According to the measurements of Vogt (1930) and Engelhard (1933), the conductivity of cuprous oxide can be represented by a formula of the type $\sigma = Ae^{-b/T}$, with kb about 0·3 eV. The Hall coefficient at 0° C. is about 9×10^{-14} in Gaussian units, which shows that the current is carried by positive holes and that their number is of the order of 10^{12} per c.c.

In one of the typical specimens used by Vogt, the number of positive holes was given approximately by $1·1 \times 10^{18} e^{-b/T}$ per c.c. with $kb = 0·33$ eV. In this specimen σ was 6×10^{-6} ohm^{-1}cm.$^{-1}$ at 0° C., i.e. $5·5 \times 10^6$ e.s.u., $R\sigma$ was 5×10^{-7} Gauss^{-1} and l was about 3×10^{-7} cm.

The thermoelectric power in a copper/cuprous oxide thermocouple was found by Vogt to be 1·3 mV. per degree at room temperature and to increase with decreasing temperature more or less according to equation (5·42·1). The positive sign is consistent with the current being due to positive holes. On the other hand, Anderson and Greenwood (1952) found that $d\Theta/dT$ remained constant at about 1·5 mV/degree between 155 and 355° C.

Similar experiments were carried out by Dünwald and Wagner at elevated temperatures, mainly to determine the effect of oxygen concentration upon the conductivity (§ 5·311) and to prove that cuprous oxide is a deficit conductor. They employed a differential thermocouple formed by two cuprous oxide wires with junctions at 900 and 1000° C., one of the wires being in nitrogen containing 1 % of oxygen and the second being in air. They found that $d\Theta_{12}/dT = 2·96 \times 10^{-5}$ V./degree and that $\sigma_1/\sigma_2 = 1/1·52$. The value for $d\Theta_{12}/dT$ calculated from (5·42·2) with the sign changed is therefore $3·6 \times 10^{-5}$ V./degree, which is in reasonable agreement with the observed value.

5·431. The absorption of light by cuprous oxide has been investigated by Schönwald (1932) and Engelhard (1933) in an attempt to disentangle the energy levels involved in the conduction phenomena. In addition to the absorption band of pure cuprous oxide, which has a maximum at $0·63\mu$, there is a weak absorption band in the infra-red round about 2μ. This band is quite broad and extends from about 4μ to 1μ, with its maximum at 2μ (i.e. from $h\nu = 0·3$ eV. to $1·2$ eV., the maximum being at $0·6$ eV.). It is due to the same impurity atoms as cause the 'dark conductivity', and absorption in this band is accompanied by photoconductivity.

It is tempting to identify the threshold energy $h\nu = 0.3$ eV. of the absorption band with the value $kb = 0.3$ eV. found by Vogt from the temperature dependence of the conductivity. But if equation (5·21·4) is correct, then $kb = \frac{1}{2}\Delta E$, where ΔE is the dissociation energy required to produce a free hole, and there is therefore a discrepancy amounting to a factor of 2. An alternative hypothesis is to identify the maximum of the absorption band with kb, and since $h\nu = 0.6$ eV. at the maximum, this would remove the discrepancy. There is, however, the further question whether kb is equal to $\frac{1}{2}\Delta E$ or to ΔE or to something intermediate between the two. According to the calculations in § 5·211, kb is equal to ΔE if the number of atoms to which an electron (or a positive hole) can attach itself is very much greater than the total number of electrons (or positive holes) which are capable of becoming free. Now in any crystal in thermal equilibrium there must be lattice imperfections quite apart from the imperfections due to an excess of atoms of one kind or another. For although a perfect lattice has lower energy than a disordered one, the presence of imperfections results in an increase in the entropy, and, except when $T = 0$, a crystal containing imperfections has lower free energy than a perfect crystal. There is, therefore, the possibility that the conditions envisaged in § 5·211 are fulfilled, in which case kb would be equal to ΔE. On the other hand, the measurements of Dünwald and Wagner on the variation of σ with the oxygen pressure are only easily explicable if the free holes and the lattice defects to which they can attach themselves are present in equivalent numbers, and this leads to $kb = \frac{1}{2}\Delta E$. The situation is therefore very obscure and none of the arguments for or against either hypothesis is conclusive.

It is probable that the apparent agreement between the values of $h\nu$ and of kb is entirely fortuitous, since, as pointed out by de Boer and van Geel (1935), the thermal dissociation energy and the photoelectric excitation energy are different quantities. When an electron is removed from an atom by the action of light, the lattice, by the Franck-Condon principle, does not change, since the ions do not have time to move during the short time in which the process takes place. The lattice is therefore not in equilibrium and the ions surrounding the impurity will in time move into new positions with lower energy. Since the thermal dissociation energy is the energy difference between two equilibrium configurations it must be less than the photoelectric excitation energy, which is the energy difference between two configurations only one of which is an equilibrium one. It therefore appears that in general there is no simple relation between the thermal and photoelectric data.

5·432. If a semi-conductor is very carefully purified, then at a sufficiently high temperature it should be possible to observe the intrinsic conductivity provided that it is not masked by electrolytic conduction. Jusé and Kurtschatow (1932) found that the conductivity of various specimens of cuprous oxide could be represented by the formula $\sigma = A_1 e^{-b_1/T} + A_2 e^{-b_2/T}$, with $A_1 = 100\,\text{ohm}^{-1}\,\text{cm.}^{-1}$ approximately and $kb_1 = 0.72\,\text{eV.}$, while kb_2, and more particularly A_2, varied from specimen to specimen. A_2 was of the order of $1\,\text{ohm}^{-1}\,\text{cm.}^{-1}$, while kb_2 varied between 0.13 and $0.134\,\text{eV.}$ They interpret the first term in σ as the intrinsic conductivity and the second term as that due to excess oxygen. On the other hand, Dünwald and Wagner, whose measurements were carried out at higher temperatures than those of Jusé and Kurtschatow, found no evidence of the intrinsic conductivity. In the more recent and more carefully controlled experiments of Anderson and Greenwood (1952) the existence of two contributions to the conductivity above 355° C. was confirmed. They found, however, that $d\Theta/dT$ was constant and of the order of $1.5\,m\text{V/degree}$ up to 355° C., while it decreased linearly with $1/T$ between 355 and 1000° C. but did not change sign. These results are difficult to reconcile with equation (5·42·1) for the extrinsic range and with equation (8·611·5) for the intrinsic range. It must therefore be concluded that all the details of the conductivity mechanisms in cuprous oxide have not so far been satisfactorily disentangled, and that in particular the anomalies in the thermoelectric power require further investigation.

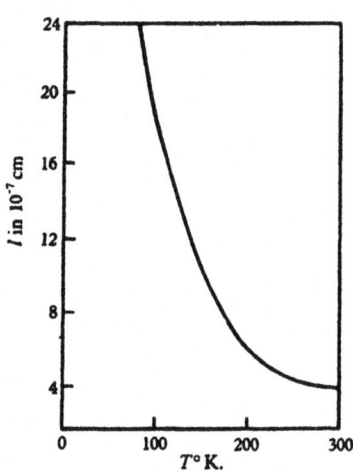

Fig. V 2. The free path in cuprous oxide.

5·433. The collision mechanisms in a semi-conductor are discussed in §§ 9·36 and 9·361. At room temperatures the scattering in a polar semi-conductor is mainly due to the polarization waves of the lattice. The mean free path l, defined by $\sigma = \frac{1}{3}n\varepsilon^2 l/(2\pi mkT)^{\frac{1}{2}}$, is then proportional to $T^{\frac{1}{2}}(e^{h\nu_l/kT} - 1)$, where ν_l is the frequency of the polarization waves, provided that $h\nu_l > kT$. The free path therefore increases rapidly as T decreases. At somewhat lower temperatures, the main cause of the scattering is the acoustical vibrations of the lattice and then l is proportional to $1/T$. At very low temperatures, depending upon the amount of impurity present, the most effective scatterers are the ions which are responsible for the electrons and holes. In this case l is proportional to T^2.

In cuprous oxide, l has been measured by Engelhard down to about 100° K. His results are shown in fig. V 2, and it is clear that the scattering is essentially due to the lattice vibrations, but whether to the polarization waves or to the acoustical waves is by no means certain.

ELEMENTAL SEMI-CONDUCTORS

5·5. The elements silicon, germanium, grey tin, selenium and tellurium were long considered to be semi-conductors before the nature of semi-conductors had been established. They were later held to be insulators, but recent work has established that they are intrinsic semi-conductors. It has so far proved impossible to prepare really homogeneous specimens of selenium and tellurium, and the electrical behaviour of these elements is largely influenced by the presence of inhomogeneous layers. Silicon and germanium, on the other hand, can be prepared in the pure state and their properties can be controlled by the addition of known amounts of other elements. It is therefore possible to obtain results with silicon and germanium which have a much clearer theoretical interpretation than those relating to semi-conducting compounds, and a large amount of experimental work has been carried out in recent years in connexion with the technically important topics of rectifiers and amplifiers (transistors). It is only possible to give a brief summary of the work here, much of which has been published in short notes. Detailed references are available in Shockley's book (1950).

5·51. *Silicon.* In the pure state, silicon has a measurable electronic conductivity above about 500° C., which can be fitted by the formula

$$\sigma = A\,e^{-\frac{1}{2}\Delta E/kT},$$

with $A = 9 \times 10^3\,\text{ohm}^{-1}\,\text{cm.}^{-1} = 8 \times 10^{15}\,\text{e.s.u.}$ and $\Delta E = 1\cdot12\,\text{eV}.$

(Pearson and Bardeen, 1949). It forms substitutional alloys with trivalent and pentavalent elements, and the alloys are conducting at much lower temperatures than pure silicon. The trivalent elements such as boron and aluminium act as acceptors of electrons and give alloys which have positive Hall coefficients at normal and low temperatures, whereas phosphorus, antimony and arsenic act as donors of electrons and give alloys with negative Hall coefficients.

The great advantage of investigating these substitutional alloys is that the conditions can be so arranged that there are, for example, only holes present, and therefore the fundamental parameters determining the motion of holes can be determined. Similarly, by carrying out measurements on specimens in which the current is entirely due to electrons, the fundamental parameters for electrons can be found. If then we have to consider cases in which both holes and electrons are present simultaneously, we have sufficient knowledge of the various parameters to interpret the measurements. (This assumes that the free path for holes is not influenced, for example, by the number of electrons, which is not entirely true if there are a large number of ionized atoms from which the electrons are derived, but it will be true as a rough approximation.)

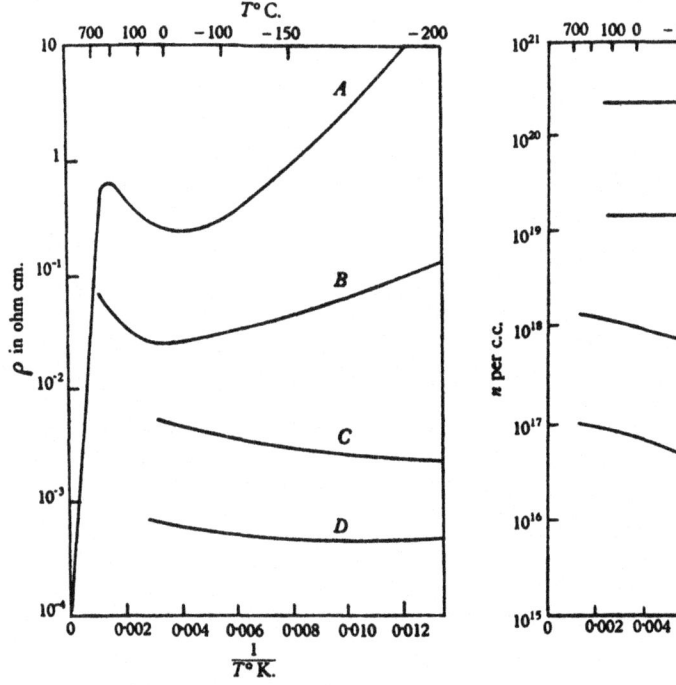

Fig. V 3. The resistivity of silicon-phosphorus alloys as a function of $1/T$. The numbers of phosphorus atoms per c.c. are as follows: A, $4·7 \times 10^{17}$; B, $2·7 \times 10^{18}$; C, $4·7 \times 10^{19}$; D, $4·7 \times 10^{20}$.

Fig. V 4. The number of electrons in silicon-phosphorus alloys as a function of $1/T$.

The resistivities of a number of silicon-phosphorus alloys are shown in fig. V 3, and in the case of the purest specimen the intrinsic conductivity is clearly predominant at high temperatures. The results can be analysed as follows. In the extrinsic range we can determine n from measurements of the Hall coefficient by the relation $R = -3\pi/(8nec)$ if the electrons are non-degenerate. If the amount of phosphorus present is so large that the electrons must be treated as degenerate, then we have to use the formula $R = -1/(nec)$ to determine n, but the difference between the formulae is scarcely significant. The values of n determined in this way are shown in fig. V 4. When n has been found, the mean free path l, defined by

$$\tfrac{3}{4}(2\pi mkT)^{\frac{1}{2}}\,\sigma/ne^2$$

if the electrons are non-degenerate, can be obtained. Corresponding measurements on silicon-boron alloys give the analogous results for the number of holes and their free paths. The free path l_h for holes varies with temperature very much like l_e, but l_e is about three times as large as l_h. At normal temperatures $l_e = 6 \times 10^{-4}T^{-1}$ cm. and $l_h = 2 \times 10^{-4}T^{-1}$ cm.

In the intrinsic range we have to use the formulae (8·611·6) and (8·62·5) for σ and R, namely,

$$\sigma = \frac{4}{3}\frac{\epsilon^2}{(2\pi kT)^{\frac{1}{2}}}\left(\frac{n_1 l_1}{m_1^{\frac{3}{2}}}+\frac{n_2 l_2}{m_2^{\frac{3}{2}}}\right),$$

$$R = -\frac{3\pi}{8ec}\frac{n_1 l_1^2/m_1 - n_2 l_2^2/m_2}{(n_1 l_1/m_1^{\frac{3}{2}}+n_2 l_2/m_2^{\frac{3}{2}})^2}.$$

Now the difference between the number of holes and the number of electrons is equal to the number of donor atoms present, which is known, and, if we assume that the free paths (or rather $l_1/m_1^{\frac{3}{2}}$, $l_2/m_2^{\frac{3}{2}}$) are the same as when only one type of carrier is present, the measurement of σ and R is sufficient to determine the number of electrons and holes separately. The results for one specimen are given in fig. V 5. It is then possible to deduce from the curves in fig. V 4 for the extrinsic range and from the curves in fig. V 5 for the

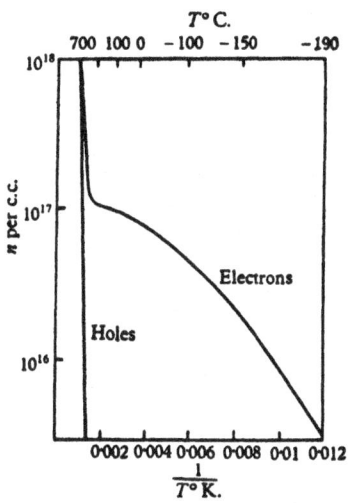

Fig. V 5. The numbers of electrons and holes as functions of $1/T$ for a silicon-phosphorus alloy containing $4·7 \times 10^{17}$ atoms of phosphorus per c.c.

intrinsic range, the values of the energies ΔE_1 and ΔE required to activate an electron from a donor atom and from the filled band of pure silicon. The formulae required are (5·21·4), and (5·22·7) with ζ determined by (5·22·2). It is found that $\Delta E = 1·12$ eV. and $\Delta E_1 = 0·054$ eV. for very dilute alloys if we assume that $m_1 = m_2 = m$.

Exactly similar results are found for the silicon-boron alloys, except that the Hall coefficient is positive at low and normal temperatures and becomes negative at high temperatures where the intrinsic conductivity is dominant. It is found that the energy required to activate an electron on to an acceptor atom is $\Delta E_2 = 0·08$ eV. for very dilute alloys.

5·511. Although the free path is inversely proportional to T at high temperatures there are considerable deviations at low temperatures (see fig. V 6), and it is found that

$$1/l = aT + b/T^2 \qquad (5\text{·}511\text{·}1)$$

approximately, and therefore that there must be two scattering mechanisms. It is shown in § 9·36 that, in a monatomic crystal, scattering by the lattice vibrations gives rise to a free path proportional to $1/T$, and we therefore attribute the first term in (5·511·1) to the effect of the lattice vibrations. It is further shown in § 9·361 that, in the scattering of the electrons by the ions from which the free electrons are derived, the time of relaxation is proportional to the cube of the velocity and that the mean free path is proportional to T^2. We therefore identify the second term in (5·511·1) with

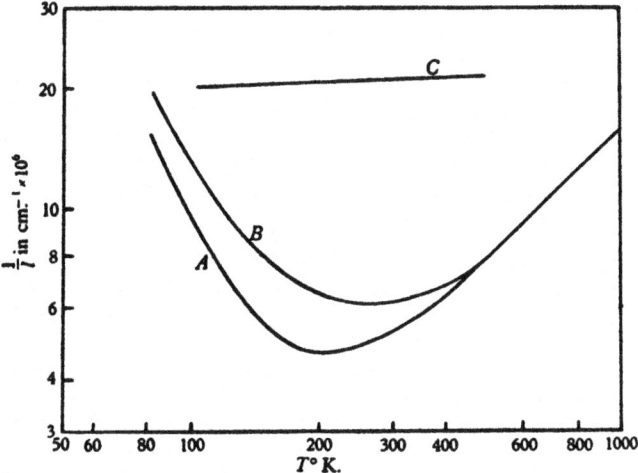

Fig. V 6. The reciprocal of the free path in silicon-phosphorus alloys as a function of T, plotted on a logarithmic scale. Specimens *A, B, C* as in fig. V 3.

the impurity scattering, which in semi-conductors must be considered as due to impurity ions and not to impurity atoms as in metals.

When τv is not independent of the velocity the formulae used in analysing the conduction data need revision (see § 8·631). In particular, the Hall coefficient is given by

$$R = -\frac{315}{512}\frac{\pi}{nec}$$

for the ionic impurity scattering of electrons. These refinements could be taken into account without difficulty, but they would not fundamentally alter the interpretation. They are, however, important in connexion with the magneto-resistance effects, as discussed in § 8·64, since these are very sensitive to the form of the scattering law.

5·512. It is found that the activation energies ΔE_1 and ΔE_2 of the donor and acceptor atoms decrease as the number of impurity atoms increases, a phenomenon which is common to all semi-conductors. For silicon-boron alloys, ΔE_2 in eV. is given by

$$\Delta E_2 = 0.08 - 4.3 \times 10^{-8} n_0^{\frac{1}{3}},$$

where n_0 is the number of acceptor atoms per unit volume, so that ΔE_2 becomes zero for $n_0 = 5.6 \times 10^{18}$ per c.c. Owing to the high dielectric constant of the medium, the orbit of an electron attached to an acceptor atom is large, and it is not surprising that the orbits of individual electrons tend to overlap when the number of impurities is quite low. Any overlap will result in ΔE_2 depending explicitly on n_0. As soon as ΔE_2 becomes zero the number of holes will be independent of the temperature.

5·52. *Germanium.* The properties of germanium and its substitutional alloys are very similar to those of silicon (see, for example, Dunlap (1950), Putley (1949) and Shockley (1950)). The energy difference between the full and the conduction band is given by $\Delta E = 0.76$ eV. approximately, while $\Delta E_1 = 0.03$ eV. in germanium-antimony alloys and $\Delta E_2 = 0.04$ eV. approximately in germanium-aluminium alloys at infinite dilution.

In many specimens the number of free electrons is such that they can no longer be treated as a non-degenerate gas (this applies to silicon crystals also), and it is necessary to generalize the preceding results so as to deal with the transition from classical to quantum statistics.

If we consider semi-conductors with donor impurities in the extrinsic range, the Fermi energy ζ is determined by equation (5·21·1), which can be written as

$$\frac{n_0}{1 + 2e^{(\Delta E_1 + \zeta)/kT}} = 4\pi \frac{(2mkT)^{\frac{3}{2}}}{h^3} F_{\frac{1}{2}}(\xi),$$

where
$$F_{\frac{1}{2}}(\xi) = \int_0^\infty \frac{x^{\frac{1}{2}} dx}{e^{x-\xi} + 1} \qquad (\xi = \zeta/kT).$$

The function $F_{\frac{1}{2}}(\xi)$ has been tabulated by McDougall and Stoner (1938), and their tables can be used to determine ζ when the approximate analytical expressions for $F_{\frac{1}{2}}(\xi)$ are inadequate. Some computed values of ξ and n are given in Table V 1 for $\Delta E_1 = 0.03$ eV. and $T = 290°$ K., and it is clear that the electrons must be considered to form a degenerate gas when the number of impurities is large. The variation of ζ/kT and n with T for $n_0 = 10^{20}$ per c.c. is shown in Table V 2.

Table V 1. *The variation of ζ and the number of free electrons with the number of impurities for $\Delta E_1 = 0.03$ eV. and $T = 290°$ K.*

n_0 per c.c.	10^{18}	10^{19}	10^{20}	10^{21}	10^{22}
ζ/kT	-3.4	-1.6	-0.2	1.2	2.7
n per c.c.	7.9×10^{17}	4.5×10^{18}	1.6×10^{19}	4.3×10^{19}	9.5×10^{19}

Table V 2. *The variation of ζ and the number of free electrons with*
temperature ($n_0 = 10^{20}$ per c.c., $\Delta E_1 = 0·03$ eV.)

T in °K.	10	50	100	150	200	250
ζ/kT	$-14·5$	$-1·8$	$-0·50$	$-0·22$	$-0·15$	$-0·17$
n per c.c.	$7·7 \times 10^{10}$	$2·7 \times 10^{17}$	$2·5 \times 10^{18}$	$5·7 \times 10^{18}$	$9·3 \times 10^{18}$	$1·3 \times 10^{19}$
T in °K	290	350	400	450	500	550
ζ/kT	$-0·20$	$-0·27$	$-0·33$	$-0·40$	$-0·46$	$-0·53$
n per c.c.	$1·6 \times 10^{19}$	2×10^{19}	$2·3 \times 10^{19}$	$2·5 \times 10^{19}$	$2·8 \times 10^{19}$	$4·5 \times 10^{19}$

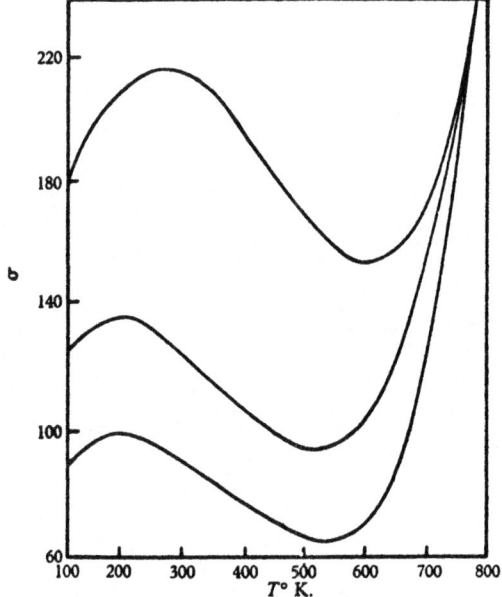

Fig. V 7. The conductivity of various specimens of germanium as a function
of T. ($1/\sigma$ is measured in ohm cm.)

It will be seen that at low temperatures the number of free electrons is
small and that they form a non-degenerate gas. As the temperature is
increased the number of free electrons increases exponentially until there
are so many free electrons that ζ/kT becomes small and the electron gas
must be treated as degenerate. As the temperature is increased still further,
ζ/kT reaches a maximum and then decreases steadily, so that at sufficiently
high temperatures the electron distribution function is classical once more.
This occurs when the majority of the impurity atoms are ionized, since then
the slow increase in the number of free electrons with increasing temperature is insufficient to outweigh the effect of the high temperature.

At still higher temperatures it is necessary to take the intrinsic conductivity into account and to calculate ζ from (5·22·1). Ultimately we must

have $\zeta = -\frac{1}{2}\Delta E$, where ΔE is the band separation in pure germanium, and the number of free electrons will increase with temperature like $e^{-\frac{1}{2}\Delta E/kT}$.

If we do not assume that the electrons are either degenerate or non-degenerate, then according to (1·9·4) the conductivity must be determined from the exact formula

$$\sigma = -\frac{16\sqrt{2}\,\pi m^{\frac{1}{2}}\epsilon^2}{3h^3}\int \tau E^{\frac{3}{2}}\frac{\partial f_0}{\partial E}\,dE,$$

if carriers of only one sign are present. In a monatomic semi-conductor τv is independent of v unless the temperature is sufficiently low for the impurity scattering to be important. At moderate and high temperatures, therefore, σ can be written as

$$\sigma = -\frac{16\pi m l\epsilon^2}{3h^3}\int_0^\infty E\,\frac{d}{dE}\frac{1}{e^{(E-\zeta)/kT}+1}\,dE = \frac{16\pi m l\epsilon^2}{3h^3}kT\log\left(1+e^{\zeta/kT}\right),$$

where l is proportional to $1/T$. If the variation of ζ/kT with T is as given in Table V 2, σ will increase rapidly at first as T increases from a low value and will reach a maximum at the temperature where ζ/kT is a maximum. At higher temperatures σ will decrease with increasing T until the intrinsic conductivity sets in, when it will again increase rapidly.

The measured values of σ for three specimens are shown in fig. V 7, and it will be seen that the curves follow the theoretical predictions. Above 750° K. all the curves coincide due to the intrinsic conductivity becoming dominant.

REFERENCES

Anderson, J. S. and Greenwood, N. N. (1952). The semi-conducting properties of cuprous oxide. *Proc. Roy. Soc.* A, **215**, 353.

v. Baumbach, H. H. and Wagner, C. (1933). The electrical conductivity of zinc oxide and cadmium oxide. *Z. Phys. Chem.* B, **22**, 199.

de Boer, J. H. and van Geel, W. C. (1935). The long wave limit of the inner photo-electric effect in semi-conductors. *Physica*, **2**, 286.

de Boer, J. H. and Verwey, E. J. W. (1937). Semi-conductors with partially and completely filled 3d-bands. *Proc. Phys. Soc.* **49** (extra part), 59.

Dunlap, W. C. (1950). Some properties of high resistivity p-type germanium. *Phys. Rev.* **79**, 286.

Dünwald, H. and Wagner, C. (1933). The electrical properties of cuprous oxide. *Z. Phys. Chem.* B, **22**, 212.

Engelhard, E. (1933). The dark and photoelectric conductivities in cuprous oxide. *Ann. Phys., Lpz.* (5), **17**, 501.

Fritsch, O. (1935). Electrical measurement on zinc oxide. *Ann. Phys., Lpz.* (5), **22**, 375.

Grondahl, L. O. (1933). The copper-cuprous oxide rectifier and photoelectric cell. *Rev. Mod. Phys.* **5**, 141.

Gudden, B. (1930). Electrical conductivity in semi-conductors. *S.B. phys.-med. Soz. Erlangen,* **62**, 289.

Gudden, B. (1934). Elektrische Leitfähigkeit elektronischer Halbleiter. *Ergebn. exakt. Naturw.* **13** (Berlin).

Henisch, H. K. (1949). *Metal rectifiers* (Oxford).

Henisch, H. K. (1951). *Semi-conducting materials* (London).

Jusé, W. and Kurtschatow, B. W. (1932). The electrical conductivity of cuprous oxide. *Phys. Z. Sowjet.* **2**, 453.

Klaiber, F. (1929). Hall effect and conductivity of silver sulphide. *Ann. Phys., Lpz.* (5), **3**, 229.

McDougall, J. and Stoner, E. C. (1938). The computation of Fermi-Dirac functions. *Philos. Trans.* A, **237**, 67.

Pearson, G. L. and Bardeen, J. (1949). Electrical properties of pure silicon and silicon alloys containing boron and phosphorus. *Phys. Rev.* **75**, 865.

Putley, E. H. (1949). The electrical conductivity of germanium. *Proc. Phys. Soc.* A, **62**, 284.

Schönwald, B. (1932). The measurement of photoelectric currents in semi-conductors. *Ann. Phys., Lpz.* (5), **15**, 395.

Shockley, W. (1950). *Electrons and holes in semi-conductors* (New York).

Torrey, H. C. and Whitmer, C. A. (1948). *Crystal rectifiers* (New York).

Vogt, W. (1930). Electrical measurements on cuprous oxide. *Ann. Phys., Lpz.* (5), **7**, 183.

Wagner, C. (1933). The nature of the electrical conductivity of α-silver sulphide. *Z. Phys. Chem.* B, **21**, 42.

Wilson, A. H. (1931a). The theory of electronic semi-conductors. *Proc. Roy. Soc.* A, **133**, 458.

Wilson, A. H. (1931b). The theory of electronic semi-conductors. II. *Proc. Roy. Soc.* A, **134**, 277.

Chapter VI

THE THERMAL AND MAGNETIC
PROPERTIES OF METALS

The Lattice Specific Heat

6·1. For small amplitudes of vibration (i.e. for temperatures considerably less than the melting-point) the motion of the atoms in a crystal can be considered to be simple-harmonic. Since the motions of the various atoms are coupled together, the normal modes are oscillations of the whole assembly of atoms. If atoms of only one mass are present, the normal oscillations are the ordinary sound waves in the crystal, which can be taken to be standing waves (in a finite crystal) or travelling waves (in a cyclical or an infinite crystal) as convenience may dictate. The sound waves can be characterized by a wave vector $\mathbf{q}$, and for each value of $\mathbf{q}$ there are three independent waves, which may be classified as one longitudinal and two transverse. In general, the frequency ν of a wave of given type is a complicated function of $\mathbf{q}$, but the approximate behaviour of ν can be obtained by considering the crystal to be an elastic continuum in which the velocity of propagation is independent of the frequency. The displacement being of the form $e^{2\pi i \nu t - i \mathbf{q} \cdot \mathbf{r}}$, the wave-length λ is $2\pi/|\mathbf{q}|$, and, if u_i ($i = 1, 2, 3$) is the velocity of the sound waves ($i = 1$ for waves of longitudinal type, $i = 2, 3$ for waves of transverse type), the usual relation $\lambda\nu = u_i$ gives $\nu = u_i |\mathbf{q}|/2\pi$. This cannot, however, be universally true, since in a crystal lattice there is a lower limit to the wave-length (and therefore an upper limit to $\mathbf{q}$), whereas in a continuum, waves of any length can be present. We consider therefore a cyclical crystal containing $(2G + 1)^3$ unit cells, which we take to be simple cubic of side a. The values of q_1, q_2, q_3 are then restricted to the set

$$2\pi(0, \pm 1, \pm 2, ..., \pm G)/\{(2G + 1)a\},$$

i.e. there are $(2G + 1)^3$ in all, making the total number of vibrations, both longitudinal and transverse, equal to $3(2G + 1)^3$, which is the same as the number of degrees of freedom of the $(2G + 1)^3$ atoms, as of course it must be.

For infinite G, the values of q_1, q_2, q_3 lie within the cube

$$-\pi \leqslant a(q_1, q_2, q_3) \leqslant \pi.$$

It is usually convenient, and often sufficiently accurate, to consider the wave vector, instead of lying inside the cube, to lie within a sphere of equal volume, the radius of which is

$$q_0 = \left(\frac{3}{4\pi}\right)^{\frac{1}{3}} \frac{2\pi}{a}. \tag{6·1·1}$$

If we further assume for simplicity that the relation between ν and $\mathbf{q}$ is that given above for an elastic continuum, the maximum frequency for waves of the ith type is

$$\nu_{Di} = u_i \frac{q_0}{2\pi} = \left(\frac{3}{4\pi}\right)^{\frac{1}{3}} \frac{u_i}{a}. \qquad (6·1·2)$$

6·11. The energy content of a harmonic oscillation of frequency ν is given by Planck's original formula

$$\frac{h\nu}{e^{h\nu/kT} - 1}.$$

Let $g(\nu)\,d\nu$ be the number of normal modes of vibration per unit volume of a crystal with frequencies lying between ν and $\nu + d\nu$. Then the energy of the crystal per unit volume due to the thermal vibrations is

$$U = \int_0^\infty \frac{h\nu g(\nu)}{e^{h\nu/kT} - 1}\,d\nu, \qquad (6·11·1)$$

where $g(\nu)$ is zero when ν is greater than the maximum frequency of vibration. If we adopt the approximation discussed in the preceding section of treating the lattice as a continuum but with a maximum frequency, then†

$$g(\nu) = 4\pi \left(\frac{1}{u_1^3} + \frac{1}{u_2^3} + \frac{1}{u_3^3}\right) \nu^2, \qquad (6·11·2)$$

and

$$U = \sum_{i=1}^3 \frac{4\pi}{u_i^3} \int_0^{\nu_{Di}} \frac{h\nu^3\,d\nu}{e^{h\nu/kT} - 1}. \qquad (6·11·3)$$

It is usual to ignore the differences between the separate u_i's and ν_{Di}'s and to replace them by average values u and ν_D. Then, on eliminating u_i by means of $(6·1·2)$, and replacing $1/a^3$ by N, the number of atoms per unit volume, U becomes

$$U = \frac{9N}{\nu_D^3} \int_0^{\nu_D} \frac{h\nu^3\,d\nu}{e^{h\nu/kT} - 1}. \qquad (6·11·4)$$

The approximate theory given above is due to Debye (1912) and the characteristic temperature Θ given by $h\nu_D = k\Theta$ is known as the Debye temperature. With this notation, $(6·11·4)$ becomes

$$U = \frac{9NkT^4}{\Theta^3} \int_0^{\Theta/T} \frac{x^3\,dx}{e^x - 1} = 3NkT D\left(\frac{\Theta}{T}\right), \qquad (6·11·5)$$

where

$$D(x) = \frac{3}{x^3} \int_0^x \frac{t^3\,dt}{e^t - 1}. \qquad (6·11·6)$$

† Denote the values of q_1, q_2, q_3 for a cyclical crystal by $2\pi(\kappa_1, \kappa_2, \kappa_3)/\{(2G+1)\,a\}$. The values of κ_1, κ_2, κ_3 are then (positive or negative) integers, and the number of vibrations in the range $\Delta\kappa_1\Delta\kappa_2\Delta\kappa_3$ is clearly $\Delta\kappa_1\Delta\kappa_2\Delta\kappa_3 = (2G+1)^3\,a^3 dq_1 dq_2 dq_3/(2\pi)^3$. The number of vibrations per unit volume is therefore $dq_1 dq_2 dq_3/(2\pi)^3$. Expressed in terms of ν, this becomes $4\pi\nu^2 d\nu/u^3$ for waves of type i.

For small values of x, the denominator in the integral (6·11·6) can be expanded in ascending powers of x, giving

$$D(x) = 1 - \tfrac{3}{8}x + \tfrac{1}{20}x^2 + \dots \tag{6·11·7}$$

For large values of x, the upper limit in the integral can be replaced by infinity with an error of order e^{-x}. Then, since

$$\int_0^\infty \frac{t^3\,dt}{e^t - 1} = \sum_{n=1}^\infty \int_0^\infty t^3 e^{-nt}\,dt = 3! \sum_{n=1}^\infty \frac{1}{n^4} = \frac{\pi^4}{15},$$

we have
$$D(x) \sim \tfrac{1}{5}\pi^4 x^{-3} + O(e^{-x}) \tag{6·11·8}$$

for large x. Therefore

$$U \sim 3NkT \quad (T \gg \Theta),$$

and
$$U \sim \tfrac{3}{5}\pi^4 NkT^4/\Theta^3 \quad (\Theta \gg T),$$

so that
$$C_v \sim 3Nk \quad (T \gg \Theta), \tag{6·11·9}$$

and
$$C_v \sim \tfrac{12}{5}\pi^4 Nk(T/\Theta)^3 \quad (\Theta \gg T). \tag{6·11·10}$$

The general expression for C_v is

$$C_v = 9Nk(T/\Theta)^3 \mathscr{J}_4(\Theta/T), \tag{6·11·11}$$

where
$$\mathscr{J}_n\!\left(\frac{\Theta}{T}\right) = \int_0^{\Theta/T} \frac{z^n\,dz}{(e^z - 1)(1 - e^{-z})}, \tag{6·11·12}$$

as is most easily seen by differentiating (6·11·4).

6·12. The Debye theory gives a simple and excellent account of the specific heat of solids in spite of the crude nature of some of the approximations made. The high-temperature form of C_v is a consequence of the fact that at sufficiently high temperatures the energy content of an oscillator is $3kT$, no matter what its frequency may be. The low-temperature form, on the other hand, depends upon $g(\nu)$ being proportional to ν^2 for sufficiently small ν; this must be true, since for sufficiently long waves a crystal lattice must behave like a continuum. For intermediate temperatures, however, the specific heat must depend upon the exact form of $g(\nu)$, and deviations from Debye's formula are bound to occur.

The calculation of $g(\nu)$ for more complicated models was carried out simultaneously with the work of Debye and has also received considerable attention in recent years (Born and von Kármán, 1912; Blackman, 1935, 1941). The principles involved are most easily appreciated by considering a linear chain of atoms executing longitudinal vibrations.

6·13. *A simple linear chain.* Consider an infinite linear chain of equal atoms of mass m at a normal distance a apart, and suppose that the forces on the atoms are elastic forces and are only exerted by each atom on its two neighbours. Let x_g be the displacement of the atom g along the chain. Then the equations of motion are

$$m\,d^2x_g/dt^2 = -\lambda\{(x_g - x_{g-1}) + (x_g - x_{g+1})\}, \qquad (6\cdot13\cdot1)$$

where λ is the elasticity constant. To solve this put $x_g = \alpha(t)\,e^{iqga}$. Then

$$m\,d^2\alpha/dt^2 = -(4\lambda\sin^2\tfrac{1}{2}qa)\,\alpha,$$

and $\alpha \propto \cos(2\pi\nu_q t + \delta)$, where δ is the phase and where

$$2\pi\nu_q = 2(\lambda/m)^{\frac{1}{2}}\,|\sin\tfrac{1}{2}qa\,|. \qquad (6\cdot13\cdot2)$$

The behaviour of ν_q as a function of q is shown in fig. VI 1. It is a periodic function of q with period $2\pi/a$, and to obtain each vibration once only it is necessary to restrict the range of q, which can be done in two equivalent ways.

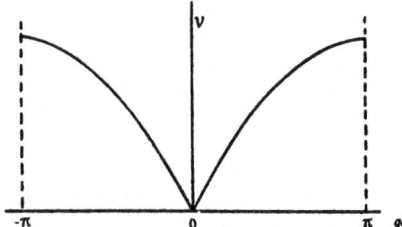

Fig. VI 1. The frequency spectrum for a simple linear chain.

In the solution given above no account has been taken of the requirement that x_g must be real. To introduce the reality condition we write

$$x_g = \alpha(t)\,e^{iqga} + \alpha(t)^*\,e^{-iqga}, \qquad (6\cdot13\cdot3)$$

so that

$$x_g = (A\cos qga + B\sin qga)\cos(2\pi\nu_q t + \delta), \qquad (6\cdot13\cdot4)$$

where A and B are real independent constants. For each value of q there are therefore two possible vibrations, but, on the other hand, the q vibration is the same as the $-q$ vibration, so that only positive values of q must be considered. In other words, we obtain all frequencies by restricting q to the range $0 \leqslant qa \leqslant \pi$, but there are two vibrations with a given frequency.

An alternative description of the vibrations is in terms of travelling waves. An obvious solution of (6·13·1) is

$$x_g = C\cos(2\pi\nu_q t - qga + \delta), \qquad (6\cdot13\cdot5)$$

with ν_q given by (6·13·2). This represents a travelling wave, and the oscillations with q and $-q$ are different, since they represent waves travelling in opposite directions. To obtain all modes of vibration once only, we have therefore to consider the range of q to be $-\pi \leqslant qa \leqslant \pi$, but to each value of q there now corresponds one vibration, so that the total number of independent vibrations is the same whether standing or travelling waves are used. (Note that in either method of describing the vibrations, $q = 0$ gives rise to one vibration only.)

Normal coordinates. The kinetic energy $\frac{1}{2}m\Sigma\dot{x}_g^2$ is a sum of squares, but the potential energy is not. The two quadratic forms can be reduced simultaneously to their principal axes in the usual way by introducing the normal coordinates found in the preceding section. Consider a cyclical lattice containing $2G + 1$ atoms, and put

$$x_g = \frac{1}{\sqrt{2}\,(2G+1)^{\frac{1}{2}}}\sum_q (\alpha_q e^{iqga} + \alpha_q^* e^{-iqga}). \qquad (6\cdot13\cdot6)$$

The values of q are then given by

$$aq = 2\pi(0, 1, 2, ..., G)/(2G+1). \qquad (6\cdot13\cdot7)$$

With $(6\cdot13\cdot6)$ the kinetic energy is

$$T = \frac{m}{4(2G+1)}\sum_{q,\,q'}\sum_g (\dot{\alpha}_q\dot{\alpha}_{q'}\,e^{i(q+q')ga} + \dot{\alpha}_q\dot{\alpha}_{q'}^*\,e^{i(q-q')ga}$$
$$+ \dot{\alpha}_q^*\dot{\alpha}_{q'}\,e^{i(q'-q)ga} + \dot{\alpha}_q^*\dot{\alpha}_{q'}^*\,e^{-i(q+q')ga}). \quad (6\cdot13\cdot8)$$

Now
$$\sum_{g=0}^{2G} e^{i(q+q')ga} = \frac{1 - e^{i(q+q')(2G+1)a}}{1 - e^{i(q+q')a}},$$

and since the values of q and q' are given by $(6\cdot13\cdot7)$, this is zero unless $q = q' = 0$. Similarly, the last term in $(6\cdot13\cdot8)$ vanishes unless $q = q' = 0$, while the second and third are non-zero if $q = q'$. Hence T reduces to

$$T = \tfrac{1}{4}m(\dot{\alpha}_0^2 + \dot{\alpha}_0^{*2} + 2\sum_q \dot{\alpha}_q\dot{\alpha}_q^*). \qquad (6\cdot13\cdot9)$$

If we put $\alpha_0 = \frac{1}{2^{\frac{1}{2}}}A_0$, $B_0 = 0$, and $\alpha_q = A_q + iB_q$ $(q \neq 0)$, T becomes

$$T = \tfrac{1}{2}m\sum_q \left\{ \left(\frac{dA_q}{dt}\right)^2 + \left(\frac{dB_q}{dt}\right)^2 \right\}, \qquad (6\cdot13\cdot10)$$

there being $2G + 1$ terms in all.

The potential energy is treated similarly. We have

$$V = \tfrac{1}{2}\lambda \sum_{g=0}^{2G} (x_g - x_{g-1})^2$$

$$= \frac{\lambda}{4(2G+1)}\sum_{q,\,q'}\sum_g \{\alpha_q\alpha_{q'}\,e^{i(q+q')ga}\,(1 - e^{-iqa})\,(1 - e^{-iq'a})$$
$$+ \alpha_q\alpha_{q'}^*\,e^{i(q-q')ga}\,(1 - e^{-iqa})\,(1 - e^{iq'a})$$
$$+ \alpha_q^*\alpha_{q'}\,e^{i(q'-q)ga}\,(1 - e^{iqa})\,(1 - e^{-iq'a})$$
$$+ \alpha_q^*\alpha_{q'}^*\,e^{-i(q+q')ga}\,(1 - e^{iqa})\,(1 - e^{iq'a})\},$$

and the only non-vanishing terms give

$$V = 2\lambda \sum_q \alpha_q \alpha_q^* \sin^2 \tfrac{1}{2}qa = \tfrac{1}{2}m \sum_q 4\pi^2 \nu_q^2 (A_q^2 + B_q^2), \qquad (6\cdot13\cdot11)$$

where ν_q is given by (6·13·2), so that the total energy is

$$\mathscr{H} = T + V = \tfrac{1}{2}m \sum_q \left\{ \left(\frac{dA_q}{dt}\right)^2 + \left(\frac{dB_q}{dt}\right)^2 + 4\pi^2 \nu_q^2 (A_q^2 + B_q^2) \right\}, \qquad (6\cdot13\cdot12)$$

where the summation is over the $G+1$ values of q given in equation (6·13·7).

In general, the equilibrium properties of the lattice can be calculated from the energy spectrum without reference to the normal coordinates. The latter are, however, required when dealing with interaction effects (cf. § 9·2).

Fig. VI 2. A linear chain with a basis.

6·14. *A linear lattice with a basis.* As a more complicated example consider a linear chain with lattice constant $a+b$, with two atoms in each unit cell, the masses of the atoms being m and M and being placed unsymmetrically in the cell (see fig. VI 2). If we label the m atoms as the even atoms, and the M atoms as the odd atoms, the equation of motion can be written

$$\left. \begin{aligned} m\ddot{x}_{2g} &= -\lambda(x_{2g} - x_{2g+1}) - \mu(x_{2g} - x_{2g-1}), \\ M\ddot{x}_{2g+1} &= -\lambda(x_{2g+1} - x_{2g}) - \mu(x_{2g+1} - x_{2g+2}). \end{aligned} \right\} \qquad (6\cdot14\cdot1)$$

If we leave aside the reality conditions, these equations can be solved by putting $\quad x_{2g} = A \exp[igq(a+b)], \quad x_{2g+1} = B \exp[i(g+1)qa + igqb]$

and $\qquad d^2A/dt^2 = -4\pi^2\nu_q^2 A, \quad d^2B/dt^2 = -4\pi^2\nu_q^2 B.$

Then $\qquad (4\pi^2 m\nu_q^2 - \lambda - \mu) A + (\lambda e^{iqa} + \mu e^{-iqb}) B = 0,$

$$(\lambda e^{-iqa} + \mu e^{iqb}) A + (4\pi^2 M\nu_q^2 - \lambda - \mu) B = 0,$$

which are only compatible if

$$(4\pi^2 m\nu_q^2 - \lambda - \mu)(4\pi^2 M\nu_q^2 - \lambda - \mu) - \{\lambda^2 + \mu^2 + 2\lambda\mu \cos q(a+b)\} = 0.$$

The roots of this are

$$4\pi^2\nu_q^2 = [(m+M)(\lambda+\mu) \pm \sqrt{\{(\lambda+\mu)^2(m+M)^2}$$
$$- 16\lambda\mu mM \sin^2 \tfrac{1}{2}q(a+b)\}]/(2mM), \qquad (6\cdot14\cdot2)$$

so that the ν_q, q curve now consists of two branches, the general shape of which is shown in fig. VI 3. The branch relating to the minus sign before the square root is called the acoustical branch, and for this branch $\nu_q \to 0$

as $q \to 0$. The other branch is called the optical branch, and, as $q \to 0$, ν_q tends to the non-zero limit $[(m+M)(\lambda+\mu)/(4\pi^2 mM)]^{\frac{1}{2}}$. The names are appropriate, since for the acoustical branch the velocity of propagation tends to a finite limit as $q \to 0$, namely, the velocity in a continuous medium, whereas the velocity of propagation for the optical branch tends to infinity as $q \to 0$. (A finite velocity of the order of the velocity of light would be obtained if the retardation of the forces acting on the atoms were taken into account.)

The reality conditions are the same as for the simple lattice, and the arguments need not be repeated. We have the choice of standing waves,

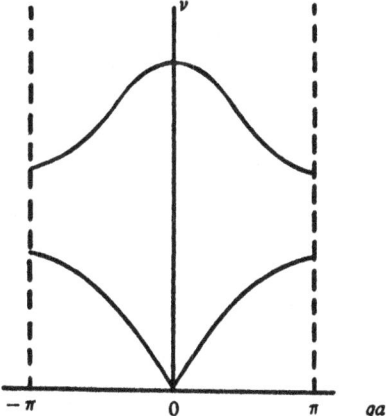

Fig. VI 3. The frequency spectrum of a composite lattice as a function of the reduced wave number.

with q lying in the range $0 \leqslant qa \leqslant \pi$ and with two normal modes to each frequency, or of travelling waves with q lying in the range $-\pi \leqslant qa \leqslant \pi$ and with one normal mode to each frequency. The only novelty is that there are two frequencies (one acoustical and one optical) for each value of q. We can, if we wish, make ν_q a single-valued function of qa for travelling waves by transferring the optical branch to the ranges $(-2\pi, -\pi)$ and $(\pi, 2\pi)$, since qa can be increased by an integral multiple of 2π without affecting any of the physical quantities. Similarly, ν_q can be made a single-valued function of qa for standing waves by assigning the range $(0, \pi)$ to the acoustical and the range $(\pi, 2\pi)$ to the optical branch (see fig. VI 4).

6·15. *Three-dimensional lattices.* The principles involved in treating the vibrations of three-dimensional lattices are the same as for one-dimensional lattices, but the details are much more complicated. The general theory is due to Born (1923); an outline only will be given here. Let the displacement of the atom at the lattice point $\mathbf{g}$ be $\mathbf{R_g}$. Then for small oscillations we can expand the potential energy of the lattice as a quadratic function of the displacements

$$\tfrac{1}{2} \sum_{\mathbf{g,h}} \sum_{k,l} V_{\mathbf{gh}}^{kl} R_{\mathbf{g},k} R_{\mathbf{h},l},$$

where $V_{\mathbf{gh}}^{kl}$ is a constant and the $R_{\mathbf{g},k}$ $(k = 1, 2, 3)$ are the components of the displacement. If $M_{\mathbf{g}}$ is the mass of the atom $\mathbf{g}$, the equations of motion are

$$M_{\mathbf{g}} d^2 R_{\mathbf{g},k} / dt^2 = - \sum_{\mathbf{h}} \sum_{l} V_{\mathbf{gh}}^{kl} R_{\mathbf{h},l},$$

which can be solved by putting

$$\mathbf{R}_{\mathbf{g}} = \alpha_{\mathbf{q}} \mathbf{e}_{\mathbf{q}} \, e^{i(\mathbf{q} \cdot \mathbf{g}_a - 2\pi\nu_{\mathbf{q}} t)},$$

where $\mathbf{e}_{\mathbf{q}}$ is a unit vector defining the direction of vibration. For a given value of $\mathbf{q}$ there are $3s$ values of ν, where s is the number of atoms in a unit

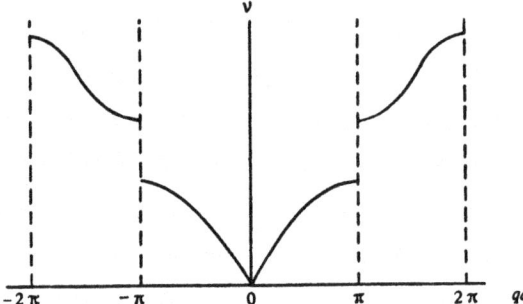

Fig. VI 4. The frequency spectrum of fig. VI 3 as a single-valued function of the wave number.

cell. Of these $3s$ modes of vibration there are 3 acoustical modes (one of which may be termed longitudinal and two transverse), while there are $3(s-1)$ optical modes. When all the atoms have the same mass and the lattice is of the Bravais type, the optical modes are absent and the problem is considerably simplified. The equations for $\mathbf{e}_{\mathbf{q}}$ and $\nu_{\mathbf{q}}$ then take the form

$$4\pi^2 M \nu_{\mathbf{q}}^2 \, e_{\mathbf{q},k} = \sum_{l=1}^{3} e_{\mathbf{q},l} \sum_{\mathbf{h}} V_{0\mathbf{h}}^{kl} e^{i\mathbf{q} \cdot \mathbf{h}_a} \quad (k = 1, 2, 3). \tag{6·15·1}$$

For these equations to be soluble for the e's, the determinant must vanish, which gives rise to a cubic equation for $\nu_{\mathbf{q}}^2$. The corresponding vectors $\mathbf{e}_{\mathbf{q}j}$ $(j = 1, 2, 3)$ define the directions of polarization of the three types of vibrations; they form an orthogonal system, as is clear from the fact that they define the directions of the principal axes of the quadratic form (in three variables) associated with the equations (6·15·1). In general the e's bear no simple relation to the axes of the crystal or to the direction of propagation $\mathbf{q}$, but for very long waves the crystal behaves like an isotropic solid, and one of the vibrations is longitudinal with e_1 parallel to q, while the other two are transverse and have equal frequencies, less than that of the longitudinal vibration.

6·16. *The specific heat of a three-dimensional lattice.* When the lattice is not of the Bravais type, Debye's approximation is obviously inadequate on account of the existence of the optical modes of vibration. If there are s atoms per unit cell, only 3 of the $3s$ vibrations are acoustical and, to a first approximation, their contribution to the internal energy can be taken to be given by (6·11·4), where N is now the number of unit cells per unit volume, not the number of atoms. For the optical modes it is usually sufficient to neglect the dependence of ν_q on q and to ascribe a single frequency to each optical branch. We may therefore consider each optical branch to have N vibrations per unit volume of frequency ν_i^0, and, if we apply Debye's approximation to the acoustical modes, the total contribution of the thermal vibrations to the energy per unit volume is

$$U = 3NkTD\left(\frac{\Theta}{T}\right) + N \sum_{i=4}^{3s} \frac{h\nu_i^0}{e^{h\nu_i^0/kT} - 1}. \tag{6·16·1}$$

The second set of terms are usually known as the Einstein terms, since they were the basis for Einstein's theory of the specific heat of solids put forward in 1907. At low temperatures the Einstein terms are negligible, but they must be included in the calculation of the specific heat at high temperatures since otherwise we should not obtain the classical value $3sNk$ for $T \gg \Theta$ and $kT \gg h\nu_i^0$.

By considering special models (Blackman, 1935, 1941) it has been found that there are divergences from Debye's formula even in temperature regions where the Einstein terms are negligible. This is to be expected *a priori* for temperatures of the order of Θ, but the anomalies are in fact most marked at much lower temperatures where (6·11·10) might have been thought to hold with considerable accuracy. The density of levels $g(\nu)$ is always proportional to ν^2 for sufficiently small ν, so that Debye's formula must ultimately hold at sufficiently low temperatures, but the shape of the curve is such as to introduce deviations from the Debye law at low temperatures. The usual way of analysing these anomalies is as follows. The specific heat at a given temperature T is inserted in Debye's formula, and the value of Θ is calculated from it. If Debye's formula held exactly, Θ would be constant, but in practice it is a slowly varying function of T, and the departure from constancy is a measure of the inadequacy of Debye's approximations.

Detailed calculations have been carried out for rock-salt (Kellermann, 1941) and for diamond (Smith, 1948), both of which have composite lattices, and for silver (Leighton, 1948), which is face-centred cubic. The results are of the same general type in all cases, and those for rock-salt are shown in figs. VI 5 and VI 6. The frequency distribution is shown in fig. VI 5, split up into the contributions from the four different types of waves, the longitudinal acoustic, the transverse acoustic, the longitudinal optical and

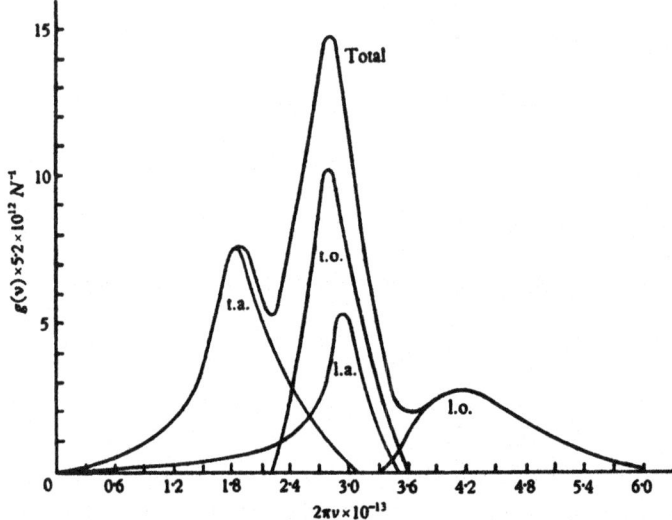

Fig. VI 5. The frequency spectrum for rock salt; total curve and four partial curves, transverse acoustic (t.a.), transverse optical (t.o.), longitudinal acoustic (l.a.) and longitudinal optical (l.o.).

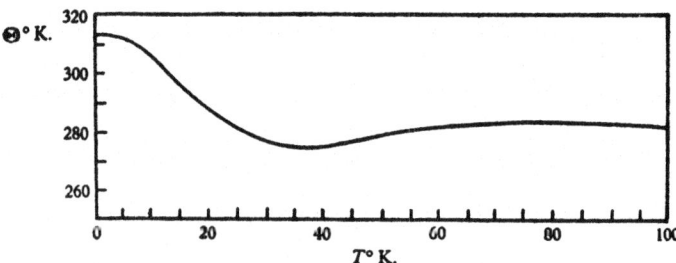

Fig. VI 6. The apparent Debye Θ as a function of T for rock salt.

the transverse optical, the transverse waves being doubly degenerate. Fig. VI 6 shows the calculated temperature dependence of Θ. The observed values lie very close to the theoretical curve, and it will be seen that variations of up to 10 % in Θ are to be expected, and that the variations extend to very low temperatures.

6·17. Since the value of Θ depends upon the velocity of sound, it should be possible to calculate Θ from the compressibility β and Poisson's ratio σ, which determine the velocities u_l and u_t of the longitudinal and transverse sound waves according to the formulae

$$u_l^2 = \frac{3(1-\sigma)}{(1+\sigma)\,\beta\rho}, \quad u_t^2 = \frac{3(1-2\sigma)}{2(1+\sigma)\,\beta\rho},$$

where ρ is the density. Values of Θ can then be obtained from the relation

$$\frac{1}{\Theta_D^3} = \left(\frac{k}{h\nu_D}\right)^3 = \frac{4\pi}{9N}\frac{k^3}{h^3}\left(\frac{1}{u_l^3}+\frac{2}{u_t^3}\right). \tag{6·17·1}$$

Some of the values obtained from the elastic constants are shown in Table VI 1, and it is seen that they agree reasonably well with those calculated from the thermal data. (These have been collected together by Dr M. Blackman.) As is shown later, Θ can be calculated from the electrical conductivity, but in view of the different ways in which the lattice vibrations enter into the various calculations, it cannot be expected that there will be agreement between the values of Θ calculated from the different phenomena.

Table VI 1. *Characteristic temperatures* Θ

From specific heat data for the temperature range $\frac{1}{3}\Theta$ to $\frac{2}{3}\Theta$

Li	430	Fe	355	Sn	160
Be	900	Co	385	Ta	245
Na	160	Ni	320	W	315
Mg	330	Cu	310	Ir	285
Al	410	Zn	240	Pt	225
K	99	Mo	360	Au	185
Ca	220	Ag	220	Hg	96
Cr	405	Cd	165	Pb	88

From the elastic constants at room temperature

Al	394	Cd	189
Cu	342	W	384
Zn	306	Au	158
Ag	212	Hg	69

Estimated from the melting-point†

Rb	59	Sb	140
Sr	148	Cs	43
Pd	270	Ba	116
In	100	Bi	80

The values for the ferromagnetic metals are very uncertain owing to the difficulty in separating the magnetic specific heat from that due to the lattice.

† These values, deduced from Lindemann's formula, are unlikely to be accurate, but they are given here, as the best available values, for use later in the text.

THE ELECTRONIC SPECIFIC HEAT

6·2. It was shown in § 1·82 that the free electrons in a metal contribute to the specific heat, but so long as the electron gas is degenerate the contribution is small. We now consider this matter in more detail.

If $n(E)$ is the density of states of the free electrons (for one direction of spin) the internal energy per unit volume is

$$U = 2 \int_0^\infty E f_0(E)\, n(E)\, dE = -2 \int_0^\infty \frac{\partial f_0}{\partial E}\, dE \int_0^E E' n(E')\, dE'.$$

Evaluation of this by (1·7·2) gives

$$U = 2 \int_0^\zeta E n(E)\, dE + \frac{\pi^2}{3} (kT)^2 \left(\frac{\partial}{\partial E} \{ E n(E) \} \right)_{E=\zeta}. \tag{6·2·1}$$

The specific heat (at constant volume) per unit volume is therefore, to the first order in kT/ζ,

$$C_v = \frac{\partial U}{\partial T} = 2\zeta n(\zeta) \frac{d\zeta}{dT} + \frac{2\pi^2}{3} k^2 T \left(n(\zeta) + \zeta \frac{dn(\zeta)}{d\zeta} \right). \tag{6·2·2}$$

To evaluate this, we require ζ as a function of T expressed in terms of $n(E)$, as a generalization of (1·81·3). In the general case ζ is determined by the condition

$$n = 2 \int f_0(E)\, n(E)\, dE = -2 \int \frac{\partial f_0}{\partial E}\, dE \int_0^E n(E')\, dE',$$

which, for small kT/ζ, reduces to

$$n = 2 \int_0^\zeta n(E)\, dE + \frac{\pi^2}{3} k^2 T^2 \frac{dn(\zeta)}{d\zeta}. \tag{6·2·3}$$

We define ζ_0 by the relation $n = 2 \int_0^{\zeta_0} n(E)\, dE$, and the second approximation is found by writing

$$\int_0^\zeta n(E)\, dE = \int_0^{\zeta_0} n(E)\, dE + (\zeta - \zeta_0)\, n(\zeta_0),$$

which gives

$$\zeta = \zeta_0 - \frac{\pi^2}{6} k^2 T^2 \frac{1}{n(\zeta)} \frac{dn(\zeta)}{d\zeta} \tag{6·2·4}$$

and

$$\frac{d\zeta}{dT} = -\frac{\pi^2}{3} k^2 T \frac{1}{n(\zeta)} \frac{dn(\zeta)}{d\zeta}. \tag{6·2·5}$$

Equations (6·2·1) and (6·2·2) now reduce to

$$U = 2 \int_0^{\zeta_0} E n(E)\, dE + \tfrac{1}{3} (\pi kT)^2\, n(\zeta), \tag{6·2·6}$$

and

$$C_v = \tfrac{2}{3} \pi^2 k^2 T n(\zeta). \tag{6·2·7}$$

(To the order of approximation used, it is immaterial whether we put ζ or ζ_0 in the temperature-dependent terms.) The physical meaning of (6·2·7) is that the only electrons that contribute to the specific heat are those in an energy range of order kT centred round the Fermi energy ζ. Each of these contributes k to the specific heat while the remainder contribute nothing.

6·21. It has been convenient at various places in the preceding chapters to point out that there is a strong similarity between the properties of electrons and of vacant levels in an energy band. So far as the specific heat is concerned, there is complete symmetry between electrons and 'holes', as is shown by the following transformations.

We define the distribution function for holes by putting $E' = -E$, $\zeta' = -\zeta$ and

$$f_0^+(E') = 1 - f_0(E) = \frac{1}{e^{(E'-\zeta')/kT} + 1}. \tag{6·21·1}$$

Then, if N is the total number of energy levels per unit volume in a band or in a number of bands, we have

$$N = 2 \int_{-\infty}^{\infty} \mathfrak{n}(E) \, dE.$$

Hence, if n is the number of electrons and n^+ is the number of holes per unit volume,

$$n^+ = N - n = 2 \int_{-\infty}^{\infty} \{1 - f_0(E)\} \mathfrak{n}(E) \, dE = 2 \int_{-\infty}^{\infty} f_0^+(E') \mathfrak{n}(-E') \, dE', \tag{6·21·2}$$

which is of exactly the same form as the relation between n and $f_0(E)$. Also

$$U = 2 \int_{-\infty}^{\infty} E f_0(E) \mathfrak{n}(E) \, dE = 2 \int_{-\infty}^{\infty} E\{1 - f_0^+(E')\} \mathfrak{n}(E) \, dE$$

$$= 2 \int_{-\infty}^{\infty} E'\{f_0^+(E') - 1\} \mathfrak{n}(-E') \, dE'. \tag{6·21·3}$$

The expression for U is therefore the same (apart from a constant) whether expressed in terms of E and $f_0(E)$ or in terms of E' and $f_0^+(E')$, and the alternative expressions for C_v are identical in form, being

$$C_v = 2 \int_{-\infty}^{\infty} ET \frac{\partial}{\partial T} \frac{E - \zeta}{T} \frac{\partial f_0}{\partial E} \mathfrak{n}(E) \, dE$$

$$= 2 \int_{-\infty}^{\infty} E'T \frac{\partial}{\partial T} \frac{E' - \zeta'}{T} \frac{\partial f_0^+}{\partial E'} \mathfrak{n}(-E') \, dE'. \tag{6·21·4}$$

In view of the above results it is immaterial whether we consider the specific heat to be due to n electrons or to n^+ holes, and we therefore use whichever description is most appropriate to the physical conditions which occur in any particular metal.

6·22. In a normal band of standard form $n(E)$ is given by (2·71·3) and ζ by (1·81·2), so that

$$C_v = \frac{4\pi^3 m^* k^2}{3h^2} \left(\frac{3n}{\pi}\right)^{\frac{1}{3}} T.$$ (6·22·1)

The same formula applies to an inverted band of standard form if n now denotes the number of vacant levels in the band. More generally, for two overlapping bands of standard form

$$C_v = \frac{4\pi^3 k^2}{3h^2} \left\{ \varpi_1 m_1 \left(\frac{3n_1}{\pi\varpi_1}\right)^{\frac{1}{3}} + \varpi_2 m_2 \left(\frac{3n_2}{\pi\varpi_2}\right)^{\frac{1}{3}} \right\} T,$$ (6·22·2)

where m_1, m_2 are the effective masses, ϖ_1, ϖ_2 are the weights of the bands if they are degenerate, and n_1 and n_2 are the total number of occupied and vacant levels per unit volume in the two bands respectively, the number of levels in each band of weight unity being n_1/ϖ_1 and n_2/ϖ_2. In general it is sufficient to write

$$C_v = \frac{4\pi^3 k^2}{3h^2} \varpi^{\frac{2}{3}} m^* \left(\frac{3n}{\pi}\right)^{\frac{1}{3}} T,$$ (6·22·3)

if it is borne in mind that then n is an average of n_1 and n_2, and that $\varpi^{\frac{2}{3}} m^*$ is an average of $\varpi_1^{\frac{2}{3}} m_1$ and $\varpi_2^{\frac{2}{3}} m_2$.

On account of its smallness, we can only hope to observe the electronic specific heat either at very low temperatures where its magnitude must, in the limit as $T \to 0$, be greater than that of the lattice specific heat, or at very high temperatures where its temperature variation can be observed, the lattice specific heat being then practically constant.

6·23. *The temperature variation of the specific heat.* The proportionality of C_v to T only holds for very low temperatures, and at sufficiently high temperatures C_v must tend asymptotically to the classical value $\frac{3}{2}nk$. The exact form of C_v as a function of T, however, depends upon the functional dependence of $n(E)$ on E over the whole of the energy range, and the calculations have to be carried out numerically. If we assume that $n(E)$ is of standard form, the calculations can be simplified by using the extensive tables of Fermi-Dirac functions compiled by McDougall and Stoner (1938).

In general we have to evaluate

$$U = 2 \int_0^\infty E n(E) f_0(E) \, dE,$$

subject to the condition

$$n = 2 \int_0^\infty n(E) f_0(E) \, dE.$$

For a band of standard form, we put $x = E/kT$, $\xi = \zeta/kT$, and we then have

$$U = nkT F_{\frac{3}{2}}(\xi)/F_{\frac{1}{2}}(\xi), \tag{6·23·1}$$

where

$$F_s(\xi) = \int_0^\infty \frac{x^s\,dx}{e^{x-\xi}+1}. \tag{6·23·2}$$

Now for $T = 0$ we have $\zeta = \zeta_0$, and $f_0(E) = 1$ for $E < \zeta_0$ and $f_0(E) = 0$ for $E > \zeta_0$, so that

$$n = A \int_0^\infty \frac{E^{\frac{1}{2}}\,dE}{e^{(E-\zeta)/kT}+1} = A \int_0^{\zeta_0} E^{\frac{1}{2}}\,dE = \tfrac{2}{3}A\zeta_0^{\frac{3}{2}},$$

where A is a constant. Hence the condition for the constancy of n is

$$F_{\frac{1}{2}}(\xi) = \tfrac{2}{3}(\zeta_0/kT)^{\frac{3}{2}}, \tag{6·23·3}$$

which is an implicit equation for ξ. The elimination of ξ between (6·23·1) and (6·23·3) gives U as a function of T.

The limiting form of (6·23·1) at very low temperatures has already been evaluated to a first approximation. To obtain a second approximation we proceed as follows. From (6·23·1) and (6·23·3) we have

$$U = \tfrac{3}{2}nkT(kT/\zeta_0)^{\frac{3}{2}} F_{\frac{3}{2}}(\xi), \tag{6·23·4}$$

which by (1·7·2) is

$$U = \tfrac{3}{5}nkT \left(\frac{kT}{\zeta_0}\right)^{\frac{3}{2}} \xi^{\frac{5}{2}}\left(1 + \frac{15c_2}{2\xi^2} - \frac{15c_4}{8\xi^4} + \dots\right). \tag{6·23·5}$$

Also (6·23·3) gives

$$\left(\frac{\zeta_0}{kT}\right)^{\frac{3}{2}} = \xi^{\frac{3}{2}}\left(1 + \frac{3c_2}{2\xi^2} + \frac{9}{8}\frac{c_4}{\xi^4} + \dots\right), \tag{6·23·6}$$

and on inverting the series we find

$$\xi = \frac{\zeta_0}{kT}\left\{1 - c_2\left(\frac{kT}{\zeta_0}\right)^2 - \tfrac{3}{4}(c_2^2 + c_4)\left(\frac{kT}{\zeta_0}\right)^4 - \dots\right\}, \tag{6·23·7}$$

so that, on substituting into (6·23·5), U becomes

$$U = \tfrac{3}{5}n\zeta_0\left\{1 + 5c_2\left(\frac{kT}{\zeta_0}\right)^2 - \tfrac{15}{4}(c_2^2 + c_4)\left(\frac{kT}{\zeta_0}\right)^4 - \dots\right\}$$

$$= \tfrac{3}{5}n\zeta_0\left\{1 + \frac{5\pi^2}{12}\left(\frac{kT}{\zeta_0}\right)^2 - \frac{\pi^4}{16}\left(\frac{kT}{\zeta_0}\right)^4 - \dots\right\}. \tag{6·23·8}$$

Hence

$$C_v = \tfrac{1}{2}\pi^2 nk\frac{kT}{\zeta_0}\left\{1 - \frac{3\pi^2}{10}\left(\frac{kT}{\zeta_0}\right)^2 - \dots\right\}. \tag{6·23·9}$$

At very high temperatures such that $kT \gg \zeta_0$, ξ is large and negative, and the first approximation to $F_s(\xi)$ is $\Gamma(s+1)\,e^\xi$ (see Appendix (A 41·3)). Equation (6·23·1) then reduces immediately to $U = \tfrac{3}{2}nkT$, which is the classical value of U.

It is possible to obtain further approximations to the integrals, but numerical calculations are required to cover the whole temperature range. The results are shown in fig. VI 7, where C_v is given as a function of T/T_0.

SURVEY OF THE EXPERIMENTAL DATA ON SPECIFIC HEATS

6·3. The electronic specific heat of silver was first observed by Keesom and Kok in 1934, and a large number of measurements have been made in recent years of the specific heats of metals near the absolute zero of sufficient accuracy to enable the specific heat of the conduction electrons to be disentangled from the specific heat of the lattice vibrations. In some cases the specific heats of the electrons have been measured directly by calorimetric methods, while for the superconducting metals they have been obtained from the entropy differences between the normal and superconducting states on the assumption that the specific heat in the

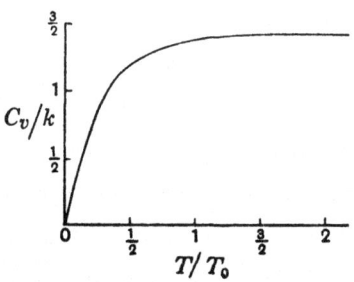

Fig. VI 7. The specific heat per electron of a gas of free electrons.

superconducting state has no linear term. The results for metals of groups one to four are shown in Table VI 2, which gives the values of $(C_v)_A$, the specific heat per gram atom, and the values of $n(\zeta_0)/n_a$, where n_a is the number of atoms per unit volume, deduced from equation (6·2·7). Values are also given for $\varpi^{\frac{2}{3}}m^*/m$, where m^* is the effective mass of the electrons and ϖ is the weight factor of the electronic bands, it being assumed that n in equation (6·22·3) is equal to the number of valency electrons. It will be seen that the values of m^*/m required to explain the experimental results are reasonable.

Table VI 2. *Specific heats of metals at low temperatures*
in units of $10^{-4}T$ *joule/degree/g. atom*

	$(C_v)_A$	$n_a \times 10^{-22}$	$n(\zeta_0)/n_a$ in (eV.)$^{-1}$	$\varpi^{\frac{2}{3}}m^*/m$
Cu	7·4	8·5	0·16	1·5
Ag	6·7	5·9	0·14	1·0
Zn	6·3	6·7	0·13	0·85
Hg	15·6	4·3	0·33	1·5
Al	14·7	6·1	0·3	1·6
Tl	16	3·6	0·33	1·2
Sn	14·7	4·3	0·3	1·1
Pb	30	3·3	0·63	1·9

The values for Cu, Ag, Zn and Al are calorimetric ones, while the others are obtained from the entropy differences between the superconducting and normal states. Note that 1 joule = 0·24 calorie.

The measurement of electronic specific heats is especially easy for the transition elements, since the density of the states is very high when the valency electrons form incomplete d- or f-shells. The experimental results are given in Table VI 3, and the values of $n(\zeta_0)/n_a$ are derived from equation (6·2·7). For the ferromagnetic metals the formula $C_v = \frac{1}{3} 2^{\frac{2}{3}} \pi^2 k^2 T n(\zeta)$ is used instead of (6·2·7) (cf. equation (7·5·7)). The assumptions under which this formula is derived probably hold for nickel though not for iron and cobalt, but the values of $n(\zeta)$ deduced can only be wrong by a factor of order 1. Some calculated values of $(C_v)_A$ are given in the table, based upon the band structures calculated by the methods of Chapter III (see figs. III 13, 15 and 16, pp. 83, 87 and 88). It will be seen that, though the calculated values of $(C_v)_A$ are large, they are very much smaller than the observed values, except for nickel, where the agreement is reasonable. Therefore, while theory and experiment are in good agreement for normal metals there is an enormous discrepancy in the results for the transition metals. The inclusion of the exchange forces (§ 3·442) would make matters much worse, and the theory of abnormally large electronic specific heats is in an unsatisfactory state.

Table VI 3. *Specific heats of transition metals at low temperatures in units of $10^{-4}T$ joule/degree/g. atom*

	$(C_v)_A$ observed	$(C_v)_A$ calculated	$n_a \times 10^{-22}$	$n(\zeta_0)/n_a$ in (eV.)$^{-1}$
W	220	20	6·3	4·5
Nb	250	—	5·6	5·1
Ta	{ 82 { 60	27	5·7	— 1·2
Mn	178	—	8·3	3·7
Fe	50	19	8·5	1·3
Co	50	—	8·9	1·3
Ni	73	46	9·2	1·9
Pd	130	—	6·9	2·7
Pt	67	—	6·7	1·4

All the observed specific heats in the table are obtained from calorimetric measurements with two exceptions (Nb and the higher value for Ta). For the sources of the experimental data see Silvidi and Daunt (1950).

6·31. It is not easy to measure the temperature variation of the electronic specific heat except in the case of nickel which, on account of its being ferromagnetic, is discussed in Chapter VII. In the first place the measurements give C_p, and it is necessary to know, or estimate, the compressibility and the coefficient of thermal expansion in order to obtain C_v from C_p. Secondly, the Debye theory is only an approximation and does not take into account the anharmonic lattice forces which determine, for example, the thermal expansion of a solid, and the theory of the effect of the anharmonic forces is in a rudimentary state. It is, however, known that they can give either a positive or a negative contribution to the specific heat.

Some rough values of $(C_v)_A - (C_D)_A$, the excess of the specific heat over the Debye values, at 1000° K. are given in Table VI 4, and the low-temperature values of the electronic specific heats are also shown (Stoner, 1938). It will be seen, by reference to the table and fig. VI 7, that if $(C_v)_A - (C_D)_A$ is considered to be an electronic specific heat, the experimental values at 1000° K. are too high for copper and much too low for the transition metals. It must, however, be borne in mind that the temperature variation of the specific heat depends upon the behaviour of $n(E)$ over the whole energy range, and that the curve in fig. VI 7 is based upon the assumption that the energy band is of standard form. On the other hand, a high electronic specific heat necessarily entails a low degeneracy temperature, and whatever the energy band structure, it is difficult to reconcile the high- and low-temperature values of the specific heats of the transition metals; both the absolute magnitudes and the temperature variations are anomalous. It is perhaps worth noting that the high-temperature values of the electronic specific heat for tungsten and tantalum are in agreement with the values calculated from the theoretical energy-level systems. If this is significant it leaves the low-temperature values unexplained.

Table VI 4. *Specific heats at high and low temperatures*
in joules/degree/g. atom

	Cu	Ag	W	Ta	Pd	Pt
$10^3 (C_v)_A/T$ $(T \to 0)$	0·7	0·6	22	6	13	7
$(C_v)_A - (C_D)_A$ $(T = 1000°\,K.)$	1·5	0·8	2·1	2·9	2·9	2·2

THE SPIN PARAMAGNETISM

6·4. The problem of calculating the susceptibility of a substance composed of molecules having a permanent magnetic moment μ_0 was first considered by Langevin using classical statistics. He derived the formula

$$\chi = \tfrac{1}{3} N \mu_0^2 / kT, \tag{6·4·1}$$

where N is the number of molecules per unit volume. Although this formula fits well the experimental results for gases and for salts of the rare earths, the susceptibility of metals is small and nearly independent of the temperature. This can readily be explained by using Fermi-Dirac statistics instead of classical statistics for the electron gas.

If E is the energy of an electron when $H = 0$, the energy of an electron with magnetic moment $\pm \mu_0$ in a field H is $E \mp \mu_0 H$. The magnetic moment per unit volume of an electron gas is then

$$M = \mu_0 \int \{f_0(E - \mu_0 H) - f_0(E + \mu_0 H)\} \, \mathfrak{n}(E) \, dE, \qquad (6\cdot4\cdot2)$$

where the dependence of the parameter ζ on H must be taken into account. Now ζ is determined by the condition that

$$n = \int \{f_0(E - \mu_0 H) + f_0(E + \mu_0 H)\} \, \mathfrak{n}(E) \, dE = 2 \int f_0(E) \, \mathfrak{n}(E) \, dE + O(H^2), \qquad (6\cdot4\cdot3)$$

so that any variation in ζ due to the magnetic field is at least of order H^2 and can be neglected in calculating M to order H. Equation $(6\cdot4\cdot2)$ now gives

$$M = -2H\mu_0^2 \int \frac{\partial f_0}{\partial E} \mathfrak{n}(E) \, dE + O(H^2) \qquad (6\cdot4\cdot4)$$

$$= 2H\mu_0^2 \left(\mathfrak{n}(\zeta) + \frac{\pi^2}{6} k^2 T^2 \frac{d^2 \mathfrak{n}(\zeta)}{d\zeta^2} \right) + O(H^2). \qquad (6\cdot4\cdot5)$$

The first approximation to the susceptibility is therefore

$$\chi = 2\mu_0^2 \mathfrak{n}(\zeta). \qquad (6\cdot4\cdot6)$$

The second approximation is obtained by substituting for ζ from $(6\cdot2\cdot4)$ into $(6\cdot4\cdot5)$, which gives

$$\chi = 2\mu_0^2 \mathfrak{n}(\zeta_0) \left[1 + \frac{\pi^2}{6} k^2 T^2 \left\{ \frac{1}{\mathfrak{n}} \frac{d^2 \mathfrak{n}}{dE^2} - \left(\frac{1}{\mathfrak{n}} \frac{d\mathfrak{n}}{dE} \right)^2 \right\}_{E = \zeta_0} \right]. \qquad (6\cdot4\cdot7)$$

6·41. It was shown in § 6·21 that it is immaterial whether the electronic specific heat is considered to be due to electrons or to holes. A similar dual description is possible for the spin paramagnetism. If we use the same notation as in § 6·21, then when a magnetic field H is present, the total number of energy levels per unit volume in a band or number of bands is

$$N = 2 \int_{-\infty}^{\infty} \mathfrak{n}(E) \, dE,$$

and the number n^+ of holes per unit volume is

$$n^+ = N - n = \int_{-\infty}^{\infty} \{f_0^+(E' - \mu_0 H) + f_0^+(E' + \mu_0 H)\} \, \mathfrak{n}(-E') \, dE'. \qquad (6\cdot41\cdot1)$$

Also if we change the variable of integration in $(6\cdot4\cdot2)$ from E to E', we find that

$$M = \mu_0 \int_{-\infty}^{\infty} \{f_0^+(E' - \mu_0 H) - f_0^+(E' + \mu_0 H)\} \, \mathfrak{n}(-E') \, dE', \qquad (6\cdot41\cdot2)$$

so that the equations determining M are identical in form whether they are expressed in terms of E, $f_0(E)$ and n or E', $f_0^+(E')$ and n^+. It is therefore sufficient to carry out the calculations in the form appropriate to electrons, and to apply the results to both electrons and holes.

6·42. *The temperature variation of the spin paramagnetism.* It is impossible to calculate the variation of χ with T unless $n(E)$ is known explicitly, and, as when discussing the electronic specific heat, we consider a band of standard form with $n(E) \propto E^{\frac{1}{2}}$. Now $\partial f_0/\partial E = -\partial f_0/\partial \zeta$, and so (6·4·4) can be written as

$$M = 2H\mu_0^2 \int \frac{\partial f_0}{\partial \zeta} n(E)\, dE. \qquad (6\cdot42\cdot1)$$

Hence, for a band of standard form,

$$M = \frac{nH\mu_0^2}{F_{\frac{3}{2}}(\xi)} \frac{\partial F_{\frac{3}{2}}(\xi)}{\partial \zeta},$$

$$\chi = \frac{n\mu_0^2}{kT} \frac{F_{\frac{3}{2}}'(\xi)}{F_{\frac{3}{2}}(\xi)} \quad (\xi = \zeta/kT), \qquad (6\cdot42\cdot2)$$

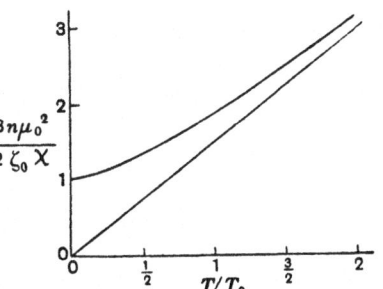

Fig. VI 8. The reciprocal of the paramagnetic susceptibility as a function of the temperature for a gas of free electrons. The straight line gives the classical value of $1/\chi$.

where the dash denotes differentiation with respect to ξ, and where ξ is given in terms of T by (6·23·3).

For high temperatures $F_{\frac{3}{2}}(\xi) = \Gamma(\frac{5}{2})\, e^{\xi}$ and (6·42·2) becomes

$$\chi = n\mu_0^2/kT \quad (T \gg T_0). \qquad (6\cdot42\cdot3)$$

The absence of the factor $\frac{1}{3}$ which occurs in (6·4·1) is due to the fact that the spin of an electron must be parallel or antiparallel to the field, whereas in the classical theory all directions of spin are possible.

From (6·42·2), (1·7·2) and (6·23·7), the low-temperature value of χ is

$$\chi = \frac{3n\mu_0^2}{2\zeta_0}\left\{1 - \frac{\pi^2}{12}\left(\frac{kT}{\zeta_0}\right)^2 + \dots\right\} = \frac{4m}{h^2}(3\pi^2 n)^{\frac{1}{3}}\mu_0^2\left\{1 - \frac{\pi^2}{12}\left(\frac{kT}{\zeta_0}\right)^2 + \dots\right\} \quad (T_0 \gg T).$$
$$(6\cdot42\cdot4)$$

For intermediate temperatures χ is most simply evaluated numerically from (6·42·2) and (6·23·3) by using McDougall and Stoner's tables (1938) (see also Stoner, 1936). The general behaviour of χ is shown in fig. VI 8. If the electrons have effective mass m^*, then m in (6·42·4) must be replaced by m^*. The other formulae remain unchanged, but it must be borne in mind that m^* appears implicitly through ζ_0.

6·43. The above calculations can be extended to calculate the variation of χ with H, but it is then necessary to take into account the dependence of

ζ upon H. For a band of standard form (6·4·2) and (6·4·3) can be most conveniently written as

$$M = \tfrac{3}{4} n \mu_0 (kT/\zeta_0)^{\frac{3}{2}} \{F_{\frac{1}{2}}(\xi+\alpha) - F_{\frac{1}{2}}(\xi-\alpha)\}, \tag{6·43·1}$$

$$1 = \tfrac{3}{4} (kT/\zeta_0)^{\frac{3}{2}} \{F_{\frac{1}{2}}(\xi+\alpha) + F_{\frac{1}{2}}(\xi-\alpha)\}, \tag{6·43·2}$$

where $\alpha = \mu_0 H/kT$ and $\xi = \zeta/kT$. Now when $\zeta_0 \gg kT$, we have $F_{\frac{1}{2}}(\xi) = \tfrac{2}{3}\xi^{\frac{3}{2}}$, and so

$$M = \tfrac{1}{2} n \mu_0 (kT/\zeta_0)^{\frac{3}{2}} \{(\xi+\alpha)^{\frac{3}{2}} - (\xi-\alpha)^{\frac{3}{2}}\},$$

$$1 = \tfrac{1}{2} (kT/\zeta_0)^{\frac{3}{2}} \{(\xi+\alpha)^{\frac{3}{2}} + (\xi-\alpha)^{\frac{3}{2}}\}.$$

Therefore

$$M = n\mu_0 \left(\frac{\zeta}{\zeta_0}\right)^{\frac{3}{2}} \left\{\frac{3}{2}\frac{\mu_0 H}{\zeta} - \frac{1}{16}\left(\frac{\mu_0 H}{\zeta}\right)^3 + \dots\right\},$$

$$1 = \left(\frac{\zeta}{\zeta_0}\right)^{\frac{3}{2}} \left\{1 + \frac{3}{8}\left(\frac{\mu_0 H}{\zeta}\right)^2 + \dots\right\}.$$

Hence ζ is given by

$$\zeta = \zeta_0 (1 - \tfrac{1}{4}\mu_0^2 H^2/\zeta_0^2 + \dots), \tag{6·43·3}$$

and M by

$$M = \frac{3n\mu_0^2}{2\zeta_0} H \left(1 - \frac{\mu_0^2 H^2}{6\zeta_0^2} + \dots\right). \tag{6·43·4}$$

Therefore

$$\chi = \frac{dM}{dH} = \frac{3n\mu_0^2}{2\zeta_0}\left(1 - \frac{\mu_0^2 H^2}{2\zeta_0^2} + \dots\right). \tag{6·43·5}$$

If we wish to include higher order terms it is necessary to consider the spin paramagnetism and the orbital diamagnetism simultaneously, since they are not additive to all orders (see § 6·7). The variation with H is too small to be observable, and the results are of no practical importance.

If the temperature is sufficiently high for the statistics to be classical, we have $F_{\frac{1}{2}}(\xi) = \Gamma(\tfrac{3}{2}) e^{\xi}$ and, dividing (6·43·1) by (6·43·2), we obtain

$$M = n\mu_0 \tanh(\mu_0 H/kT) \quad (T \gg T_0), \tag{6·43·6}$$

which is the generalization of (6·42·3).

6·44. *The effect of exchange forces on the paramagnetism.* We have so far neglected the exchange and correlation forces which arise when we take the wave function of the assembly of electrons to have the correct anti-symmetrical properties, instead of assuming that the wave function is a simple product and taking account of the Pauli principle merely by using the Fermi-Dirac statistics. To calculate the effect of these forces at $T = 0$ it is necessary to know the internal energy U per unit volume as a function of n_1 and n_2, the numbers of spins per unit volume parallel and antiparallel to an external magnetic field H. If terms which depend only on $n = n_1 + n_2$ are omitted, we can write formally

$$U = (h^2/80\pi^4 m)\{(6\pi^2 n_1)^{\frac{5}{3}} + (6\pi^2 n_2)^{\frac{5}{3}}\} + f(n_1) + f(n_2)$$
$$+ n_1 g(n_2) + n_2 g(n_1) - \mu_0(n_1 - n_2)H. \quad (6\cdot44\cdot1)$$

The first two terms are the kinetic energy of the electrons, the third and fourth give the effect of the exchange forces, the fifth and sixth give the effect of the correlation forces and the last term is the energy due to the presence of the magnetic field. According to the calculations in §3·441 which are approximately valid for perfectly free electrons,

$$f(n) = -\epsilon^2(6\pi^2 n)^{\frac{4}{3}}/(8\pi^3), \quad g(n) = -0\cdot58\,\epsilon^2/(10\cdot3r_0 + n^{-\frac{1}{3}}). \quad (6\cdot44\cdot2)$$

If we put $n_1 = \frac{1}{2}n + p$, $n_2 = \frac{1}{2}n - p$ and calculate U to order p^2 we obtain

$$U(p) = U_0 + \alpha p^2 - 2\mu_0 pH, \quad (6\cdot44\cdot3)$$

where

$$\alpha = (\tfrac{1}{3}h^2/m)(3\pi^2 n)^{-\frac{1}{3}} + f''(\tfrac{1}{2}n) - 2g'(\tfrac{1}{2}n) + \tfrac{1}{2}ng''(\tfrac{1}{2}n). \quad (6\cdot44\cdot4)$$

If α is negative, $U(p)$ is a maximum for $p = 0$ when $H = 0$, which means that there is a spontaneous magnetism and the metal is ferromagnetic. We shall discuss this case in more detail in Chapter VII, but in the meantime we shall confine ourselves to the case $\alpha > 0$. $U(p)$ then has a minimum at $p = \mu_0 H/\alpha$, and at the minimum

$$U = U_0 - \tfrac{1}{2}\chi H^2, \quad \chi = 2\mu_0^2/\alpha, \quad (6\cdot44\cdot5)$$

χ being the susceptibility.

If $f(n)$ is given by $(6\cdot44\cdot2)$ and $g(n) = 0$, then

$$\frac{1}{\chi} = \frac{1}{\chi_{i0}}\left\{1 - \frac{4m\epsilon^2}{h^2}\left(\frac{\pi}{3n}\right)^{\frac{1}{3}}\right\}, \quad (6\cdot44\cdot6)$$

where χ_{i0} is the susceptibility at $T = 0$ when the exchange forces are neglected. The effect of the exchange forces is therefore to increase the susceptibility, whereas the correlation forces, represented by the terms involving g, tend to decrease it.

The calculation of the effect of the correlation and exchange forces is somewhat more complicated when the temperature is not zero. The details are given in §7·6, the result being that $(6\cdot42\cdot2)$ is replaced by

$$\frac{1}{\chi} = \frac{1}{n\mu_0^2}\left(kT\frac{F_{\frac{1}{2}}(\xi)}{F'_{\frac{1}{2}}(\xi)} - k\theta'\right) = \frac{1}{\chi_i} - \frac{k\theta'}{n\mu_0^2} = \frac{1}{\chi_i} - \frac{3}{2\chi_{i0}}\frac{\theta'}{T_0}, \quad (6\cdot44\cdot7)$$

where

$$k\theta' = -\tfrac{1}{2}nf''(\tfrac{1}{2}n) + ng'(\tfrac{1}{2}n) - \tfrac{1}{4}n^2 g''(\tfrac{1}{2}n), \quad (6\cdot44\cdot8)$$

and χ_i is the susceptibility when the correlation and exchange forces are neglected. With this notation, the effect of the correlation and exchange forces is to introduce a term $-\frac{1}{2}k\theta' M/(n\mu_0^2)$ into the energy U per unit volume at $T = 0$, where $M = (n_1 - n_2)\mu_0$ is the magnetic moment per unit volume. The $1/\chi - T$ curve is the same as that shown in fig. VI 8, except that all the ordinates are reduced by an amount $-k\theta'/(n\mu_0^2)$.

SURVEY OF THE EXPERIMENTAL DATA ON THE MAGNETIC SUSCEPTIBILITY

6·5. There are a number of complications which must be taken into account in comparing the formulae derived above with the experimental results. In the first place the metal ions show a diamagnetic effect which can, however, be estimated by assuming that the core electrons behave exactly as in a free atom (see equation (6·6·1)). In the following discussion it is assumed that this correction has been made, and that we are therefore dealing with the susceptibility of the conduction electrons and not of the metal as a whole. In the second place the conduction electrons themselves contribute a diamagnetic effect through their translational motion, as well as a paramagnetic effect due to their spin. The theory of the diamagnetism of the conduction electrons is of a different order of mathematical difficulty from that of the spin paramagnetism. The results are therefore quoted here before the details of the calculations are given in § 6·6. If the conduction band is of standard form the diamagnetic susceptibility is given by

$$\chi_{\text{dia.}} = -\frac{4}{3}\frac{m^2\mu_0^2}{m^*h^2}(3\pi^2 n)^{\frac{1}{3}} \tag{6·5·1}$$

(see equation (6·81·2)), which is exactly one-third of the paramagnetic susceptibility if $m^* = m$. The total susceptibility of the conduction electrons is therefore given by adding (6·42·4), with m replaced by m^*, and (6·5·1), so that χ is given by

$$\chi = \frac{4m^*\mu_0^2}{h^2}(3\pi^2 n)^{\frac{1}{3}}\left(1 - \frac{m^2}{3m^{*2}}\right). \tag{6·5·2}$$

Alternatively, if $n(\zeta_0)/n_a$ is the density of states per electron-volt per atom (for one direction of spin), the susceptibility χ_A per gram atom is given by

$$\chi_A = 64 \times 10^{-6}\left(1 - \frac{m^2}{3m^{*2}}\right)\frac{n(\zeta_0)}{n_a}. \tag{6·5·3}$$

6·51. *Monovalent metals.* For the monovalent metals it is reasonable to put $m^* = m$. The results calculated in this way are compared with the observed results in Table VI 5, and it will be seen that the agreement is reasonably good, though in general the calculated susceptibilities are too low. An attempt to improve the agreement has been made by Sampson and Seitz (1940), who calculated the effect of the exchange and correlation forces on the paramagnetic susceptibility, by means of equations (6·44·5),

(6·44·4) and (6·44·2), assuming that the diamagnetic susceptibility is unchanged. Their results for sodium are given in Table VI 6, where the observed value of χ has been corrected for the diamagnetism of the ions and of the conduction electrons to give χ_{spin} (observed), i.e. it is assumed to be 1·5 times the figure in the fourth column of Table VI 5. The theoretical value of $1/\chi_{spin}$ is obtained by calculating the three contributions to α given by the three different types of terms in (6·44·4), the first being due to the kinetic energy of the electrons, the second due to the exchange forces and the third due to the correlation forces. It will be seen that the agreement between the observed and calculated values is improved by the inclusion of the exchange and correlation forces, but in view of the crude approximations made in deriving the theoretical formulae, no great weight can be given to the figures.

Table VI 5. *Volume susceptibilities of the monovalent metals in units of* 10^{-7}

	$n \times 10^{-21}$	Total χ observed	χ due to ions	χ due to valency electrons	χ calculated
Li	48·3	(2·8)	− 0·56	(3·36)	5·3
Na	26·2	5·0 6·5	− 2·3	7·3 8·8	4·2
K	13·8	3·6 4·9	− 3·0	6·6 7·9	3·5
Rb	11·3	1·1 3·3	− 4·2	5·3 7·5	3·3
Cs	9·0	−2·0 4·4	− 5·2	3·2 9·6	3·1
Cu	85	− 7·6	− 20	12·4	6·5
Ag	59	− 21	−(25 to 30)	(4 to 9)	5·9
Au	59	− 29	− 43	14	5·9

Table VI. 6. *Reciprocal of the spin paramagnetic susceptibility of sodium in units of* 10^5

	$1/\chi_{spin}$ (theoretical)			
$1/\chi_{spin}$ (observed)	Due to kinetic energy terms	Due to exchange force terms	Due to correlation terms	Total
7·6 to 9·2	16	−10	3	9

6·52. *Divalent metals.* The alkaline earth metals are moderately paramagnetic, and the observed susceptibilities are appreciably higher than the theoretical values given by (6·5·2) with $m^* = m$ and with $n = 2n_a$. The results are given in Table VI 7, and it will be seen that, especially for calcium, the calculated values are much too low. This is only to be expected since the band structure of calcium (see fig. III 14) is such that $n(\zeta)$ is much higher than that given by the free-electron model. Further, with such a system of energy levels it is possible for the distribution of electrons between the two bands to vary significantly with temperature. The paramagnetic susceptibility could then increase with increasing temperature. Such a variation of χ has been observed for barium, but numerical calculations of the order of magnitude to be expected are lacking.

Table VI 7. *Volume susceptibilities of Mg and Ca in units of* 10^{-7}

	$n_a \times 10^{-21}$	Total χ observed	χ due to ions	χ due to valency electrons	χ calculated
Mg	43·5	9·5	−2·5	12	6·5
Ca	23·3	17	−4	21	5

The hexagonal metals zinc and cadmium have negative susceptibilities which depend upon the orientation of the field relative to the crystal axes. At room temperature, the total average susceptibility $\chi = \frac{1}{3}\chi_{\parallel} + \frac{2}{3}\chi_{\perp}$ is nearly equal to that of the ions, so that the contribution of the valency electrons is small (see Table VI 8). This means that, although the spin paramagnetism is of the normal order of magnitude, the diamagnetism of the valency electrons is some two or three times the value for free electrons. Most of the interest therefore centres round the diamagnetism, which behaves anomalously (Marcus, 1949). For zinc, $-\chi_{\perp}$ increases linearly from 9×10^{-7} at room temperature to $10·5 \times 10^{-7}$ at 100° K. and is constant at lower temperatures, but $-\chi_{\parallel}$ has a flat minimum at about 180° K., increases to a maximum of about 15×10^{-7} at 105° K. and has a further minimum at 65° K. and a maximum at 40° K. For cadmium, on the other hand, $-\chi_{\perp}$ is effectively constant, while $-\chi_{\parallel}$ increases steadily by a factor of about 3 as the temperature decreases from room temperature to 14° K. No explanation has yet been put forward to account for the strong temperature dependence of $\chi_{\parallel}$ or for the difference in behaviour between zinc and cadmium. In addition these metals exhibit complex de Haas-van Alphen effects at low temperatures (see § 6·7).

Table VI 8. *Volume susceptibilities of Zn and Cd in units of* 10^{-7} *at* 18° C.

	$\chi_{\parallel}$	$\chi_{\perp}$	χ	χ due	χ due to
	observed		observed	to ions	valency electrons
Zn	−12	−9	−10	−14	4
Cd	−21	−12	−15	−16	1

6·53. Transition metals. Many of the transition metals which are not ferromagnetic are so strongly paramagnetic that the effect of the exchange forces must be considerable. The expressions (6·44·2) for the exchange and correlation forces are not likely to be valid for the d-electrons of the transition metals, and in the present state of the theory it seems best to treat the quantity θ' which occurs in (6·44·7) as an empirical parameter to be determined from the observations. On this basis, Wohlfarth (1949) has discussed the paramagnetism of palladium and platinum, particularly the former.

The susceptibility of alloys of palladium and the noble metals is shown as a function of composition in fig. VI 9. The susceptibility of pure palladium is high, due to the large density of the $4d$ states. The addition of a monovalent metal decreases the susceptibility, and when 60 % has been added the alloys have a diamagnetic susceptibility of normal magnitude. These results can be interpreted as showing that in pure palladium there is about 0·6 electron per atom in $5s$ states and the same number of vacant $4d$ states,

and that when a noble metal is alloyed with palladium the valency electrons of the noble metal go into one of the $4d$ states until all the $4d$ levels are filled (the number of occupied $5s$ states remaining constant at 0·6 electron per atom). When more than 60 % of the noble metal has been added the number of electrons in $5s$ states increases linearly with the concentration of the noble metal. We may therefore assume that $n/n_a = 0·6$, where n is the number of vacant $4d$ levels and n_a is the number of atoms, both per unit volume.

If we neglect the small effect of the $5s$ electrons, and assume that the $4d$ band in palladium is of standard form, then the low-temperature value of the electronic specific heat in palladium given in Table VI.3 gives $T_0 = 2100°$ K.,

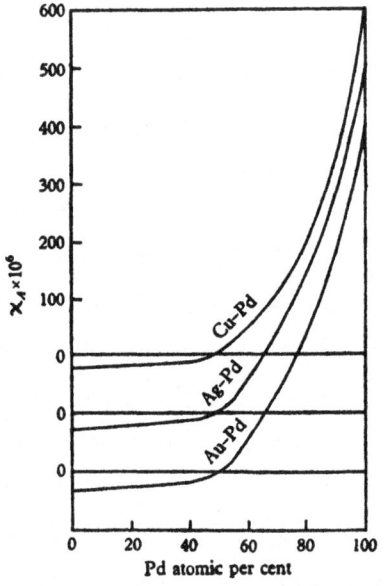

Fig. VI 9. The susceptibilities of palladium alloys.

where T_0 is the degeneracy temperature of the $4d$ electrons. Now the paramagnetic susceptibility of palladium is much too large to be consistent with (6·4·6) and the value of $n(\zeta)$ given by the specific heat; and the effect of the exchange forces must therefore be taken into account. If then we write (6·44·7) for small T as

$$n\mu_0^2/\chi_0 = \tfrac{2}{3}kT_0 - k\theta', \qquad (6·53·1)$$

the observed value of 10^4 c.g.s. units for $1/\chi_0$ gives $\theta' = 1200°$ K. approximately.

The behaviour of $1/\chi$ against T (fig. VI 10) is qualitatively in accordance with that shown in fig. VI 8, but it is not possible to fit the experimental results by an expression of the type (6·44·7) with a single value of θ', the observed susceptibility decreasing less fast than is predicted theoretically. In view of the discrepancy between the observed and calculated specific heats at high temperatures it is not surprising that the temperature variation of χ does not follow the law given by theory for a band of standard form. Hoare and Matthews (1952) have suggested that n/n_a should be treated as an adjustable parameter, in which case a much better fit can be obtained by choosing $n/n_a = 0.29$, $T_0 = 1020°$ K., $\theta' = 554°$ K. However, Hoare and Matthews found that χ behaved anomalously at low temperatures since it has a flat maximum at 80° K. No reasonable explanation of this anomaly has so far been offered, and it was ignored by Hoare and Matthews in fitting their data to (6·44·7).

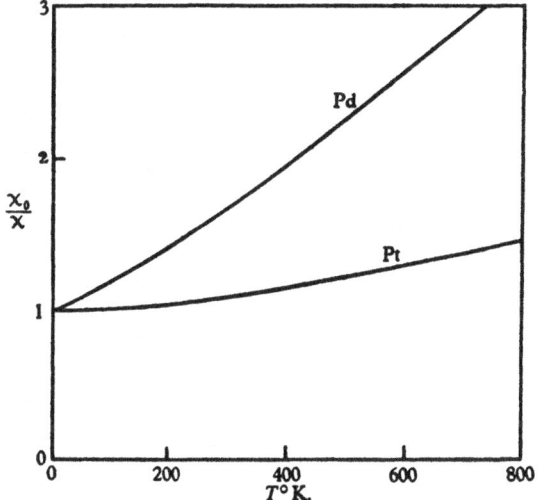

Fig. VI 10. The temperature variation of $1/\chi$ for palladium and platinum.

The results for platinum are similar but less definite, the number of s-electrons, based on the behaviour of the alloys, lying somewhere between 0·3 and 0·6 per atom. The values of the parameters adopted by Hoare and Matthews are $n/n_a = 0.25$, $T_0 = 1750°$ K., $\theta' = 770°$ K. For platinum, χ behaves normally and decreases steadily as T increases.

6·54. Highly diamagnetic metals. Metals of the bismuth group are highly diamagnetic. To explain the observed susceptibility of bismuth on the basis of equation (6·5·2) it is necessary to assume that the effective mass of the free electrons is very small, m^*/m being of the order of 2.5×10^{-3}. The number of free electrons per atom is also small, being about 10^{-3} (see § 8·56). The temperature variation of χ is large, and at sufficiently low temperatures there is a considerable and complicated field variation. These phenomena are discussed in more detail in §§ 6·66 and 6·7.

THE DIAMAGNETISM OF FREE ELECTRONS

6·6. The diamagnetic susceptibility of a free atom or molecule is given by

$$\chi = - (\epsilon^2/6mc^2) \sum_i \overline{r_i^2}, \tag{6·6·1}$$

where $\overline{r_i^2}$ denotes the mean square distance of the ith electron from the centroid of the atom or molecule. This formula also applies to the core electrons in a metal, but it cannot hold without modification for the conduction electrons, since there is no obvious definition of $\overline{r^2}$ for electrons which are free to move over the whole of the metal.

In classical electrodynamics a diamagnetic effect could be expected to occur for free electrons, attributable to the helical path of the electrons in a magnetic field. Detailed analysis, however, shows that the diamagnetic effect is identically zero, the positive contribution to the diamagnetism due to the electrons in the interior of the metal being exactly counterbalanced by the negative contribution of those electrons which hit the boundary of the metal. In quantum theory, on the other hand, not only are the paths of the electrons helical, but the radii of the helices are quantized, and on account of this the diamagnetic effect is not zero, as was first shown by Landau (1930). In order to calculate the diamagnetism we have to construct the free-energy function, which is equivalent to calculating the partition function $Z = \Sigma e^{-E_i/kT}$, where the summation is over all energy states; and since E_i depends upon the magnetic field, we should expect Z to do so too. However, the zero result obtained in the classical theory shows that it is extremely dangerous to rely upon qualitative arguments when dealing with diamagnetism, and it is therefore necessary to carry out the calculations rigorously in order to obtain formulae whose validity is beyond doubt.

In general the diamagnetic susceptibility of metals is small (of the same order as that of a collection of free atoms) and independent of the temperature. For some metals, however, and for bismuth in particular, the diamagnetic susceptibility is large and increases as the temperature is lowered, while at very low temperatures it depends strongly on the magnetic field and approximately like $\sin(a + b/H)$, a phenomenon known as the de Haas-van Alphen effect after its discoverers.

Landau's method was to find the energy levels by solving the Schrödinger equation in terms of Hermite polynomials and then to obtain the partition function by summation. We shall adopt here a more powerful method based upon the use of the density matrix, which, for perfectly free electrons, can be obtained by direct solution of the Schrödinger equation (Sondheimer and Wilson, 1951).

6·61. Consider the expression $e^{-\mathscr{H}/kT}\psi$, where $\mathscr{H}$ is the Hamiltonian operator. This is to be interpreted as

$$e^{-\mathscr{H}/kT}\psi = \sum_{n=0}^{\infty} \frac{1}{n!}\frac{(-1)^n}{(kT)^n}\mathscr{H}^n\psi,$$

where $\mathscr{H}^n\psi = (\mathscr{H}(\mathscr{H}\dots(\mathscr{H}(\mathscr{H}\psi))\dots))$. Now if ψ_i is the ith eigenfunction of $\mathscr{H}$, we have $\mathscr{H}\psi_i = E_i\psi_i$ and $\mathscr{H}^n\psi_i = E_i^n\psi_i$, so that

$$e^{-\mathscr{H}/kT}\psi_i = \sum_{n=0}^{\infty}\frac{1}{n!}\left(-\frac{E_i}{kT}\right)^n\psi_i = e^{-E_i/kT}\psi_i.$$

Thus, if ψ_i is normalized,

$$\sum_i \int \psi_i^* e^{-\mathscr{H}/kT}\psi_i\,d\tau = \sum_i e^{-E_i/kT} = Z. \tag{6·61·1}$$

If we write $\qquad \Psi'(\mathbf{r}',\mathbf{r},\gamma) = \sum_i \psi_i(\mathbf{r}')^* e^{-\gamma\mathscr{H}}\psi_i(\mathbf{r}), \tag{6·61·2}$

then $\qquad\qquad Z = \int \Psi(\mathbf{r},\mathbf{r},\gamma)\,d\tau \quad (\gamma = 1/kT). \tag{6·61·3}$

The quantity $\Psi'(\mathbf{r}',\mathbf{r},\gamma)$ is called the **density matrix**. It satisfies the Schrödinger equation

$$-\partial\Psi/\partial\gamma = \mathscr{H}(\mathbf{r})\,\Psi, \tag{6·61·4}$$

as is obvious from the definition (6·61·2), since $\mathscr{H}(\mathbf{r})$ only acts on $\psi_i(\mathbf{r})$. Further, since the ψ_i's form a complete orthogonal set, we have

$$\Psi'(\mathbf{r}',\mathbf{r},0) = \delta(\mathbf{r}-\mathbf{r}'). \tag{6·61·5}$$

The function $\Psi'(\mathbf{r}',\mathbf{r},\gamma)$ can be defined without reference to the eigenfunctions ψ_i, and it is an invariant for orthogonal transformations. More generally, if A is any operator and $\phi_i\ (i=1,2,\dots)$ is any complete orthogonal set of functions, the quantity

$$(\mathbf{r}'\,|\,A\,|\,\mathbf{r}) = \sum_i \phi_i(\mathbf{r}')^* A\phi_i(\mathbf{r}) \tag{6·61·6}$$

is an invariant for orthogonal transformations of the ϕ's.† We can therefore

---

† Let $\phi_i(q)$ and $\psi_i(q)$ be any two complete sets of normalized orthogonal functions so that

$$\psi_i(q) = \sum_j \phi_j(q)\int \phi_j(q')^*\,\psi_i(q')\,dq', \tag{A}$$

$$\phi_k(q) = \sum_i \psi_i(q)\int \psi_i(q')^*\,\phi_k(q')\,dq'. \tag{B}$$

Then, using (A) we have

$$(q'\,|\,A\,|\,q) = \sum_i \psi_i(q')^*\,A\psi_i(q)$$

$$= \sum_{i,k,l}\phi_k(q')^*\,A\phi_l(q)\int \phi_k(q'')\,\psi_i(q'')^*\,dq''\int \psi_i(q''')\,\phi_l(q''')^*\,dq'''$$

$$= \sum_{k,l}\phi_k(q')^*\,A\phi_l(q)\int \phi_k(q'')\,\phi_l(q'')^*\,dq'',$$

obtain Ψ as the solution of the equation (6·61·4) that satisfies the initial condition (6·61·5), and it is not necessary to know the eigenfunctions and eigenvalues of the Hamiltonian operator.

When a magnetic field is present, the vector potential $\mathbf{A}$ appears explicitly in the Schrödinger equation and introduces a certain degree of arbitrariness, the eigenfunctions being indeterminate to the extent of an unobservable phase factor of modulus unity. The Schrödinger equation is, however, gauge invariant under the transformation

$$\psi \to \psi\, e^{2\pi i \epsilon \Lambda/hc}, \quad \mathbf{A} \to \mathbf{A} - \operatorname{grad} \Lambda, \quad V \to V - \epsilon \partial \Lambda / \partial(ct), \quad (6·61·7)$$

and such a transformation leaves all the physical properties of a system unchanged.

6·62. In the particular case of a free electron in a uniform magnetic field, every point in space is equivalent to every other point, and none of the observable properties of the system can depend upon the coordinates. Hence $\Psi(\mathbf{r}, \mathbf{r}, \gamma)$ must be independent of $\mathbf{r}$. $\Psi(\mathbf{r}', \mathbf{r}, \gamma)$, on the other hand, is a covariant and not an invariant under the gauge transformation (6·61·7). If we change the origin by the translation $\mathbf{r} \to \mathbf{r} + \mathbf{a}$, the vector potential $\mathbf{A}$ is increased by $\frac{1}{2}\mathbf{H} \times \mathbf{a} = \frac{1}{2}\operatorname{grad}\mathbf{r}.(\mathbf{H} \times \mathbf{a})$, and the Schrödinger equation is invariant if every wave function is multiplied by $\exp\{-\pi i \epsilon \mathbf{r}.(\mathbf{H} \times \mathbf{a})/hc\}$. Hence

$$\Psi(\mathbf{r}', \mathbf{r}, \gamma) \to \exp\{\pi i \epsilon(\mathbf{r}' - \mathbf{r}).(\mathbf{H} \times \mathbf{a})/hc\}\,\Psi(\mathbf{r}', \mathbf{r}, \gamma),$$

which means that $\Psi(\mathbf{r}', \mathbf{r}, \gamma)$ is of the form

$$\Psi(\mathbf{r}', \mathbf{r}, \gamma) = \exp\{-\pi i \epsilon \mathbf{H}.(\mathbf{r}' \times \mathbf{r})/hc\}\,F(\mathbf{r} - \mathbf{r}'). \quad (6·62·1)$$

The function F will depend differently on the components of $\mathbf{r} - \mathbf{r}'$ along and perpendicular to the magnetic field.

6·63. *The density matrix for a free electron is a uniform magnetic field.* If we take the magnetic field to be along the z-axis, the most convenient form for the vector potential is $\mathbf{A} = (-\frac{1}{2}Hy, \frac{1}{2}Hx, 0)$, and the equation to be solved for $\Psi(\mathbf{r}', \mathbf{r}, \gamma)$ is then

$$\frac{\partial \Psi}{\partial \gamma} = \left[\frac{h^2}{8\pi^2 m}\nabla^2 - \frac{\epsilon h H}{4\pi i m c}\left(x\frac{\partial}{\partial y} - y\frac{\partial}{\partial x}\right) - \frac{\epsilon^2 H^2}{8mc^2}(x^2 + y^2)\right]\Psi. \quad (6·63·1)$$

by (B). Hence

$$(q'\,|\,A\,|\,q) = \sum_{k,l} \phi_k(q')^* A \phi_l(q)\,\delta_{kl} = \sum_k \phi_k(q')^* A \phi_k(q),$$

so that $(q'\,|\,A\,|\,q)$ is independent of the choice of the orthogonal set of functions. Also, for any function $f(q)$,

$$f(q) = \sum_i \phi_i(q) \int \phi_i(q')^* f(q')\, dq',$$

which is equivalent to $(q'\,|\,1\,|\,q) = \delta(q - q')$.

When $H = 0$, this becomes

$$\partial \Psi / \partial \gamma = (h^2 / 8\pi^2 m) \nabla^2 \Psi,$$

which is the equation of heat conduction. The solution in this case is well known and corresponds to the temperature in an infinite medium at the point (x, y, z) at time γ due to an instantaneous unit point source of heat at (x', y', z') at time $\gamma = 0$. The required solution is (see, for example, Carslaw and Jaeger, 1947, p. 217)

$$\Psi(\mathbf{r}', \mathbf{r}, \gamma)_{H=0} = \left(\frac{2\pi m}{h^2 \gamma}\right)^{\frac{3}{2}} \exp\left[-\frac{2\pi^2 m}{h^2 \gamma} \{(x - x')^2 + (y - y')^2 + (z - z')^2\} \right]. \tag{6·63·2}$$

The form of the generalized expression valid for $H \neq 0$ is suggested by (6·62·1) and (6·63·2). We try to determine a solution of the type

$$\Psi(\mathbf{r}', \mathbf{r}, \gamma) = f(\gamma) \exp\left[-\frac{\pi i \epsilon H}{hc} (x'y - y'x) \right.$$
$$\left. - g(\gamma)\{(x - x')^2 + (y - y')^2\} - \frac{2\pi^2 m}{h^2 \gamma} (z - z')^2 \right]. \tag{6·63·3}$$

On substituting this into (6·63·1) we find the following equations for f and g:

$$\frac{df}{d\gamma} = -\frac{h^2}{2\pi^2 m} fg - \frac{f}{2\gamma}, \quad \frac{dg}{d\gamma} = \frac{\epsilon^2 H^2}{8mc^2} - \frac{h^2}{2\pi^2 m} g^2. \tag{6·63·4}$$

The second equation gives

$$g(\gamma) = \frac{\pi \epsilon H}{2hc} \coth\frac{\epsilon h H \gamma}{4\pi mc}, \tag{6·63·5}$$

and the first then gives

$$f(\gamma) = \left(\frac{\pi m}{2\gamma}\right)^{\frac{1}{2}} \frac{\epsilon H}{h^2 c} \operatorname{cosech}\frac{\epsilon h H \gamma}{4\pi mc}, \tag{6·63·6}$$

the arbitrary constant in f having been chosen so that Ψ reduces to (6·63·2) when $H = 0$. The required solution is therefore

$$\Psi(\mathbf{r}', \mathbf{r}, \gamma) = \left(\frac{2\pi m}{h^2 \gamma}\right)^{\frac{3}{2}} \frac{\mu_0 H \gamma}{\sinh \mu_0 H \gamma} \exp\left[-\frac{2\pi^2 m}{h^2 \gamma} [2i\mu_0 H\gamma(x'y - y'x) \right.$$
$$\left. + \mu_0 H\gamma \coth \mu_0 H\gamma\{(x - x')^2 + (y - y')^2\} + (z - z')^2] \right], \tag{6·63·7}$$

where $\mu_0 = \epsilon h / (4\pi mc)$. To obtain the classical partition function per unit volume Z we put $\mathbf{r} = \mathbf{r}'$, and we then have

$$Z = \Psi(\mathbf{r}, \mathbf{r}, \gamma) = \left(\frac{2\pi m}{h^2 \gamma}\right)^{\frac{3}{2}} \frac{\mu_0 H \gamma}{\sinh \mu_0 H \gamma}. \tag{6·63·8}$$

Since the free energy F per unit volume is $-nkT \log Z$, and since the magnetic moment M per unit volume is $-\partial F/\partial H$, we have

$$M = nkT \frac{\partial \log Z}{\partial H} = -n\mu_0 \left(\coth \frac{\mu_0 H}{kT} - \frac{kT}{\mu_0 H} \right). \qquad (6·63·9)$$

This is a monotonic function of $\mu_0 H/kT$, decreasing steadily from zero to $-n\mu_0$. For small values of $\mu_0 H/kT$, it gives

$$\chi = -\tfrac{1}{3} n\mu_0^2/kT,$$

which is just one-third of the paramagnetic susceptibility, though this ratio does not hold for all values of $\mu_0 H/kT$.

6·64. The generalization to Fermi-Dirac statistics. In a real metal we must employ Fermi-Dirac and not Boltzmann statistics. We must therefore calculate the free energy per unit volume from the relation (Appendix (A 2·8))

$$F - n\zeta = -2kT \sum_i \log(1 + e^{(\zeta - E_i)/kT}) = \sum_i f(E_i), \qquad (6·64·1)$$

where

$$f(E) = -2kT \log(1 + e^{(\zeta - E)/kT}). \qquad (6·64·2)$$

The function F can be obtained from the classical partition function Z by the theory of Laplace transforms in the following way. We define two functions $z(E)$ and $\phi(\gamma)$ by the equations

$$\frac{Z(\gamma)}{\gamma^2} = \int_0^\infty z(E) e^{-\gamma E} dE, \quad \phi(\gamma) = \int_0^\infty f(E) e^{-\gamma E} dE. \qquad (6·64·3)$$

Then these formulae can be inverted (Titchmarsh, 1937, p. 6) to give

$$z(E) = \frac{1}{2\pi i} \int_{c-i\infty}^{c+i\infty} e^{Es} \frac{Z(s)}{s^2} ds, \quad f(E) = \frac{1}{2\pi i} \int_{c-i\infty}^{c+i\infty} e^{Es} \phi(s) ds \qquad (6·64·4)$$

for $E > 0$, where the constant c is such that all the singularities of the integrands lie to the left of the contour of integration. Hence

$$F - n\zeta = \sum_i f(E_i) = \frac{1}{2\pi i} \int_{c-i\infty}^{c+i\infty} \sum_i e^{E_i s} \phi(s) ds = \frac{1}{2\pi i} \int_{c-i\infty}^{c+i\infty} \frac{Z(-s)}{s^2} s^2 \phi(s) ds. \qquad (6·64·5)$$

The last integral is a resultant, and since the transform of $s^2 \phi(s)$ is $\partial^2 f/\partial E^2$, we have (Titchmarsh, 1937, p. 53)

$$F - n\zeta = \int_0^\infty z(E) \frac{\partial^2 f}{\partial E^2} dE = 2 \int_0^\infty z(E) \frac{\partial f_0}{\partial E} dE, \qquad (6·64·6)$$

where $f_0(E)$ is, as usual, the Fermi function, by (6·64·2). The calculation of F is therefore reduced to the determination of the function $z(E)$.

6·65. Evaluation of the contour integrals. In the particular case under consideration the function $z(E)$ is essentially a Riemann ζ-function, and it can therefore be calculated by standard complex variable methods. It is given by inserting (6·63·8) into (6·64·4) and is

$$\frac{h^3}{(2\pi m)^{\frac{3}{2}}(\mu_0 H)^{\frac{3}{2}}}z(E) = \frac{1}{2\pi i}\int_{c-i\infty}^{c+i\infty}\frac{e^{Es}\,ds}{(\mu_0 Hs)^{\frac{3}{2}}\sinh\mu_0 Hs}$$

$$= \frac{1}{2\pi i}\int_{c-i\infty}^{c+i\infty}\frac{2e^{Es-\mu_0 Hs}\,ds}{(\mu_0 Hs)^{\frac{3}{2}}(1-e^{-2\mu_0 Hs})}. \qquad (6\cdot65\cdot1)$$

If we could evaluate the integral simply by deforming the contour into a loop starting from $-\infty$, encircling the origin in the positive direction and ending at $-\infty$, we should obtain a diamagnetic moment which was a monotonic function of $\mu_0 H/kT$. However, the integrand has poles at the points $\mu_0 Hs = \pm r\pi i$ $(r=1,2,...)$ as well as a branch point at the origin, and the poles give rise to trigonometrical terms in the expression for $z(E)$ when it is reduced to a real form. The trigonometrical terms have their physical origin in the fact that the orbits of a free electron in a magnetic field are quantized, thus giving rise to a quasi-periodic variation of the density of states as a function of the energy. When the distribution of the energies of the individual electrons is given by the Boltzmann law, these variations are

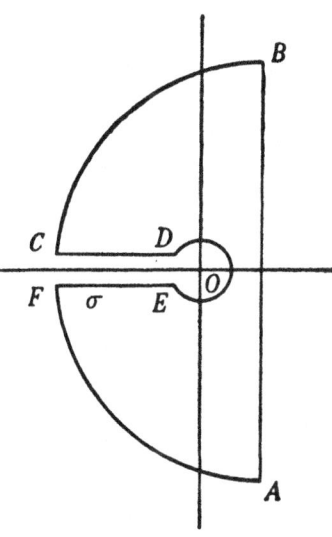

Fig. VI 11.

smoothed out, but this must be considered to be the exceptional case, and for general distributions the oscillations remain.

To obtain $z(E)$ in real form, consider the integral (6·65·1) taken round the closed contour $ABCDEFA$ shown in fig. VI 11, where there is a cut in the complex s-plane along the negative real axis, and where BC and FA are portions of the circle $\mu_0 H |s| = (N+\frac{1}{2})\pi$, N being a large number. The integral is now single-valued and the contour does not pass through any of the poles of the integrand. Further, the integrals over the paths BC and FA tend to zero as $N \to \infty$. Hence the integral over AB can be replaced by the integral over the loop $FEDC$ plus $2\pi i$ times the sum of the residues. We therefore write (6·65·1) as

$$\frac{1}{2\pi i}\int_{-\infty}^{(0+)}\frac{2e^{Es-\mu_0 Hs}\,ds}{(\mu_0 Hs)^{\frac{3}{2}}(1-e^{-2\mu_0 Hs})} + \sum_{r=1}^{\infty}\{R(r\pi i)+R(-r\pi i)\}, \qquad (6\cdot65\cdot2)$$

where the integral is taken over a loop starting at $-\infty$ on the real axis, encircling the origin in the positive direction and returning to the starting point, but not containing any of the points $\mu_0 Hs = \pm r\pi i$. The contribution $R(r\pi i)$ from the pole at $\mu_0 Hs = r\pi i$ is

$$-\frac{1}{\mu_0 H(r\pi)^{\frac{3}{2}}} \exp\left(\frac{r\pi E}{\mu_0 H} + r\pi - \tfrac{1}{4}\pi\right) i,$$

so that

$$R(r\pi i) + R(-r\pi i) = \frac{(-1)^{r+1}}{\mu_0 H(r\pi)^{\frac{3}{2}}} 2 \cos\left(\frac{r\pi E}{\mu_0 H} - \tfrac{1}{4}\pi\right). \qquad (6\cdot65\cdot3)$$

The contour integral in (6·65·2) can be transformed into a real integral as follows. We write

$$\int_{-\infty}^{(0+)} \frac{e^{Es}\, ds}{(\mu_0 Hs)^{\frac{3}{2}} \sinh \mu_0 Hs} = \int_{-\infty}^{(0+)} \left\{\frac{1}{(\mu_0 Hs)^{\frac{5}{2}}} - \frac{1}{6(\mu_0 Hs)^{\frac{3}{2}}}\right\} e^{Es}\, ds$$

$$-\int_{-\infty}^{(0+)} \left\{\frac{1}{(\mu_0 Hs)^{\frac{5}{2}}} - \frac{1}{6(\mu_0 Hs)^{\frac{3}{2}}} - \frac{1}{(\mu_0 Hs)^{\frac{3}{2}} \sinh \mu_0 Hs}\right\} e^{Es}\, ds.$$

The first integral can be evaluated by means of Hankel's formula

$$\frac{1}{\Gamma(z)} = \frac{1}{2\pi i} \int_{-\infty}^{(0+)} t^{-z} e^t\, dt,$$

while the second can be transformed into a real infinite integral since the integrand is of order $s^{\frac{1}{2}}$ near the origin.

When all the terms are collected together, the complete expression for $z(E)$ is

$$z(E) = \frac{(2\pi m)^{\frac{3}{2}}}{h^3} \left\{\frac{8}{15\sqrt{\pi}} E^{\frac{5}{2}} - \frac{1}{3\sqrt{\pi}} (\mu_0 H)^2 E^{\frac{1}{2}}\right.$$

$$+ \frac{1}{\pi} (\mu_0 H)^{\frac{5}{2}} \int_0^\infty \left(\frac{1}{y^{\frac{5}{2}}} - \frac{1}{6y^{\frac{3}{2}}} - \frac{1}{y^{\frac{3}{2}} \sinh y}\right) e^{-yE/\mu_0 H}\, dy$$

$$\left. - 2(\mu_0 H)^{\frac{3}{2}} \sum_{r=1}^\infty \frac{(-1)^r}{(r\pi)^{\frac{3}{2}}} \cos\left(\frac{r\pi E}{\mu_0 H} - \tfrac{1}{4}\pi\right)\right\}. \qquad (6\cdot65\cdot4)$$

We can obtain the free energy for temperatures such that $kT \ll \zeta$ by using the asymptotic formula for Fermi integrals. The part of the integral coming from the trigonometrical terms must, however, be evaluated to a higher degree of accuracy than usual since the integrand is a rapidly varying function of E. Write

$$\cos\left(\tfrac{1}{4}\pi - r\pi E/\mu_0 H\right) = \mathscr{R}[\exp\left(-\tfrac{1}{4}\pi i + r\pi i \zeta/\mu_0 H\right) \exp\{r\pi i(E-\zeta)/\mu_0 H\}],$$

where $\mathscr{R}$ denotes the real part, and consider the integral

$$\int_0^\infty e^{r\pi i(E-\zeta)/\mu_0 H} \frac{\partial f_0}{\partial E}\, dE.$$

Put $\eta = (E - \zeta)/kT$ and replace the lower limit of integration $-\zeta/kT$ by $-\infty$ as usual, the error being $O(e^{-\zeta/kT})$. The integral is then

$$-\frac{1}{4}\int_{-\infty}^{\infty} \frac{e^{r\pi ikT\eta/\mu_0 H}}{\cosh^2 \frac{1}{2}\eta}\,d\eta,$$

which can readily be evaluated by contour integration, its value being

$$-\frac{r\pi^2 kT/\mu_0 H}{\sinh(r\pi^2 kT/\mu_0 H)}.$$

We therefore have

$$F - n\zeta = 2\frac{(2\pi m)^{\frac{3}{2}}}{h^3}\int_0^{\infty}\left[\frac{8}{15\sqrt{\pi}}E^{\frac{5}{2}} - \frac{1}{3\sqrt{\pi}}(\mu_0 H)^2 E^{\frac{1}{2}}\right.$$
$$\left. + \frac{1}{\pi}(\mu_0 H)^{\frac{3}{2}}\int_0^{\infty}\left(\frac{1}{y^{\frac{1}{2}}} - \frac{1}{6y^{\frac{3}{2}}} - \frac{1}{y^{\frac{1}{2}}\sinh y}\right)e^{-\nu E/\mu_0 H}\,dy\right]\frac{\partial f_0}{\partial E}\,dE$$
$$+ \frac{8\sqrt{2}\,\pi m^{\frac{3}{2}}}{h^3}kT(\mu_0 H)^{\frac{3}{2}}\sum_{r=1}^{\infty}\frac{(-1)^r\cos(\frac{1}{4}\pi - r\pi\zeta/\mu_0 H)}{r^{\frac{3}{2}}\sinh(r\pi^2 kT/\mu_0 H)}. \quad (6\cdot65\cdot5)$$

6·66. The normal diamagnetism. The normal diamagnetism is determined by the integrals in (6·65·5), and if we neglect the dependence of χ upon H we have

$$\chi = -\frac{1}{H}\frac{\partial F}{\partial H} = \frac{4(2\pi m)^{\frac{3}{2}}\mu_0^2}{3\sqrt{\pi}\,h^3}\int_0^{\infty}E^{\frac{1}{2}}\frac{\partial f_0}{\partial E}\,dE.$$

Hence, with the notation of § 6·42,

$$\chi = -\frac{n\mu_0^2}{3kT}\frac{F'_{\frac{1}{2}}(\xi)}{F_{\frac{1}{2}}(\xi)}, \quad (6\cdot66\cdot1)$$

which is exactly one-third of the paramagnetic spin susceptibility for all values of the temperature (cf. equation (6·42·2)).

The variation of χ with H is negligible in practice, but for the sake of mathematical completeness we derive the exact formula here. In this case it is necessary to include the spin paramagnetism explicitly from the beginning, since although the spin and translatory motion of a free electron give additive contributions to the susceptibility to the first order in H, this is no longer true for all values of the magnetic field. We therefore have to recalculate the classical partition function Z. Since the effect of the spin is to split each energy level E_i into two with energies $E_i \pm \mu_0 H$, the partition function is

$$Z = \frac{1}{2}(e^{\mu_0 H/kT} + e^{-\mu_0 H/kT})\left(\frac{2\pi m}{h^2\gamma}\right)^{\frac{3}{2}}\frac{\mu_0 H\gamma}{\sinh\mu_0 H\gamma}$$
$$= (2\pi m/h^2\gamma)^{\frac{3}{2}}\mu_0 H\gamma\coth\mu_0 H\gamma. \quad (6\cdot66\cdot2)$$

We can follow the calculation of the preceding section exactly and obtain F in a form similar to (6·65·5), but for our present purpose it is convenient to proceed slightly differently. We use the formula

$$\coth x = \frac{1}{x} + 2x \sum_{n=1}^{\infty} \frac{1}{n^2\pi^2 + x^2}$$

and write

$$\frac{1}{x^{\frac{1}{2}}} \coth x = \frac{1}{x^{\frac{3}{2}}} + 2 \sum_{n=1}^{\infty} \frac{1}{n^2\pi^2} \left(\frac{1}{x^{\frac{1}{2}}} - \frac{x^{\frac{1}{2}}}{n^2\pi^2 + x^2} \right)$$

$$= \frac{1}{x^{\frac{3}{2}}} + \frac{1}{3x^{\frac{1}{2}}} - 2 \sum_{n=1}^{\infty} \frac{x^{\frac{1}{2}}}{n^2\pi^2(n^2\pi^2 + x^2)}.$$

If we now calculate F by the method of the preceding section and omit the trigonometrical terms we obtain

$$F - n\zeta = 2\frac{(2\pi m)^{\frac{3}{2}}}{h^3} \int_0^{\infty} \left[\frac{8}{15\sqrt{\pi}} E^{\frac{5}{2}} + \frac{2}{3\sqrt{\pi}} (\mu_0 H)^2 E^{\frac{1}{2}} \right.$$

$$\left. + \frac{2}{\pi} (\mu_0 H)^{\frac{5}{2}} \sum_{n=1}^{\infty} \int_0^{\infty} \frac{x^{\frac{1}{2}}}{n^2\pi^2(n^2\pi^2 + x^2)} e^{-xE/\mu_0 H} dx \right] \frac{\partial f_0}{\partial E} dE. \quad (6\cdot66\cdot3)$$

The integrals are now in a form in which they can readily be evaluated, either exactly by numerical integration, or approximately by analytical methods.

A more important matter than the field variation of the susceptibility is the effect of the binding of the conduction electrons. The results given in § 6·8 show that, at least for the zero-order approximation for χ, the effect of the binding of the electrons can be taken into account by using an effective mass m^* for the electrons if the energy band is of standard form. Since the spin of the electron is unaffected by the value of the effective mass, the total susceptibility is obtained by adding (6·42·2) and (6·66·1) with μ_0 in the latter replaced by μ_0^*, where $\mu_0^* = \epsilon h/(4\pi m^* c)$. We therefore have, for the combined paramagnetic and diamagnetic susceptibility,

$$\chi(T) = \frac{n\mu_0^2}{kT} \frac{F'_{\frac{3}{2}}(\xi)}{F_{\frac{1}{2}}(\xi)} \left(1 - \frac{m^2}{3m^{*2}} \right), \quad \chi(0) = \frac{4m^*\mu_0^2}{h^2} (3\pi^2 n)^{\frac{1}{3}} \left(1 - \frac{m^2}{3m^{*2}} \right).$$

$$(6\cdot66\cdot4)$$

According to this result the temperature variation of χ should be the same for all metals (expressed as a function of T/T_0), but χ should be positive or negative according as m^* is greater or less than $m/\sqrt{3}$. A large diamagnetic susceptibility can only occur, therefore, if the effective mass is small, and this means that the surface of the Fermi distribution must lie in regions of the energy surfaces which have very large curvatures, that is, that it must approach very closely the boundary of a Brillouin zone. This condition is satisfied in the bismuth group of metals and in the γ-alloys, all of which are highly diamagnetic, and also, to a less extent, in the divalent metals, zinc, cadmium and mercury, which are moderately diamagnetic.

For polycrystalline bismuth, χ is about -155×10^{-7} at $85°\,\mathrm{K}$. The electrons nearly fill a Brillouin zone, there being about 10^{-3} vacant levels per atom in the zone and the same number n of electrons in the next zone. The various estimates of n differ appreciably (see §§ 8·56 and 8·561), but since only $n^{\frac{1}{3}}$ occurs in χ the uncertainty is not important. If we assume that the holes and the free electrons contribute equally to the diamagnetism, equation (6·66·4) gives $m^*/m = 2·5 \times 10^{-3}$. This is of the same order as the value required to explain the electrical properties of bismuth (§ 8·561).

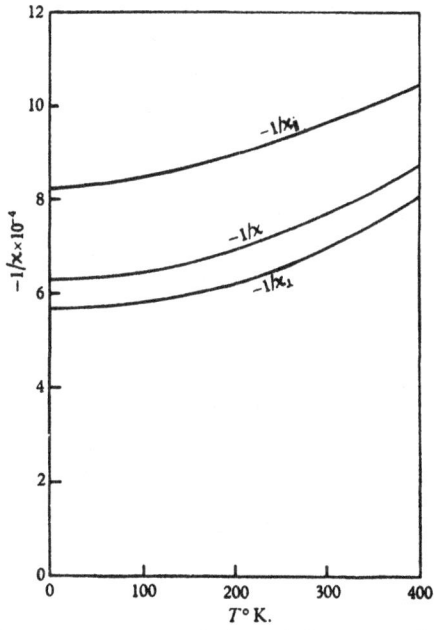

Fig. VI 12. The temperature variation of the susceptibility of bismuth. χ_1 and $\chi_\perp$ refer to single crystals, χ to polycrystalline material.

The temperature variation of χ for bismuth is quite large, as shown by the curves in fig. VI 12. The general shape of the $1/\chi - T$ curves is similar to that of the theoretical curve of fig. VI 8, and would indicate a degeneracy temperature T_0 of the order of $1000°\,\mathrm{K}$. The small value of the effective mass, however, means that the theoretical value of T_0 is extremely high, of the order of 10^5, so that the temperature variation of χ is much too large to be reconciled with that given by equation (6·66·4).

THE DE HAAS-VAN ALPHEN EFFECT

6·7. A quasi-periodic variation of χ with the field was first observed by de Haas and van Alphen in 1930 for single crystals of bismuth at liquid hydrogen temperatures. The general features of the phenomena are shown in fig. VI 13. The susceptibility is a quasi-periodic function of $1/H$, which, for fixed T, is constant for sufficiently large magnetic fields. As H is decreased, maxima and minima of χ occur with increasing frequency, there being in theory an infinite number in any region which includes $H = 0$. These fluctuations only occur when H is perpendicular to the principal axis of the crystal and are different in detail, though similar in general aspect, according as H is parallel or perpendicular to a binary axis of the crystal.

The phenomena described above are in qualitative agreement with the formulae derived in § 6·65 for the free energy of an electron gas, since F contains trigonometrical terms. However, the formulae only apply to perfectly free electrons, and since they do not give even the correct normal diamagnetism for bismuth without modification, it is necessary to consider how they can be generalized before discussing them in detail.

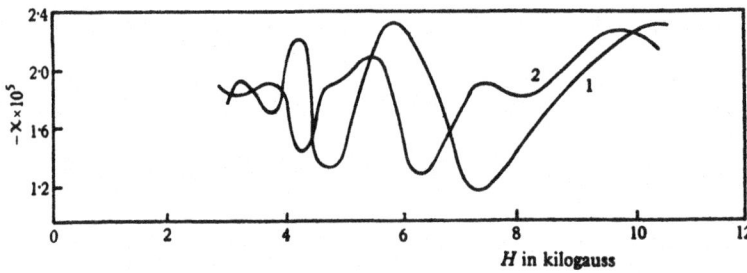

Fig. VI 13. The field dependence of $\chi_\perp$ for bismuth at 4° K. For curve 1, H is parallel to a binary axis; for curve 2, H is perpendicular to a binary axis.

In the first place we must include the effect of the electronic spin, since it is not an additive effect except for the normal susceptibility in the zero approximation. This can be done by multiplying the classical partition function Z by $\cosh(\mu_0 H/kT)$. It is the only generalization that is rigorously deducible, all the others being more or less speculative at the present stage of development of the theory. The most obvious further extension to make is the introduction of an effective mass m^*. This is justified for the normal diamagnetism by the calculations mentioned in § 6·8, but no proof exists for the complete partition function. We shall, however, replace (6·66·2) by

$$Z = \left(\frac{2\pi m^*}{h^2\gamma}\right)^{\frac{3}{2}} \mu_0^* H\gamma \frac{\cosh \mu_0 H\gamma}{\sinh \mu_0^* H\gamma}, \qquad (6\cdot7\cdot1)$$

where $\mu_0^* = \epsilon h/(4\pi m^* c)$, bearing in mind the speculative nature of the assumption involved. (Note that the factor in Z due to the electron spin involves the spin magnetic moment μ_0 and not μ_0^*.)

When the calculations of the preceding section are repeated and the small terms giving the field variation of the normal diamagnetism are omitted, the results are as follows:

$$F - n\zeta = -\frac{8\pi m^{*\frac{3}{2}}}{15h^3}(2\zeta)^{\frac{5}{2}}\left[1 + \frac{5}{8}\left(\frac{3m^{*2}}{m^2} - 1\right)\left(\frac{\mu_0^* H}{\zeta}\right)^2\right.$$

$$\left. - \frac{15kT}{4\mu_0^* H}\left(\frac{\mu_0^* H}{\zeta}\right)^{\frac{1}{2}}\sum_{r=1}^{\infty}\frac{(-1)^r\cos\left(r\pi m^*/m\right)\cos\left(\frac{1}{4}\pi - r\pi\zeta/\mu_0^* H\right)}{r^{\frac{5}{2}}\sinh\left(r\pi^2 kT/\mu_0^* H\right)}\right], \quad (6.7.2)$$

$$\chi = \frac{M}{H} = \frac{8\pi m^{*\frac{3}{2}}}{3h^3}\mu_0^{*2}(2\zeta)^{\frac{1}{2}}\left[\frac{3m^{*2}}{m^2} - 1\right.$$

$$\left. - \frac{3\pi kT}{\mu_0^* H}\left(\frac{\zeta}{\mu_0^* H}\right)^{\frac{1}{2}}\sum_{r=1}^{\infty}\frac{(-1)^{r+1}\cos\left(r\pi m^*/m\right)\sin\left(\frac{1}{4}\pi - r\pi\zeta/\mu_0^* H\right)}{r^{\frac{1}{2}}\sinh\left(r\pi^2 kT/\mu_0^* H\right)}\right], \quad (6.7.3)$$

$$n = \frac{8\pi}{3h^3}(2m^*\zeta)^{\frac{3}{2}}$$

$$\times\left[1 + \frac{3\pi kT}{2\mu_0^* H}\left(\frac{\mu_0^* H}{\zeta}\right)^{\frac{1}{2}}\sum_{r=1}^{\infty}\frac{(-1)^{r+1}\cos\left(r\pi m^*/m\right)\sin\left(\frac{1}{4}\pi - r\pi\zeta/\mu_0^* H\right)}{r^{\frac{3}{2}}\sinh\left(r\pi^2 kT/\mu_0^* H\right)}\right],$$

$$(6.7.4)$$

the last equation being obtained from the relation $\partial F/\partial\zeta = 0$ (Appendix (A 2·9)). The series are rapidly convergent, and it is usually sufficient to take a few terms, and often only one, since the denominators increase rapidly with r.

6·71. The expression (6·7·3) for χ is a function of the two parameters $\zeta/\mu_0^* H$ and $kT/\mu_0^* H$, and except for extremely high fields we have $\zeta \gg kT \gg \mu_0^* H$. Now if the oscillating terms in χ are to be appreciable in magnitude it is necessary for $\pi^2 kT/\mu_0^* H$ to be of order unity at most, i.e. H must be of the order of $10^5 Tm^*/m$ gauss. Therefore, if $m^* = m$, the de Haas-van Alphen effect will only be observable at fields of the order of 10^5–10^6 gauss. In fact, for those metals in which the effect has so far been observed (beryllium, graphite, magnesium, aluminium, zinc, gallium, cadmium, indium, tin, antimony, mercury, thallium, lead and bismuth) it occurs with fields of the order of 10^4 gauss at temperatures of the order of $4°$ K. For these metals, therefore, m^*/m must be at least as small as $0·1$.

The period of the oscillations is determined by the factor

$$\sin\left(\tfrac{1}{4}\pi - r\pi\zeta/\mu_0^* H\right).$$

Now $\zeta_0/\mu_0^* = \frac{1}{2}\pi(hc/\epsilon)(3n/\pi)^{\frac{1}{3}}$ is independent of m^* and is determined solely by n, the number of electrons present. It is found experimentally that the positions of the maxima of χ as functions of $1/H$ are equally spaced, in agreement with the theoretical formula, and that their separation is of the order of 10^{-6} gauss^{-1}. Hence n must be of the order of 10^{18}–10^{19} per c.c., i.e. the number of effective electrons per atom n/n_a must be of the order of 10^{-3}–10^{-4}.

The very small values of n/n_a and of m^*/m required to explain the de Haas-van Alphen effect are widely different from the values deduced from other phenomena. For most of the metals listed above, we should expect n/n_a and m^*/m to be only slightly less than unity, and although small values of n/n_a and m^*/m are needed to explain the electrical properties of antimony and bismuth, a more careful examination of the data reveals that the values needed to fit the de Haas-van Alphen phenomena are much smaller than those deduced from the galvanomagnetic effects.

We must therefore conclude that substantial modifications are needed to bring the theory into line with the experimental facts.

6·72. The following are the main facts in qualitative agreement with the expression (6·7·3) for χ.

(1) The amplitudes of the oscillations in χ become much more marked at low temperatures and high fields, and they vary with T and H like $e^{-\lambda T/H}$.

(2) The positions of the maxima when plotted against $1/H$ are equidistant and independent of the temperature.

(3) For a given temperature there is a value of H above which no oscillations take place.

The main differences, on the other hand, between the experimental data and (6·7·3), apart from the necessity of choosing small values of n/n_a and m^*/m, are that (6·7·3) only refers to an isotropic medium, whereas the de Haas-van Alphen effect is highly anisotropic, and that the observed susceptibility is always negative. According to the theoretical expression (6·7·3), χ is constant and equal to the steady diamagnetic susceptibility for large values of H, if we exclude the first term which is the normal spin susceptibility. As H decreases, the oscillations set in and χ changes sign; in the limit of $H = 0$, χ tends to the steady value once more. The observed oscillations in χ, however, have phases which do not agree with the predictions of (6·7·3), and their amplitudes do not seem to be large enough to make χ positive. (This last statement may only apply to bismuth, which has a large steady diamagnetism. For metals with a more normal susceptibility only the difference of the susceptibility in two directions seems to have been measured with any accuracy, and the absolute values of χ are by no means certain.) In view of this and of the extremely small values of n/n_a and m^*/m required to explain the amplitudes and periods of the oscillations, it has been suggested that the theory ought to be generalized so as to include two sets of electrons, one of which has parameters of a normal order of magnitude and is responsible for the diamagnetism, while the other is responsible for the de Haas-van Alphen phenomena. The theory has also been generalized by Blackman (1938) to apply to anisotropic media. The generalization is a formal one, in that it is assumed that the requisite formulae can be obtained by a linear transformation in **k** space, whereby the surfaces $E = \frac{1}{2}\hbar^2 |\mathbf{k}|^2/m$ become quadrics. It has not, however, been possible so far to prove that the effect of the binding of the valency electrons can be treated in this simple way.

Several further generalizations have been suggested, which are discussed by Shoenberg (1952) in an article which reviews the extensive experimental work carried out up to date by himself and a number of other workers, and compares the results with the various tentative theoretical formulae. Perhaps the most important generalization is the calculation of the effect of the collisions of the electrons with the impurities, which leads to a collision broadening of the quantized energy levels (Dingle, 1952). This results in the envelope of the oscillations being proportional to $e^{-\lambda(T+T')/H}$, where T' is a parameter describing the collision broadening. The details are too involved to be given here, and the discussion is therefore confined to some typical cases. Further information can be found in Shoenberg's article.

For bismuth the simplest assumption which is compatible with the crystal symmetry but which does not involve rotational symmetry round the trigonal axis is that the electrons occupy three separate zones, in each of which the energy surfaces are ellipsoidal. If we take rectangular axes such that O3 is along the trigonal axis and O1 along one of the three binary axes, then we can write

$$E = \frac{1}{2}\hbar^2(\alpha_{11} k_1^2 + \alpha_{22} k_2^2 + 2\alpha_{23} k_2 k_3 + \alpha_{33} k_3^2)/m$$

for one zone, while the energies for the other zones are obtained by rotations of $\pm \frac{2}{3}\pi$ round O3. The components of the mass tensor m_{ij} are given by

$$\frac{m_{11}}{m} = \frac{1}{\alpha_{11}}, \quad \frac{m_{22}}{m} = \frac{\alpha_{33}}{\alpha_{22}\alpha_{33} - \alpha_{23}^2}, \quad \frac{m_{33}}{m} = \frac{\alpha_{22}}{\alpha_{22}\alpha_{33} - \alpha_{23}^2}, \quad \frac{m_{23}}{m} = \frac{-\alpha_{23}}{\alpha_{22}\alpha_{33} - \alpha_{23}^2}.$$

Then the susceptibilities in the various directions are given by the sums of expressions of the type (6·7·3), but with various effective masses which are as follows:

$\chi_{\parallel}$: $\qquad\qquad\qquad\qquad\qquad \| M \|^{\frac{1}{2}}/m_{33}^{\frac{1}{2}};$

$\chi_{\perp}$, parallel to a binary axis: $\qquad \| M \|^{\frac{1}{2}}/m_{11}^{\frac{1}{2}}, \quad \| M \|^{\frac{1}{2}}/(\frac{1}{4}m_{11} + \frac{3}{4}m_{22})^{\frac{1}{2}};$

$\chi_{\perp}$, perpendicular to a binary axis: $\| M \|^{\frac{1}{2}}/m_{22}^{\frac{1}{2}}, \quad \| M \|^{\frac{1}{2}}/(\frac{3}{4}m_{11} + \frac{1}{4}m_{22})^{\frac{1}{2}};$

where $\qquad\qquad \| M \| = m_{11}(m_{22}m_{33} - m_{23}^2).$

In order for there to be no perceptible de Haas-van Alphen effect for $\chi_{\parallel}$, it is necessary for $\| M \|^{\frac{1}{2}}/m_{33}^{\frac{1}{2}}$ to be considerably greater than at least one of each pair of the effective masses for $\chi_\perp$, and this latter mass must be very much less than m. The following values are found to fit the results reasonably well:

$$m_{11}/m = 2\cdot4 \times 10^{-3}, \quad m_{22}/m = 2\cdot5, \quad m_{33}/m = 0\cdot05,$$

$$m_{23}/m = -0\cdot25, \quad \| M \|^{\frac{1}{3}}/m = 5\cdot2 \times 10^{-2}.$$

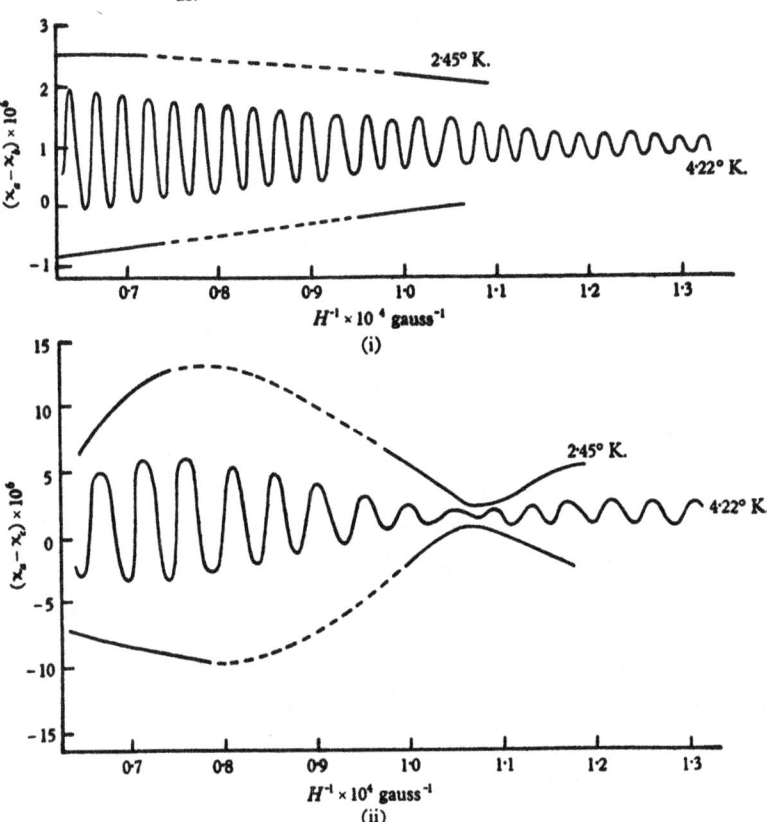

Fig. VI 14. The field dependence of the differences of the principal susceptibilities of gallium at 4·22° K. and 2·45° K. The full curves are shown for $T = 4\cdot22°$ K., but only the envelopes for $T = 2\cdot45°$ K. The curves in (i) are for H in the ab plane, with H making an angle 78·9° with the a axis. The curves in (ii) are for H in the ac plane, with H making an angle 79·8° with the a axis.

Further, ζ has to be chosen to be 0·018 eV., which corresponds to a degeneracy temperature $T_0 = 210°$ K. and to an effective number n of electrons equal to $1\cdot5 \times 10^{-5}$ per atom, where n is defined by

$$\zeta = \frac{h^2}{8 \| M \|^{\frac{1}{3}}} \left(\frac{3n}{\pi}\right)^{\frac{2}{3}}.$$

Similar results can be obtained for the other metals. For example, some experimental curves are given in fig. VI 14 for gallium, which is orthorhombic with three unequal axes $a = 4\cdot52$ A., $b = 4\cdot51$ A., $c = 7\cdot64$ A. (The method of measurement employed (Shoenberg, 1952) is such that it only gives the difference between two susceptibilities.) It will be seen that $\chi_a - \chi_b$ is given quite well by one term of (6·7·3), while for $\chi_a - \chi_c$ two terms are needed to reproduce the beats. Three energy ellipsoids are required which have their principal axes along the crystal axes in order to satisfy the symmetry relations of the crystal, and rough values of their parameters are given in Table VI 9.

Table VI 9. *Mass tensor parameters of gallium*

m_{11}/m	m_{22}/m	m_{33}/m	ζ in eV.	n/n_a
0·2	0·15	0·05	0·06	3×10^{-5}
0·1	0·3	0·03	0·07	4×10^{-5}
0·2	0·02	0·4	0·04	7×10^{-5}

Although quantitative agreement can on the whole be attained for any metal by a suitable choice of the parameters, the numbers of electrons and their effective masses turn out to be so small that the theory in its numerical aspects seems to be very artificial. While, therefore, the theory gives a good description of the qualitative features of the de Haas-van Alphen phenomena, it is difficult to believe that we have yet achieved a proper understanding of the effect of the binding of the electrons on the susceptibility.

THE DIAMAGNETISM OF QUASI-BOUND ELECTRONS

6·8. It has not so far proved possible to give a complete theory of the effect of the binding upon the diamagnetism of the conduction electrons. There have been two approximate approaches to the problem along different lines, one being by Peierls (1933a), while a method based upon the properties of the density matrix $\Psi'(\mathbf{r}', \mathbf{r}, \gamma)$ was given in the first edition of this book. An attempt was made there to determine the generalized double Fourier transform of $\Psi'(\mathbf{r}', \mathbf{r}, \gamma)$, but this gives rise to difficulties which can be avoided by the direct calculation of $\Psi'(\mathbf{r}', \mathbf{r}, \gamma)$ as a power series in H (Wilson, 1953).

In view of the complexity of the theory and the difficulty in applying it to real metals on account of our lack of sufficient knowledge of the energy levels and wave functions, we shall quote the final results and discuss their implications, referring the reader to the original sources for the details of the derivation.

The most important contribution to the normal diamagnetism is given by

$$\chi = \frac{e^2}{12\pi h^2 c^2} \int \left\{ \frac{\partial^2 E}{\partial k_1^2} \frac{\partial^2 E}{\partial k_2^2} - \left(\frac{\partial^2 E}{\partial k_1 \partial k_2} \right)^2 \right\} \frac{\partial f_0}{\partial E} d\mathbf{k} \qquad (6\cdot8\cdot1)$$

when the magnetic field is parallel to the z-axis. The susceptibility is therefore determined by the Gaussian curvature of the energy surfaces. There are a number of other terms in the complete expression for χ, the most interesting being

$$\chi = \frac{\epsilon^2}{16\pi^3 mc^2} \int f_0(E)\, d\mathbf{k} \int u^* \left(\frac{\partial^2 u}{\partial k_1^2} + \frac{\partial^2 u}{\partial k_2^2} \right) d\tau_0. \tag{6·8·2}$$

This is the analogue of the atomic diamagnetism. For, if we consider very tightly bound electrons for which

$$u_{\mathbf{k}} = \sum_{\mathbf{g}} \exp\{-i\mathbf{k}.(\mathbf{r} - \mathbf{g}_a)\}\, \phi_{\mathbf{g}}$$

and neglect the overlapping of neighbouring ϕ's, we obtain formula (6·6·1) at once.

6·81. For free electrons with an effective mass m^*, (6·8·1) can be written as

$$\chi = \frac{\epsilon^2}{3\pi hc^2(2m^*)^{\frac{1}{2}}} \int_0^\infty E^{\frac{1}{2}} \frac{\partial f_0}{\partial E}\, dE, \tag{6·81·1}$$

which, to the zero order in kT/ζ, is

$$\chi = -\frac{4}{3}\frac{m^2 \mu_0^2}{m^* h^2}(3\pi^2 n)^{\frac{1}{3}}. \tag{6·81·2}$$

With the notation of §§ 6·42 and 6·66, χ for arbitrary temperatures can be written as

$$\chi = -\frac{nm^2 \mu_0^2}{3m^{*2}kT}\frac{F'_{\frac{1}{2}}(\xi)}{F_{\frac{1}{2}}(\xi)}, \tag{6·81·3}$$

with $\xi = \zeta/kT$ and $\zeta_0 = h^2(3n/\pi)^{\frac{2}{3}}/(8m^*)$. This differs from the corresponding expression (6·66·1) by the factor $(m/m^*)^2$, and ζ_0 also depends upon m^*. The same formulae hold for an inverted band of standard form, and it should be noted that, since χ depends upon the Gaussian curvature of the energy surfaces, it can be positive as well as negative, though for a nearly full or a nearly empty band χ is negative.

The expression (6·8·1) is only large compared with that for free electrons when the effective mass is small, and (6·8·1) is then the dominant term in the magnetic susceptibility. When the effective mass is large, (6·8·1) is small compared with (6·8·2), which reduces to the susceptibility of the valency electrons in a set of free atoms, but in this case the spin paramagnetism is dominant and the diamagnetism can be neglected. In intermediate cases it is sufficient to treat the electrons as perfectly free, and the exact theory of § 6·66 then applies.

REFERENCES

Blackman, M. (1935). Contributions to the theory of the specific heat of crystals. *Proc. Roy. Soc.* A, **148**, 365, 384.

Blackman, M. (1938), The diamagnetic susceptibility of bismuth. *Proc. Roy. Soc.* A, **166**, 1.

Blackman, M. (1941). The theory of the specific heat of solids. *Rep. Progr. Phys.* 8.

Born, M. (1923). *Atomtheorie des festen Zustandes* (Leipzig).

Born, M. and von Kármán, T. (1912). Vibrations in space lattices. *Phys. Z.* **13**, 297.

Carslaw, H. S. and Jaeger, J. C. (1947). *Conduction of heat in solids* (Oxford).

Debye, P. (1912). The theory of specific heats. *Ann. Phys., Lpz.* (4), **39**, 789.

Dingle, R. B. (1952). Some magnetic properties of metals. II. The influence of collisions on the magnetic behaviour of large systems. *Proc. Roy. Soc.* A, **211**, 517.

de Haas, W. J. and van Alphen, P. M. (1930). The dependence of the susceptibility of diamagnetic metals upon the field. *Proc. Acad. Sci. Amst.* **33**, 1106.

Hoare, F. E. and Matthews, J. C. (1952). The magnetic susceptibility of platinum, rhodium and palladium from 20 to 290° K. *Proc. Roy. Soc.* A, **212**, 137.

Keesom, W. H. and Kok, J. A. (1934). The specific heat of zinc and silver at liquid helium temperatures. *Physica*, **1**, 770.

Kellerman, E. W. (1941). The specific heat of the sodium chloride lattice. *Proc. Roy. Soc.* A, **178**, 17.

Landau, L. (1930). Diamagnetism of metals. *Z. Phys.* **64**, 629.

Leighton, R. B. (1948). The vibrational spectrum and specific heat of a face-centred cubic crystal. *Rev. Mod. Phys.* **20**, 165.

Marcus, J. A. (1949). Temperature dependence of the susceptibility of zinc, cadmium and gamma-brass. *Phys. Rev.* **76**, 621.

McDougall, J. and Stoner, E. C. (1938). The computation of Fermi-Dirac functions. *Philos. Trans.* A, **237**, 67.

Peierls, R. (1933a). The diamagnetism of conduction electrons. *Z. Phys.* **80**, 763.

Peierls, R. (1933b). The diamagnetism of conduction electrons; strong fields. *Z. Phys.* **81**, 186.

Sampson, J. B. and Seitz, F. (1940). Theoretical magnetic susceptibilities of metallic lithium and sodium. *Phys. Rev.* **58**, 633.

Shoenberg, D. (1952). The de Haas-van Alphen effect. *Philos. Trans.* A, **245**, 1.

Silvidi, A. A. and Daunt, J. G. (1950). Electronic specific heats in tungsten and zinc. *Phys. Rev.* **77**, 125.

Smith, H. J. M. (1948). The theory of the vibrations and the Raman spectrum of the diamond lattice. *Philos. Trans.* A, **241**, 105.

Sondheimer, E. H. and Wilson, A. H. (1951). The diamagnetism of free electrons. *Proc. Roy. Soc.* A, **210**, 173.

Stoner, E. C. (1936). Collective electron specific heat and spin paramagnetism in metals. *Proc. Roy. Soc.* A, **154**, 656.

Stoner, E. C. (1938). Collective electron energy and specific heat. *Phil. Mag.* **25**, 899.

Titchmarsh, E. C. (1937). *The theory of Fourier integrals* (Oxford).

Vogt, E. (1932). Magnetismus der metallischen Elemente. *Ergebn. exakt. Naturw.* **11**.

Wilson, A. H. (1953). The diamagnetism of quasi-bound conduction electrons. *Proc. Cambridge Phil. Soc.* **49**, 292.

Wohlfarth, E. P. (1949). Electronic properties and band structure of palladium and platinum. *Proc. Leeds Phil. Soc.* **5**, 89.

Chapter VII

FERROMAGNETISM

INTRODUCTION

7·1. There are two separate aspects of ferromagnetism which a complete theory must explain, the occurrence of spontaneous magnetization and the phenomena associated with the magnetization curve. When an apparently unmagnetized piece of iron is placed in a magnetic field of a few gauss a magnetic moment is induced which is proportional to the field. If the magnetic field is increased the magnetic moment increases rapidly and reaches an approximately constant value for fields of the order of a few hundred gauss, after which there is a slow rise in the magnetization with increasing field. These magnetization phenomena are exceedingly complex and depend upon the previous history and the state of purity of the specimen, and also, if the specimen is a single crystal, upon the relation of the inducing field to the crystal axes. They are, however, explicable on the assumption that a ferromagnet consists of a number of domains of microscopic dimensions each exhibiting the same degree of spontaneous magnetism, which depends only on the temperature. In an apparently unmagnetized ferromagnet the moments of the various domains cancel one another on a macroscopic scale, and the phenomena which occur on applying a magnetic field are due to variations in the sizes of the domains and in the orientations of their directions of magnetization. The most convincing argument for the existence of the microscopic spontaneous magnetism is that a ferromagnetic body possesses an excess specific heat compared with normal metals; that is, a ferromagnet has an excess internal energy which, to a first approximation at least, does not depend upon the apparent magnetization but which is a function of the temperature only.

The form of the magnetization curve is much simplified if the specimen is a single crystal of ellipsoidal form, so that if it is placed in a uniform magnetic field the demagnetizing coefficient can be calculated and the magnetizing field in the specimen is uniform and known. (It is assumed in what follows either that the specimen is in the form of a long rod parallel to the external field or that the necessary correction is introduced to take care of the demagnetizing effects.) For iron the cubic axes are directions of easy magnetization, and if an external field is applied along a cubic axis the magnetization increases very rapidly as the field is increased, and reaches a constant value (the saturation value appropriate to the temperature of the specimen) at a few tens of gauss (see fig. VII 1). For other directions the magnetization curve is more complex, though the saturation value in sufficiently high fields is the same as for the directions of easy magnetization.

The theory of the magnetization curve involves ideas of a different nature to those appertaining strictly to the theory of metals, and the subject will therefore not be dealt with here.† The further discussion is restricted to the theory of the spontaneous magnetization, which is defined to be the saturation value of the magnetization when the external field is applied along a direction of easy magnetization. If we assume that the

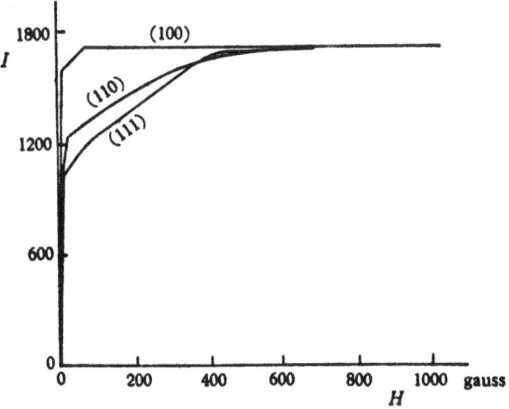

Fig. VII 1. The magnetization curves for a single crystal of iron.

magnetic moment of a ferromagnet is entirely due to the spins of the electrons, the spontaneous magnetization at $T = 0$ must be equal to the number of unpaired electrons per unit volume multiplied by μ_0, and the number of unpaired electrons per atom is known as the magneton number. At higher temperatures not all of these electrons will have their spins parallel to the field, and the spontaneous magnetization must be a decreasing function of T. The main aims in developing the theory are to explain the occurrence of the spontaneous magnetization and to calculate its variation with T.

THE FORMAL CLASSICAL THEORY

7·2. According to the calculations of § 6·43, the magnetic moment M per unit volume of an electron gas in a magnetic field H due to the electronic spin is given by $M = n\mu_0^2 H/kT$ for small values of H, where n is the number of electrons per unit volume, if the electrons have a Boltzmann distribution.

---

† The reader is referred to the review articles by E. C. Stoner (1948, 1950).

If the calculations are extended to arbitrary values of H (see equation (6·43·6)), the relationship is

$$M = n\mu_0 \tanh(\mu_0 H/kT). \qquad (7·2·1)$$

It was first suggested by Weiss in 1907 that the occurrence of ferromagnetism is due to the existence of an internal magnetic field associated with the magnetization, and that M should be determined by equation (7·2·1) and the relation

$$H = H_{\text{ext.}} + \gamma\rho M, \qquad (7·2·2)$$

where $H_{\text{ext.}}$ is the applied magnetic field, ρ is the density and γ is a constant. The spontaneous magnetization is determined by putting $H_{\text{ext.}} = 0$, and it is readily seen by considering the intersections of the curves that (7·2·1) and (7·2·2) have a non-zero solution for M provided that $T < \theta$, where $\theta = n\mu_0^2\gamma\rho/k$ is known as the Curie temperature, the temperature at which ferromagnetism disappears. At $T = 0$ the saturation magnetization is $M = n\mu_0$, so that the magneton number is n/n_a, where n_a is the number of atoms per unit volume; and in general M is given by the implicit equation

$$\frac{M}{n\mu_0} = \tanh\left(\frac{M}{n\mu_0}\frac{\theta}{T}\right). \qquad (7·2·3)$$

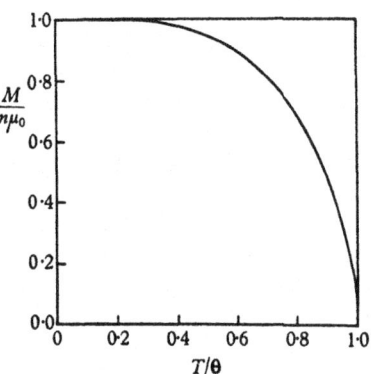

Fig. VII 2. The relative magnetization as a function of T/θ, where θ is the Curie temperature, according to Weiss' theory.

Now as $x \to \infty$, $\tanh x \sim 1 - 2e^{-2x}$, and since $M \to n\mu_0$ we have

$$M/n\mu_0 = 1 - 2e^{-2\theta/T} \quad (T \to 0). \qquad (7·2·4)$$

Also, for small x, $\tanh x = x - \tfrac{1}{3}x^3 + \ldots$, so that, for $T \to \theta - 0$,

$$M/n\mu_0 = \sqrt{3}(1 - T/\theta)^{\frac{1}{2}} \quad (T \to \theta - 0). \qquad (7·2·5)$$

The M-T curve is therefore as shown in fig. VII 2. To obtain reasonable values of θ it is necessary to assume internal fields of the order of 10^7 gauss.

7·21. The energy per unit volume associated with a magnetic moment M is $-\int H \, dM$, and so, due to the Weiss internal field, a ferromagnet has an extra internal energy of $-\tfrac{1}{2}\gamma\rho M^2$ per unit volume and a 'magnetic' specific heat

$$C_M = -\gamma\rho M \, dM/dT. \qquad (7·21·1)$$

Now, from (7·2·3),

$$\frac{d(T/\theta)}{d(M/n\mu_0)} = \frac{(T/\theta)\{1-(M/n\mu_0)^2-T/\theta\}}{(M/n\mu_0)\{1-(M/n\mu_0)^2\}},$$

so that
$$C_M = \frac{\sigma^2(1-\sigma^2)}{(T/\theta)\{(T/\theta)-(1-\sigma^2)\}} nk, \qquad (7·21·2)$$

where $\sigma = M/n\mu_0$ is the relative magnetization. The magnetic specific heat therefore increases steadily from zero at $T = 0$ to a maximum of $\frac{3}{2}nk$ at $T = \theta$, being given by

$$C_M = \frac{3}{2}(3-2\theta/T)\,nk \quad (T \to \theta-0) \qquad (7·21·3)$$

just below the Curie point. Since the magnetic energy is zero above the Curie point in the absence of an external field, there is a discontinuity of $\frac{3}{2}nk$ in the specific heat at the Curie point.

7·22. When $T > \theta$ the only solution of (7·2·3) is $M = 0$, and the metal is merely paramagnetic with M proportional to H_{ext}. Equations (7·2·1) and (7·2·2) then give

$$\frac{M}{n\mu_0} = \frac{\mu_0 H}{kT} + \ldots = \frac{\mu_0 H_{ext}}{kT} + \frac{\theta}{T}\frac{M}{n\mu_0} + \ldots,$$

so that
$$\chi = \frac{M}{H_{ext}} = \frac{n\mu_0^2}{k(T-\theta)}, \qquad (7·22·1)$$

which replaces equation (6·42·3). The graph of $n\mu_0^2/k\chi$ against T is therefore a straight line intersecting the T-axis at $T = \theta$.

7·23. The predictions of the Weiss theory outlined above are in fair agreement with the experimental facts, though there are considerable discrepancies as regards the finer details. For example, the Curie temperature obtained by extrapolating the $1/\chi - T$ curve linearly to $1/\chi = 0$ should give the same value for θ as is obtained from the disappearance of the spontaneous magnetization. In actual practice the paramagnetic and the ferromagnetic Curie points differ considerably ($\theta_p = 680°$ K. and $\theta_f = 631°$ K. for nickel) and $1/\chi$ is not a linear function of T. Also, the specific heat anomaly does not disappear abruptly at the ferromagnetic Curie point. In spite of these shortcomings the Weiss theory gives a very good qualitative picture of the phenomenon of ferromagnetism, but the origin of the large internal field remained obscure until 1928, when Heisenberg showed that in certain circumstances the exchange forces which are responsible for the valency forces between atoms can introduce a coupling between the electron spins tending to aline them.

Heisenberg's theory is based upon the Heitler-London method of approximating to the many-body problem, a method which is more suited to the discussion of molecular structure than of the properties of metals. In spite of its importance, we shall therefore omit any discussion of Heisenberg's theory and give an account of an alternative method of approach first suggested by Bloch (1929) and worked out in detail by Stoner (1938, 1939), in which it is assumed that the wave functions of the electrons concerned are unlocalized and that the energy levels form one of the conduction bands.

Collective Electron Ferromagnetism

7·3. In § 6·44 we discussed the effect of the exchange and correlation forces on the spin paramagnetism of an electron gas, and it was shown that in certain circumstances the exchange forces could give rise to a spontaneous magnetization. We now investigate this problem in more detail.

Ferromagnetism only occurs when there are holes in incomplete inner shells, presumably because it is not otherwise possible for the density of states to be sufficiently large for the exchange forces to predominate over the demagnetizing tendency of the kinetic energy of the electrons. We therefore have to deal with holes in inverted bands, but for simplicity in notation we shall consider electrons in normal bands instead. It was shown in § 6·41 in considering the spin paramagnetism that the equations for electrons and holes are identical in form, and a trivial generalization shows that the same argument applies without change to ferromagnetism.

In the problem discussed in § 6·44, namely, the influence of the exchange forces on the spin paramagnetism, the exact expressions for the exchange forces were not very important, since only one particular value of the combination of the functions occurring was required and this could be considered to be one single adjustable parameter. In the present problem, however, the functional behaviour of the exchange forces for varying degrees of occupation of the energy levels essentially determines all the finer details of the phenomena, and our knowledge of the exchange forces is much too meagre for us to base an elaborate theory upon the expressions given in (6·44·2). Stoner therefore considered an idealized model in which the magnetic energy is proportional to the square of the magnetization, not only for small intensities as in § 6·44, equation (6·44·3), but over the whole range. This choice of model enormously simplifies the calculations, which can readily be carried through to completion, but it must be borne in mind that some of the results must be specific to the model and would not hold even qualitatively for other models. There is, for example, nothing in this model corresponding to the logarithmic singularity in the Fock equation for free electrons which gives rise to the peculiar effects discussed in § 3·442. If the Fock equations are used to discuss the problem of ferromagnetism they predict a magnetized state below a critical temperature if certain conditions are satisfied. The magnetization is, however, discontinuous at the critical temperature, which would make the transition between the magnetized and unmagnetized states to be first order instead of second order (cf. Lidiard, 1951).

7·31. The calculation of the magnetic moment M is exactly the same as in § 6·43 except that the energy of an electron must now include the energy due to the spontaneous magnetism as well as the kinetic energy E and the energy $\pm \mu_0 H$ due to the magnetic field. If $\sigma = M/n\mu_0$ is the relative magnetization, Stoner's assumption is that the energy of an electron due to the spontaneous magnetization is $\pm k\theta'\sigma$, the sign depending on whether the spin of the electron is antiparallel or parallel to M. For a band of standard form we therefore have

$$M = \tfrac{1}{4}n\mu_0(kT/\zeta_0)^{\frac{3}{2}}\{F_{\frac{1}{2}}(\xi+\alpha+\alpha') - F_{\frac{1}{2}}(\xi-\alpha-\alpha')\}, \qquad (7\cdot31\cdot1)$$

$$n = \tfrac{1}{4}n(kT/\zeta_0)^{\frac{3}{2}}\{F_{\frac{1}{2}}(\xi+\alpha+\alpha') + F_{\frac{1}{2}}(\xi-\alpha-\alpha')\}, \qquad (7\cdot31\cdot2)$$

where
$$\alpha = \mu_0 H/kT, \quad \alpha' = \sigma\theta'/T, \quad \xi = \zeta/kT. \qquad (7\cdot31\cdot3)$$

These equations are the generalizations of (6·43·1) and (6·43·2). In calculating the internal energy it must be borne in mind that the energy $k\theta'\sigma$ is due to the mutual interaction of the electrons, and we must therefore halve it when summing over all the electrons, to avoid counting it twice. Therefore

$$
\begin{aligned}
U = \int\{&(E - \mu_0 H - \tfrac{1}{2}k\theta'\sigma)\, f_0(E - \mu_0 H - k\theta'\sigma) \\
&+ (E + \mu_0 H + \tfrac{1}{2}k\theta'\sigma)\, f_0(E + \mu_0 H + k\theta'\sigma)\}\, \mathfrak{n}(E)\, dE \\
= \int E\{& f_0(E - \mu_0 H - k\theta'\sigma) + f_0(E + \mu_0 H + k\theta'\sigma)\}\, \mathfrak{n}(E)\, dE \\
& - n\mu_0\sigma H - \tfrac{1}{2}nk\theta'\sigma^2, \qquad (7\cdot31\cdot4)
\end{aligned}
$$

which, for a band of standard form, becomes (cf. § 6·23)

$$U = \tfrac{3}{4}nkT(kT/\zeta_0)^{\frac{3}{2}}\{F_{\frac{3}{2}}(\xi+\alpha+\alpha') + F_{\frac{3}{2}}(\xi-\alpha-\alpha')\} - nkT\alpha\sigma - \tfrac{1}{2}nk\theta'\sigma^2. \qquad (7\cdot31\cdot5)$$

It is impossible to obtain explicit expressions for all the important quantities over the whole range of the parameters. We shall therefore only discuss the case when the parameters are such that the temperatures involved are small compared with the degeneracy temperature T_0 of the electron gas. Reference can be made to Stoner's papers for details of the other cases, and to a paper by Hunt (1953) for an extension of the theory to include a term in σ^4 and a further arbitrary parameter in U.

The Spontaneous Magnetization

7·4. The equations (7·31·1) and (7·31·2) can be written as

$$F_{\frac{1}{2}}(\xi+\alpha+\alpha') = \tfrac{2}{3}(1+\sigma)(\zeta_0/kT)^{\frac{3}{2}}, \quad F_{\frac{1}{2}}(\xi-\alpha-\alpha') = \tfrac{2}{3}(1-\sigma)(\zeta_0/kT)^{\frac{3}{2}},$$

$$(7\cdot4\cdot1)$$

and elimination of ξ gives the required relation between the physically observable quantities. If x is large we have, to the second order,

$$F_{\frac{1}{2}}(x) = \tfrac{2}{3}x^{\frac{3}{2}}\{1+\pi^2/(8x^2)+\ldots\},$$

and, if $y = \{\tfrac{3}{2}F_{\frac{1}{2}}(x)\}^{\frac{2}{3}}$, the series inverse to this is

$$x = y - \pi^2/(12y) - \ldots. \tag{7·4·2}$$

The equations (7·4·1) therefore become

$$\xi+\alpha+\alpha' = y_1 - \pi^2/(12y_1) - \ldots, \quad \xi-\alpha-\alpha' = y_2 - \pi^2/(12y_2) - \ldots, \tag{7·4·3}$$

where

$$y_1 = (1+\sigma)^{\frac{2}{3}}\zeta_0/kT, \quad y_2 = (1-\sigma)^{\frac{2}{3}}\zeta_0/kT, \tag{7·4·4}$$

and on eliminating ξ we obtain

$$2(\alpha+\alpha') = y_1 - y_2 - \pi^2(y_1^{-1} - y_2^{-1})/12 - \ldots. \tag{7·4·5}$$

7·41. When $T = 0$ and $H = 0$, equation (7·4·5) becomes

$$2\sigma_0 k\theta'/\zeta_0 = (1+\sigma_0)^{\frac{2}{3}} - (1-\sigma_0)^{\frac{2}{3}}, \tag{7·41·1}$$

where σ_0 is the relative magnetization at $T = 0$. Now the function on the right is a steadily increasing function of σ_0, and if its gradient at $\sigma_0 = 0$ is greater than $2k\theta'/\zeta_0$ the only solution of (7·41·1) is $\sigma_0 = 0$. A necessary condition for the occurrence of spontaneous magnetism is therefore

$$k\theta'/\zeta_0 > \tfrac{2}{3}. \tag{7·41·2}$$

Also, if $2k\theta'/\zeta_0 < 2^{\frac{2}{3}}$, the root of (7·41·1) is less than 1. (If the root of (7·41·1) is greater than 1 the value of σ_0 must be taken to be 1 since greater values have no physical significance.) We therefore see that, if

$$\tfrac{2}{3} < k\theta'/\zeta_0 < 2^{-\frac{1}{3}} = 0\cdot794, \tag{7·41·3}$$

the exchange forces while strong enough to give rise to a spontaneous magnetization are insufficient to produce complete alinement of the spins at $T = 0$, and the magneton number will not be equal to the number of free electrons per atom. The graph of σ_0 as a function of $k\theta'/\zeta_0$ is given in fig. VII 3.

7·42. *The Curie point.* Near the Curie point σ is small and, if $H = 0$, (7·4·5) becomes

$$\frac{k\theta'}{\zeta_0} = \frac{2}{3} + \frac{\pi^2}{18}\left(\frac{kT}{\zeta_0}\right)^2 + \tfrac{4}{81}\sigma^2 + \ldots. \tag{7·42·1}$$

The Curie point θ is the temperature at which $\sigma = 0$, and it is therefore given by

$$\frac{k\theta'}{\zeta_0} = \frac{2}{3} + \frac{\pi^2}{18}\left(\frac{k\theta}{\zeta_0}\right)^2 + \dots; \qquad \frac{k\theta}{\zeta_0} = \frac{3}{\pi}\left(\frac{2k\theta'}{\zeta_0} - \frac{4}{3}\right)^{\frac{1}{2}} + \dots, \qquad (7\cdot42\cdot2)$$

but unless $k\theta$ is small compared with ζ_0, the usefulness of these formulae is limited. Further,

$$\sigma^2 = \frac{1}{2}\left(\frac{3\pi k\theta}{2\zeta_0}\right)^2\left(1 - \frac{T^2}{\theta^2}\right) \qquad (7\cdot42\cdot3)$$

to the same order of approximation.

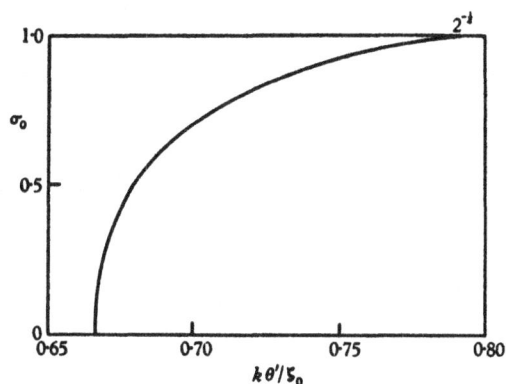

Fig. VII 3. The saturation relative magnetization $\sigma_0 = M_0/n\mu_0$
as a function of the interaction parameter $k\theta'/\zeta_0$.

7·43. *The low-temperature magnetization.* If $\sigma_0 < 1$, the low-temperature magnetization can be obtained from (7·4·4) and (7·4·5), which give

$$(1+\sigma)^{\frac{5}{3}} - (1-\sigma)^{\frac{5}{3}} - \frac{\pi^2}{12}\left(\frac{kT}{\zeta_0}\right)^2\left(\frac{1}{(1+\sigma)^{\frac{1}{3}}} - \frac{1}{(1-\sigma)^{\frac{1}{3}}}\right) - \dots$$
$$= 2\sigma k\theta'/\zeta_0 = (\sigma/\sigma_0)\{(1+\sigma_0)^{\frac{2}{3}} - (1-\sigma_0)^{\frac{2}{3}}\},$$

by (7·41·1). If we now treat $1 - \sigma/\sigma_0$ as small, we find after some manipulation

$$(\sigma/\sigma_0)^2 = 1 - \lambda(T/\theta)^2, \qquad (7\cdot43\cdot1)$$

where (see Stoner, 1938, p. 391) λ is a complicated function of σ_0 but has a value very close to unity for the important ranges of the parameters.

It will be seen from (7·4·3) and (7·4·4) that, if the approximations used are valid, the parameter ξ is given by

$$2\xi = \xi_0\{(1+\sigma_0)^{\frac{2}{3}} + (1-\sigma_0)^{\frac{2}{3}}\}$$

near $T = 0$, and hence $\xi = \xi_0/2^{\frac{1}{3}}$ if $\sigma_0 = 1$. In this case

$$\xi - \alpha' = \frac{\zeta_0}{kT}\left(\frac{1}{2^{\frac{1}{3}}} - \frac{k\theta'}{\zeta_0}\right), \qquad (7\cdot43\cdot2)$$

and $\xi - \alpha'$ is large and negative since $k\theta'/\zeta_0 > 2^{-\frac{1}{3}}$ is the condition which must be satisfied if $\sigma_0 = 1$. The approximation used so far for $F_{\frac{1}{2}}(\xi - \alpha')$ therefore breaks down and must be replaced by the appropriate expression for $\xi - \alpha' < 0$. (It is obvious that when σ_0 is nearly unity there are very few electrons with reversed spins and that the distribution function for these electrons must be the Maxwell function rather than the Fermi function.) When $\xi - \alpha'$ is large and negative we have $F_{\frac{1}{2}}(\xi - \alpha') = \frac{1}{2}\sqrt{\pi}\, e^{\xi - \alpha'}$, whereas $\xi + \alpha'$ is still given by (7·4·3), so that $\xi + \alpha' = (1 + \sigma)^{\frac{2}{3}}\zeta_0/kT$. Also (7·4·1) gives

$$\sigma = 1 - \tfrac{3}{2}(kT/\zeta_0)^{\frac{3}{2}} F_{\frac{1}{2}}(\xi - \alpha') = 1 - \tfrac{3}{4}\sqrt{\pi}\,(kT/\zeta_0)^{\frac{3}{2}} e^{\xi - \alpha'},$$

which, on substituting $\xi - \alpha' = 2^{\frac{2}{3}}(\zeta_0/kT) - 2\alpha'$, gives

$$\sigma = 1 - \tfrac{3}{4}\sqrt{\pi}\,(kT/\zeta_0)^{\frac{3}{2}} e^{-2\gamma/T} \quad (T \to 0), \tag{7·43·3}$$

where
$$\gamma = \theta' - 2^{-\frac{1}{3}}\zeta_0/k > 0. \tag{7·43·4}$$

Therefore, when $\sigma_0 = 1$ the approach to saturation is exponential as $T \to 0$.

THE SPECIFIC HEAT

7·5. The calculation of the specific heat starts from (7·31·5) and follows closely that given in § 6·23. Instead of equation (6·23·5) we have

$$U = \tfrac{3}{10}nkT\left(\frac{kT}{\zeta_0}\right)^{\frac{3}{2}}\left[(\xi + \alpha')^{\frac{5}{2}}\left(1 + \frac{15c_2}{2(\xi + \alpha')^2} - \dots\right)\right.$$
$$\left. + (\xi - \alpha')^{\frac{5}{2}}\left(1 + \frac{15c_2}{2(\xi - \alpha')^2} - \dots\right)\right] - \tfrac{1}{2}nk\theta'\sigma^2, \tag{7·5·1}$$

and instead of (6·23·6) we have

$$\left(\frac{\zeta_0}{kT}\right)^{\frac{3}{2}}(1 \pm \sigma) = (\xi \pm \alpha')^{\frac{3}{2}}\left(1 + \frac{3c_2}{2(\xi \pm \alpha')^2} + \dots\right). \tag{7·5·2}$$

Evaluating U to the second order in kT/ζ_0 and σ, thereby restricting the calculation to the neighbourhood of the Curie point, we find

$$U = \tfrac{3}{5}n\zeta_0\left\{1 + \tfrac{5}{3}\sigma^2 + \frac{5\pi^2}{12}\left(\frac{kT}{\zeta_0}\right)^2 + \dots\right\} - \tfrac{1}{2}nk\theta'\sigma^2, \tag{7·5·3}$$

and
$$C_v = \tfrac{1}{2}\pi^2nk(kT/\zeta_0) + n(\tfrac{2}{3}\zeta_0 - k\theta')\,\sigma\, d\sigma/dT,$$

which, on substituting for σ from (7·42·3), becomes

$$C_v = \tfrac{1}{2}\pi^2nk\frac{kT}{\zeta_0} + \tfrac{9}{4}\pi^2nk\frac{kT}{\zeta_0}\left(\frac{k\theta'}{\zeta_0} - \frac{2}{3}\right) + \dots. \tag{7·5·4}$$

The first term is the ordinary electronic specific heat, and the second is the magnetic specific heat which only occurs below the Curie point. There is therefore a discontinuity in the specific heat at the Curie point, of magnitude

$$\Delta C_v = \tfrac{9}{4}\pi^2 nk \frac{k\theta}{\zeta_0}\left(\frac{k\theta'}{\zeta_0} - \frac{2}{3}\right); \qquad (7\cdot5\cdot5)$$

i.e., on eliminating $k\theta'/\zeta_0$ by (7·42·2),

$$\Delta C_v = \tfrac{1}{8}\pi^4 nk(k\theta/\zeta_0)^3. \qquad (7\cdot5\cdot6)$$

At very low temperatures it can be shown by a straightforward but some-what tedious calculation that the change in magnetization does not contribute to the specific heat to the first order. Hence C_v is the sum of two separate contributions, each referring to the electrons with one direction of spin, and it is therefore given by

$$C_v = \tfrac{1}{4}\pi^2 nk\{(1+\sigma_0)^{\frac{1}{3}}$$
$$+ (1-\sigma_0)^{\frac{1}{3}}\} kT/\zeta_0 \quad (T \to 0).$$
$$(7\cdot5\cdot7)$$

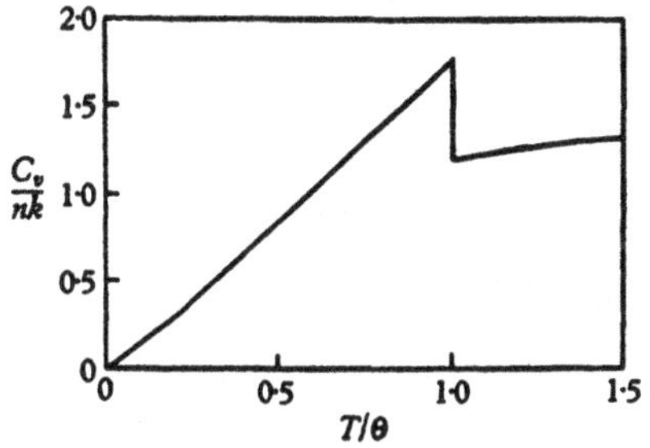

Fig. VII 4. The electronic specific heat of a ferromagnetic metal for $k\theta' = \zeta_0/2^{\frac{1}{3}}$.

The behaviour of C_v as a function of T/θ is shown in fig. VII 4 for the case $k\theta' = \zeta_0/2^{\frac{1}{3}}$.

THE SUSCEPTIBILITY ABOVE THE CURIE POINT

7·6. For temperatures above the Curie point the magnetization is small if H is small, and in (7·31·1) we can treat both α and α' as small, so that

$$M = \tfrac{3}{2}n\mu_0(\alpha + \alpha')\,(kT/\zeta_0)^{\frac{3}{2}}\,F'_{\frac{1}{4}}(\xi),$$

while (7·31·2) becomes $\qquad n = \tfrac{3}{2}n(kT/\zeta_0)^{\frac{3}{2}}\,F_{\frac{1}{4}}(\xi).$

Hence $\qquad\qquad M = \left(\dfrac{M\theta'}{T} + \dfrac{n\mu_0^2 H}{kT}\right)\dfrac{F'_{\frac{1}{4}}(\xi)}{F_{\frac{1}{4}}(\xi)}$

and $\qquad\qquad \dfrac{1}{\chi} = \dfrac{1}{n\mu_0^2}\left(kT\dfrac{F_{\frac{1}{4}}(\xi)}{F'_{\frac{1}{4}}(\xi)} - k\theta'\right). \qquad (7\cdot6\cdot1)$

In the classical limit of high temperatures, $1/\chi$ is a linear function of T, but in all other cases the $1/\chi - T$ curve is concave upwards.

In the classical limit $kT \gg \zeta_0$, where $F_{\frac{1}{4}}(\xi) = \tfrac{1}{2}\sqrt{\pi}\,e^{\xi}$, we have, without further approximation,

$$\frac{M}{n\mu_0} = \sigma = \frac{e^{\alpha+\alpha'} - e^{-(\alpha+\alpha')}}{e^{\alpha+\alpha'} + e^{-(\alpha+\alpha')}} = \tanh\frac{\mu_0 H + k\theta'\sigma}{kT}, \qquad (7\cdot6\cdot2)$$

which is the basis of Weiss's theory discussed in § 7·2.

Curves showing the variation of σ and $1/\chi$ with T are given in fig. VII 5 for various values of $k\theta'/\zeta_0$.

THE FERROMAGNETISM OF NICKEL

7·7. When copper is added to nickel the resulting alloy is less ferromagnetic, and the ferromagnetism finally disappears when about 60 °/₀ of copper has been added. This, in conjunction with the fact that the magneton number of nickel is 0·6, suggests the following electronic structure for metallic nickel. It was shown in § 3·612 that the 3d and 4s bands overlap in metallic nickel and some of the ten 3d electrons occupy 4s levels. It seems probable that all the 3d levels corresponding to one direction of the spin of the electrons are occupied at $T = 0$, thus accounting for five of the 3d electrons, while 4·4 electrons per atom occupy the lowest of the 3d levels corresponding to the other direction of the spin, and the remaining 0·6 occupy 4s levels and have paired spins. The exchange forces act on both the

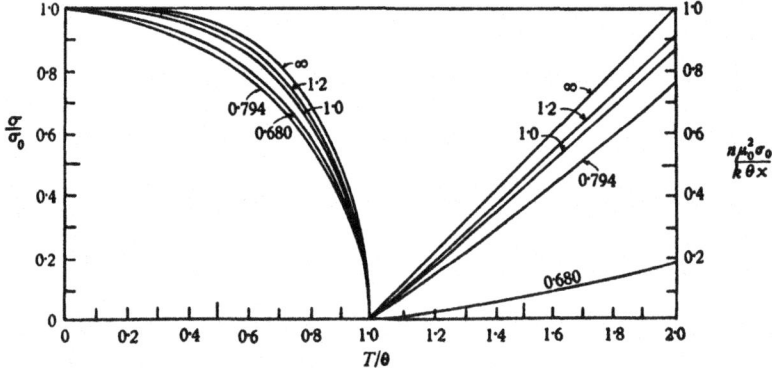

Fig. VII 5. σ/σ_0 below the Curie point and the reciprocal of the susceptibility above the Curie point as functions of T/θ. The numbers on the curves give the values of the interaction parameter $k\theta'/\zeta_0$.

4s and the 3d electrons, but since the 3d band is narrow and the 4s band is wide $(\zeta_0)_{3d}$ is small enough for the exchange forces to aline the spins in the 3d band, whereas they are too weak to aline the spins of the 4s electrons.

If we ignore the 4s electrons the preceding theory can be applied to the 3d electrons, the vacant levels in the inverted 3d band being formally equivalent to electrons in a normal band. The formulae given here are insufficient to cover the whole temperature range involved and the analysis of the data requires the use of the exact expressions, which have been tabulated numerically by Stoner. The most complete analysis has been carried out by Wohlfarth (1949), who introduced a number of refinements into the calculation. In particular, he considered both the d and the s bands and calculated the effect of the transference of electrons from one band to the other due to a change in temperature. The results for the parameters of the d band are as follows.

The value $17\cdot4 \times 10^{-4}T$ cal./degree/g. atom $(70 \times 10^{-4}T$ joule/degree/g. atom) for the low-temperature specific heat, with $n = 5\cdot5 \times 10^{22}$ per c.c. and $\sigma_0 = 1$, gives $T_0 = 2400°$ K. (by equation (7·5·7)). However, the variation of the susceptibility above the Curie point indicates a value of $1600°$ K. for T_0, and Wohlfarth therefore chose $T_0 = 2000°$ K. as a compromise. The observed value of the Curie point $(\theta = 631°$ K.) then gives $\theta' = 1200°$ K.

7·71. The theoretical results obtained by using the values of the parameters given above are in reasonable agreement with the observations, though the fit is by no means perfect. In the first place there is qualitative agreement as regards the variation of σ with T (it could hardly be otherwise).

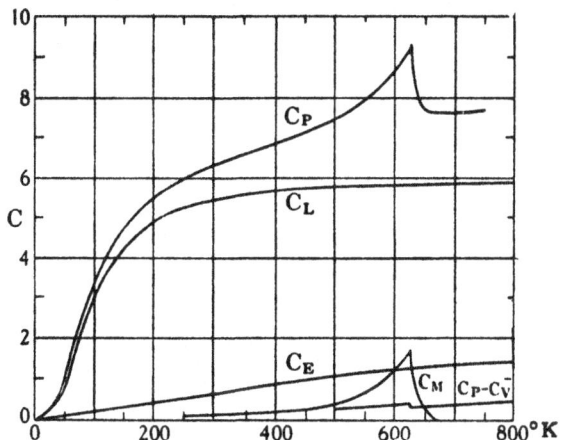

Fig. VII 6. The specific heat of nickel in cal./degree/gram atom. C_P is the observed specific heat; C_L is the lattice specific heat; C_E is the electronic and C_M is the magnetic specific heat. The last three are at constant volume.

but in general the experimentally determined magnetization decreases more rapidly with T than the theoretical σ both near $T = 0$ and more particularly near the Curie point.

The excess of the specific heat of nickel over the Debye value, as determined by Sykes and Wilkinson (1938), is shown in fig. VII 6. The value of the excess specific heat at the Curie point is approximately 3 cal./degree/g. atom (12·6 joules/degree/g. atom), and as the Curie point is passed the specific heat drops fairly rapidly to about 1·2 cal./degree/g. atom (5·0 joules/degree/g. atom) though the tail extends over about 30°. This is in good agreement with the calculated electronic specific heat, starting from the observed low-temperature value. At higher temperatures, however, the results indicate that the observed specific heat lies above the calculated curve, and Wohlfarth has attributed this to the excitation of electrons from the 3d to the 4s band. The results of his calculations are shown in fig. VII 7, the curve C_d referring to a single d band and the curve C_{d+s} to two overlapping d and s bands. The experimental results of Persoz (1940) agree well with the C_{d+s} curve if we take $T_0 = 2300°$ K.

7·72. The susceptibility of nickel above the Curie point is shown in fig. VII 8. Near the Curie point, the $1/\chi - T$ curve is concave upwards as predicted by the theory, and is then practically straight from $T = 800$ to

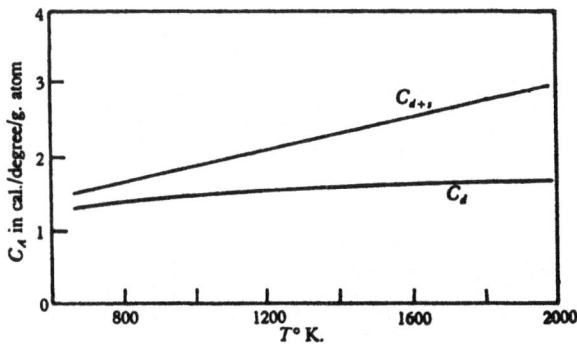

Fig. VII 7. The calculated electronic specific heat in nickel above the Curie point due to a single d-band and to overlapping s- and d-bands.

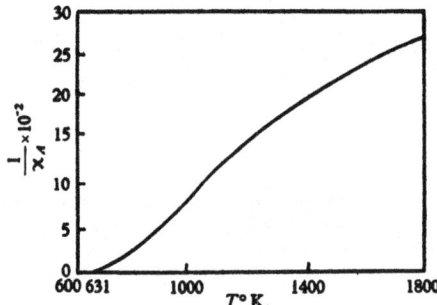

Fig. VII 8. Paramagnetic susceptibility of nickel above the Curie point.

1150° K. Above this temperature it is concave downwards. The results in the lower temperature range can be fitted reasonably by (7·6·1) with the parameters given at the end of § 7·7, but this formula cannot give a curve concave downwards. Wohlfarth attributes the anomalous character of the $1/\chi$ curve in the high-temperature region to the increase in the number of holes in the d band as the temperature rises, an effect which tends to increase the contribution of the d band to the susceptibility. His calculations show that the major part of the curvature, but not the whole, can be attributed to this cause. All these results can be improved by the introduction of another parameter into the theory (Hunt, 1953).

OTHER FERROMAGNETIC METALS

7·8. The magneton numbers of cobalt and iron are 1·7 and 2·2 respectively. If we interpret these in the same way as for nickel, we have to postulate 8·3 d-electrons per atom and 0·7 s-electron per atom for cobalt, and 7·8 d-electrons per atom and 0·2 s-electron for iron. This assignment, is, however, doubtful, since it corresponds to all the d levels with one direction of spin being full at $T = 0$, and as shown in § 7·41 the exchange forces may not be large enough to ensure complete alinement at $T = 0$. It is therefore possible that there are $5 - x$ and $3·3 - x$ d-electrons per atom with positive and negative spins and $0·7 + 2x$ s-electrons per atom

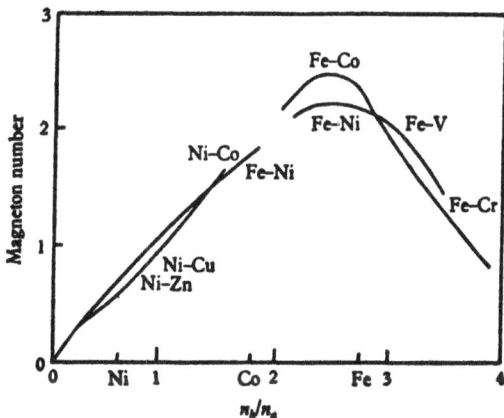

Fig. VII 9. The magneton number of ferromagnetic alloys as
a function of the number of holes in the 3d-band.

for cobalt, the corresponding numbers for iron being $5 - y$ and $2·8 - y$ d-electrons per atom and $0·2 + 2y$ s-electrons per atom. The theoretical calculations that can be made at the moment are much too crude to enable a calculation of x and y to be made, but support for the hypothesis that y at least is not zero is given by the fact that the magneton number of iron can be increased by adding cobalt or nickel which increase the number of electrons present.

The magneton numbers for some of the ferromagnetic alloys as a function of the number of electrons present outside the 3p shell are shown in fig. VII 9 (Slater, 1937), from which it is clear that one of the important factors affecting ferromagnetism is the number of electrons in the 3d band, and that there is an optimum number round about 8·5 electrons per atom. This is borne out by the behaviour of the alloys of nickel with non-ferromagnetic metals. The effect of copper has already been referred to, and zinc, aluminium and other electropositive metals show the same effect (see fig. VII 10), the magneton numbers being determined by the number of electrons, whatever their origin.

This state of affairs is similar to that in the alloys which obey the Hume-Rothery rules (Chapter IV), but the electron concentration is not the only factor of importance, as is shown by the behaviour of manganese. When manganese is added to nickel the magneton number is increased, but it reaches a maximum of about 0·75 at a concentration of about 15 % manganese, and the Mn-Ni alloys do not therefore lie on the curve of fig. VII 9. Further, the alloys of manganese and copper, though having the required electron concentration, are only paramagnetic. On the other hand, the ternary alloys of manganese, copper and aluminium embrace the strongly ferromagnetic Heusler alloys.

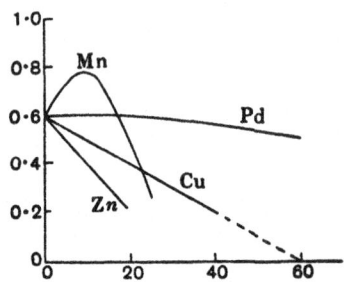

Fig. VII 10. The magneton numbers of nickel alloys as functions of the composition. The abscissae denote the atomic percentages of the second component.

It is therefore clear that although the present theory can give a good account of the properties of nickel, there are many factors which it does not take into account.

REFERENCES

Bloch, F. (1929). Remarks on the electron theory of ferromagnetism and of electrical conductivity. *Z. Phys.* **57**, 545.

Heisenberg, W. (1928). The theory of ferromagnetism. *Z. Phys.* **49**, 619.

Hunt, K. L. (1953). Collective electron ferromagnetism: a generalization of the treatment and an analysis of the experimental results. *Proc. Roy. Soc.* A, **216**, 103.

Lidiard, A. B. (1951). On the theory of free electron ferromagnetism. *Proc. Phys. Soc.* A, **64**, 814.

Persoz, B. (1940). The specific heat of metals at high temperatures. *Ann. Phys. Paris*, **14**, 237.

Slater, J. C. (1937). Electronic structure of alloys. *J. Appl. Phys.* **8**, 385.

Stoner, E. C. (1936). The specific heat of nickel. *Phil. Mag.* **22**, 81.

Stoner, E. C. (1938). Collective electron ferromagnetism. *Proc. Roy. Soc.* A, **165**, 372.

Stoner, E. C. (1939). Collective electron ferromagnetism. II. Energy and specific heat. *Proc. Roy. Soc.* A, **169**, 339.

Stoner, E. C. (1948). Ferromagnetism. *Rep. Progr. Phys.* **11**.

Stoner, E. C. (1950). Ferromagnetism: Magnetization curves. *Rep. Progr. Phys.* **13**.

Sykes, C. and Wilkinson, H. (1938). The specific heat of nickel from 100° to 600°. *Proc. Phys. Soc.* **50**, 834.

Weiss, P. (1907). The molecular field hypothesis and ferromagnetism. *J. Phys. Radium*, **6**, 661.

Wohlfarth, E. P. (1949). Collective electron ferromagnetism. III. Nickel and nickel-copper alloys. *Proc. Roy. Soc.* A, **195**, 434.

Chapter VIII

THE FORMAL THEORY OF CONDUCTION

THE FUNDAMENTAL INTEGRAL EQUATION

8·1. In Chapter I a brief description was given of the more elementary parts of the theories of conduction put forward by Lorentz and Sommerfeld. We now resume this topic, but on account of its complexity the discussion is spread over three chapters. In the present chapter it is assumed that a time of relaxation exists, and the consequences of this assumption are deduced. In the next chapter, the time of relaxation τ is calculated for a particular model, conditions being established under which τ exists for this model, and some of the simpler extensions are given. Chapter X is devoted to more sophisticated solutions of the transport equations.

We have first to set up the Boltzmann equation for the distribution function. We assume that the state of a metal can be described by one-electron wave functions, the exclusion principle only being taken into account through the limitation it imposes on the number of electrons in each state. The wave vector $\mathbf{k}$ is used to define the states of the electrons, and the number of electrons per unit volume whose wave vectors lie in the interval $d\mathbf{k} = (dk_1, dk_2, dk_3)$ is

$$(1/4\pi^3) f(\mathbf{k}, \mathbf{r}) \, d\mathbf{k}. \qquad (8\cdot1\cdot1)$$

In equilibrium $f(\mathbf{k}, \mathbf{r})$ is the Fermi function $f_0(E)$, but when there are external fields present the distribution function has to be determined as the solution of the Boltzmann equation, which is derived as follows.

The rate of change of the distribution function due to a constant electric and magnetic field was shown in § 2·85 to be the same, for short times, as that given by classical mechanics. If, in addition, the metal is not uniform, due either to the composition being variable or to the existence of a temperature gradient, the distribution function is a function of $\mathbf{r}$, and the complete Boltzmann equation is (compare § 1·3)

$$-\frac{2\pi e}{h}\left(\mathscr{E} + \frac{1}{c}\mathbf{v} \times \mathbf{H}\right).\operatorname{grad}_\mathbf{k} f + \mathbf{v}.\operatorname{grad}_\mathbf{r} f = \left[\frac{\partial f}{\partial t}\right]_{\text{coll.}}, \qquad (8\cdot1\cdot2)$$

where $[\partial f/\partial t]_{\text{coll.}}$ is the rate of change of f due to interactions with the metal ions.

An expression for $[\partial f/\partial t]_{\text{coll.}}$ can be found in the following way (Nordheim, 1931). We suppose that $\mathscr{W}(\mathbf{k}, \mathbf{k}')$ is the probability per unit time that an electron makes a transition from the state $\mathbf{k}$ to the state $\mathbf{k}'$. To take into account the effect of the exclusion principle we must multiply $\mathscr{W}(\mathbf{k},\mathbf{k}')$ by the number of electrons in the initial state $\mathbf{k}$ and the number of vacant levels in the final state $\mathbf{k}'$. Hence the net change in f is

$$\left[\frac{\partial f}{\partial t}\right]_{\text{coll.}} = \int [\mathscr{W}(\mathbf{k}',\mathbf{k})f(\mathbf{k}')\{1-f(\mathbf{k})\} - \mathscr{W}(\mathbf{k},\mathbf{k}')f(\mathbf{k})\{1-f(\mathbf{k}')\}]\,d\mathbf{k}'.$$
(8·1·3)

In the equilibrium state this expression must vanish. Substituting, therefore, the Fermi function $f_0(E)$ for f in (8·1·3), we see that the function $W(\mathbf{k},\mathbf{k}';\mathbf{r})$ defined by

$$W(\mathbf{k},\mathbf{k}';\mathbf{r}) = e^{E_{\mathbf{k}'}/kT}\mathscr{W}(\mathbf{k},\mathbf{k}') = e^{E_{\mathbf{k}}/kT}\mathscr{W}(\mathbf{k}',\mathbf{k})$$
(8·1·4)

is a symmetrical function of $\mathbf{k}$ and $\mathbf{k}'$. If we assume that the transition probabilities are the same in the equilibrium state as when a current is present, we have

$$\left[\frac{\partial f}{\partial t}\right]_{\text{coll.}} = \int W(\mathbf{k},\mathbf{k}';\mathbf{r})\,[e^{-E_{\mathbf{k}}/kT}f(\mathbf{k}')\{1-f(\mathbf{k})\} - e^{-E_{\mathbf{k}'}/kT}f(\mathbf{k})\{1-f(\mathbf{k}')\}]\,d\mathbf{k}'.$$
(8·1·5)

We shall examine this assumption in § 9·9; in the meantime we shall assume its validity.

An alternative form for $[\partial f/\partial t]_{\text{coll.}}$ is obtained by writing

$$V(\mathbf{k},\mathbf{k}') = \mathscr{W}(\mathbf{k},\mathbf{k}')f_0(\mathbf{k})\{1-f_0(\mathbf{k}')\} = V(\mathbf{k}',\mathbf{k})$$
(8·1·6)

and
$$f = f_0 - \Phi\,\partial f_0/\partial E = f_0 + \Phi f_0(1-f_0)/kT.$$
(8·1·7)

Then, if terms involving Φ^2 are neglected, (8·1·3) becomes

$$\left[\frac{\partial f}{\partial t}\right]_{\text{coll.}} = -\frac{1}{kT}\int V(\mathbf{k},\mathbf{k}')\{\Phi(\mathbf{k}) - \Phi(\mathbf{k}')\}\,d\mathbf{k}'.$$
(8·1·8)

Since the only physical problems which we consider are those in which electric fields and temperature gradients are present, the neglect of terms involving Φ^2 is equivalent to neglecting squares and products of the electric fields and temperature gradients, and this is always justified.

The formal theory of conduction is concerned with the solution of equations (8·1·2) and (8·1·5) or (8·1·8) when the function $\mathscr{W}(\mathbf{k},\mathbf{k}')$ is supposed to be known.

It can be shown (Kohler, 1941), without solving the Boltzmann equation explicitly, that the electric and heat current densities are given by

$$J_i = \sum_{j=1}^{3} S_{ij}^{(1)}(\mathbf{H})\left(\mathscr{E}_j + \frac{T}{\epsilon}\frac{\partial}{\partial x_j}\frac{\zeta}{T}\right) + \sum_{j=1}^{3} S_{ij}^{(2)}(\mathbf{H})\frac{1}{T}\frac{\partial T}{\partial x_j},$$
(8·1·9)

$$w_i = \sum_{j=1}^{3} S_{ij}^{(3)}(\mathbf{H})\left(\mathscr{E}_j + \frac{T}{\epsilon}\frac{\partial}{\partial x_j}\frac{\zeta}{T}\right) + \sum_{j=1}^{3} S_{ij}^{(4)}(\mathbf{H})\frac{1}{T}\frac{\partial T}{\partial x_j},$$
(8·1·10)

where the symmetry relation $V(\mathbf{k}, \mathbf{k}') = V(\mathbf{k}', \mathbf{k})$ gives rise to the symmetry relations

$$S_{ij}^{(1)}(\mathbf{H}) = S_{ji}^{(1)}(-\mathbf{H}), \quad S_{ij}^{(2)}(\mathbf{H}) = -S_{ji}^{(3)}(-\mathbf{H}), \quad S_{ij}^{(4)}(\mathbf{H}) = S_{ji}^{(4)}(-\mathbf{H}).$$
(8·1·11)

The equations (8·1·9) and (8·1·10) can be inverted to give

$$\mathscr{E}_i = \sum_{j=1}^{3} \left(\rho_{ij}(\mathbf{H}) J_j - \frac{1}{\epsilon} \mathfrak{S}_{ij}(\mathbf{H}) \frac{\partial T}{\partial x_j} \right) - \frac{1}{\epsilon} \frac{\partial \zeta}{\partial x_i}, \qquad (8·1·12)$$

$$w_i = \sum_{j=1}^{3} \left(\Pi_{ij}(\mathbf{H}) J_j - \kappa_{ij} \frac{\partial T}{\partial x_j} \right) - \frac{\zeta}{\epsilon} J_i, \qquad (8·1·13)$$

where

$$\rho_{ij}(\mathbf{H}) = \rho_{ji}(-\mathbf{H}), \quad T\mathfrak{S}_{ij}(\mathbf{H}) = -\epsilon \Pi_{ji}(-\mathbf{H}), \quad \kappa_{ij}(\mathbf{H}) = \kappa_{ji}(-\mathbf{H}).$$
(8·1·14)

The whole of the formal theory of conductivity is contained in equations (8·1·9)–(8·1·11) or (8·1·12)–(8·1·14), but in such a general form that it is difficult to see the significance of the various tensors or to evaluate them for particular models. We shall therefore normally limit the discussion, especially of the more complicated phenomena, to polycrystalline metals, which we assume to behave like isotropic media. (This is only rigorously true for phenomena which can be characterized by second-order tensors. For more complex phenomena such as the magneto-resistance effects, polycrystalline and isotropic media are not equivalent.) Many of the results can be extended without difficulty to the general case of anisotropic media, but this only leads to extra mathematical complications without introducing any new physical ideas, and the reader is referred to the original papers for details of these formal extensions.

8·11. *The time of relaxation.* The assumption that $[\partial f/\partial t]_{\text{coll.}}$ can be taken to be of the form

$$[\partial f/\partial t]_{\text{coll.}} = -(f - f_0)/\tau \qquad (8·11·1)$$

has already been introduced and discussed in § 1·31. We shall make this assumption in the calculations of the present chapter, reserving for Chapter X a discussion of the extra complications which arise when less restrictive assumptions are made. It is, of course, always possible to define a time of relaxation in terms of the electrical conductivity σ from the relation

$$\tau(\zeta) = m\sigma/(n\epsilon^2),$$

or in terms of the thermal conductivity κ from the relation

$$\tau(\zeta) = 3m\kappa/(n\pi^2 k^2 T),$$

which hold for a metal with perfectly free electrons (equations (1·9·6) and (1·9·7) or their generalizations (8·21·1) and (8·3·8)). In general, however, these relations give different values of τ, so that a universal definition of τ

is impossible which applies to all conduction phenomena simultaneously. It is then necessary to solve directly the full integral equation for the distribution function and to deduce σ, κ, etc., from the solution. After this has been done, we may, if we like, define τ by one of the relations given above, but it is then necessary to add to the term 'the time of relaxation' the qualification 'associated with the electrical conductivity' or 'associated with the thermal conductivity', and the concept tends to lose its usefulness.

The Boltzmann equation, obtained by combining (8·1·2) and (8·11·1), is

$$-\frac{2\pi\epsilon}{h}\left(\mathscr{E}+\frac{1}{c}\mathbf{v}\times\mathbf{H}\right).\operatorname{grad}_{\mathbf{k}}f+\mathbf{v}.\operatorname{grad}_{\mathbf{r}}f=-\frac{f-f_0}{\tau}. \qquad (8·11·2)$$

No assumptions need be made about τ and the energy levels except that $\tau(\mathbf{k})$ and $E(\mathbf{k})$ must have the same symmetry as the crystal. It is, however, convenient to apply the general theory to particular models so as to have the end-results in a form in which they can be applied, with all due reservations and approximations, to real metals. The model which we shall mostly use is that in which the energy levels form two overlapping spherically symmetrical bands, though it is often sufficient to consider the case in which only one band is present.

THE ELECTRICAL CONDUCTIVITY

8·2. To find the electrical conductivity in a uniform metal we put $H = 0$, $\operatorname{grad}_{\mathbf{r}}f = 0$ in (8·11·2). The electric field in a metal is always small, and we can neglect terms in $\mathscr{E}^2$. The solution of (8·11·2) is then obtained by putting $f = f_0$ on the left of the equation, so that

$$f = f_0 + (2\pi\epsilon/h)\tau\mathscr{E}.\operatorname{grad}_{\mathbf{k}}f_0 = f_0 + \epsilon\tau\mathbf{v}.\mathscr{E}\,\partial f_0/\partial E. \qquad (8·2·1)$$

Further, the electric current density $\mathbf{J}$ is given by

$$\mathbf{J} = -\frac{\epsilon}{4\pi^3}\int\mathbf{v}f\,d\mathbf{k} = -\frac{\epsilon^2}{4\pi^3}\int\tau\mathbf{v}(\mathbf{v}.\mathscr{E})\frac{\partial f_0}{\partial E}d\mathbf{k}. \qquad (8·2·2)$$

Writing this in tensor notation, we see that

$$J_i = \sigma_{ij}\mathscr{E}_j,$$

where the conductivity tensor σ_{ij} is given by

$$\sigma_{ij} = -\frac{\epsilon^2}{4\pi^3}\int\tau v_i v_j\frac{\partial f_0}{\partial E}d\mathbf{k}. \qquad (8·2·3)$$

One integration can be carried out by considering the $\mathbf{k}$ space to be divided up by the family of level surfaces $E(\mathbf{k}) = \text{constant}$. If $d\mathbf{k}$ refers to an element of $\mathbf{k}$ space contained between the surfaces E and $E + dE$, then

$$d\mathbf{k} = dS\,dE/(dE/dn),$$

where dS is an element of surface and dE/dn is the normal derivative of E. Now $dE/dn = |\operatorname{grad}_k E|$, and so

$$dk = dS\, dE / |\operatorname{grad}_k E|. \qquad (8\cdot2\cdot4)$$

This gives
$$\sigma_{ij} = -\frac{\epsilon^2}{4\pi^3}\int \frac{\partial f_0}{\partial E}\, dE \int \frac{\tau v_i v_j\, dS}{|\operatorname{grad}_k E|}, \qquad (8\cdot2\cdot5)$$

and the integral with respect to E can be evaluated to any desired order by the use of the asymptotic formula (1·7·2). The first approximation is

$$\sigma_{ij} = \frac{\epsilon^2}{4\pi^3}\int_{E-\zeta} \frac{\tau v_i v_j\, dS}{|\operatorname{grad}_k E|}. \qquad (8\cdot2\cdot6)$$

This is as far as we can carry the general theory; to proceed further special models must be considered.

The tensor σ_{ij} has at most two independent components; for cubic metals, and for isotropic, i.e. polycrystalline metals, it is a scalar. To show this, it is necessary to consider how the conductivity is measured. If an electric current $\mathbf{J}$ is set up in a metal crystal, the electric field $\mathscr{E}$ required to maintain it is, in general, not parallel to $\mathbf{J}$. The conductivity σ in the direction of $\mathbf{J}$ is defined to be the ratio of J to the component of the field in the direction of $\mathbf{J}$, i.e.

$$\sigma = J^2/\mathbf{J}\cdot\mathscr{E}. \qquad (8\cdot2\cdot7)$$

Now every metal has an axis of symmetry, and so, if J makes an angle θ with the axis of symmetry, we have $J_{\shortparallel} = J\cos\theta$, $J_{\perp} = J\sin\theta$, and $J_{\shortparallel} = \sigma_{\shortparallel}\mathscr{E}_{\shortparallel}$, $J_{\perp} = \sigma_{\perp}\mathscr{E}_{\perp}$. Hence

$$\mathbf{J}\cdot\mathscr{E} = \frac{J_{\shortparallel}^2}{\sigma_{\shortparallel}} + \frac{J_{\perp}^2}{\sigma_{\perp}} = J^2\left(\frac{\cos^2\theta}{\sigma_{\shortparallel}} + \frac{\sin^2\theta}{\sigma_{\perp}}\right),$$

so that
$$\frac{1}{\sigma} = \frac{\cos^2\theta}{\sigma_{\shortparallel}} + \frac{\sin^2\theta}{\sigma_{\perp}}. \qquad (8\cdot2\cdot8)$$

The conductivity can therefore be characterized by the two conductivities $\sigma_{\shortparallel}$ and $\sigma_{\perp}$, parallel to and perpendicular to the axis of symmetry. If there is more than one axis of symmetry, as in cubic metals, σ is independent of the orientation and is a scalar.

8.21. For cubic and isotropic metals, we have $\sigma = \sigma_{ii}$, and we can replace v_i^2 in (8·2·6) by $\frac{1}{3}(v_1^2 + v_2^2 + v_3^2) = \frac{1}{3}v^2$, so that

$$\sigma = \frac{\epsilon^2}{12\pi^3}\int_{E-\zeta} \frac{\tau v^2\, dS}{|\operatorname{grad}_k E|}. \qquad (8\cdot21\cdot1)$$

Now according to equations (2·71·1) and (8·2·4) the density of states is given by

$$\mathfrak{n}(E) = \frac{1}{8\pi^3}\int \frac{dS}{|\operatorname{grad}_k E|}, \qquad (8\cdot21\cdot2)$$

and if the surface $E = \zeta$ is not a single surface but consists of portions of surfaces belonging to different Brillouin zones, we can rewrite (8·21·1) in the form

$$\sigma = \tfrac{2}{3}\epsilon^2 \sum_s (\mathfrak{n}_s \overline{\tau_s v_s^2})_{E=\zeta}, \tag{8·21·3}$$

where the bar denotes the mean value and s is used to enumerate the different surfaces involved.

For a model metal consisting of a single band of standard form, we have, by (1·81·2) and (2·71·3),

$$\tfrac{1}{2}mv^2 = E, \qquad \zeta \mathfrak{n}(\zeta) = \tfrac{3}{2}n.$$

Hence

$$\sigma = n\epsilon^2 \tau(\zeta)/m. \tag{8·21·4}$$

This formula is a reasonable approximation for monovalent metals, but for multivalent metals it is more logical to consider the electrons to be distributed over two bands. In this case we have

$$\sigma = \epsilon^2 \left(\frac{n_1 \tau_1}{m_1} + \frac{n_2 \tau_2}{m_2} \right)_\zeta, \tag{8·21·5}$$

where the symbols have their customary meanings, namely, that n_1, τ_1 and m_1 are the number, time of relaxation and effective mass of the electrons in the normal band 1, while n_2, τ_2 and m_2 are the corresponding quantities for the positive holes in the inverted band 2.

It should be noted that, so far as the electrical conductivity is concerned, it makes no difference whether one of the bands is inverted or not. For the electrons in an inverted band behave as if they had positive charge, but since σ depends upon the square of the charge, the sign of the charge does not affect σ. It is, however, of great importance in any phenomenon, such as the Hall effect or the thermoelectric effect, depending upon the first power of the electric charge.

8·22. It was shown in § 3·612 that the energy levels of the transition metals can be well described in terms of two overlapping bands of standard form, the lower d band being inverted. It is not, however, immediately obvious that such a model should be applied to other metals, and some discussion of the applicability of the model is required. It is impossible in the present state of the theory to give exact quantitative expressions for any of the properties of metals, and we are therefore limited to a discussion of the factors which dominate a particular phenomenon and to deductions concerning the order of magnitude of the effect. The various conduction phenomena depend primarily upon the differential geometry of the energy surfaces, particularly in the neighbourhood of the energy ζ, rather than on the 'large-scale' configuration of the surfaces, so that we are always concerned with the derivatives of E. In any exact theory all the derivatives should be taken into account, but, on account of the complexity of the calculations, it is usual to include only the lowest derivatives which give a non-zero result for any particular phenomenon. For example, the electrical conductivity σ depends essentially upon the first derivatives of E, or rather upon their squares. A simple model therefore suffices to provide the dominant term in the general expression for σ, and the discussion of a more complicated model (and the inclusion of the effect of higher derivatives of E) would only affect the numerical values. Taken by themselves, there is little to choose between the formulae (8·21·4) and (8·21·5) unless one has some independent method of estimating the parameters. So far as σ is concerned (8·21·5) can quite well be replaced by (8·21·4) with n, τ and m as averages of n_1, n_2, τ_1, τ_2 and m_1, m_2.

There are certain conduction phenomena which depend linearly upon the second derivatives of E, and for these the free electron model is entirely inadequate. The simplest model which can then be used is one which allows the second derivatives to be either positive or negative. Such a model is that comprising two overlapping bands of standard form, one being inverted. There are other phenomena which vanish if all the fastest electrons have the same velocity; once again the two-band model is the simplest which gives a non-zero result. It should be emphasized that in general it is only the differential geometry of the energy surfaces that matters. When we introduce an inverted band the important point is that there are portions of the energy surfaces for which the second derivatives of E are negative. Whether these portions form a sphere in k space is immaterial. They could just as well form isolated portions of a sphere or have a complicated macroscopic configuration, but such refinements would only alter the exact numerical values, which cannot be calculated in any case, and would not add anything to the understanding of the factors which determine the phenomena. Provided that derivatives of E up to the second are sufficient to give the leading term for a particular effect, the two-band model can be used to describe any isotropic metal. The actual energy surface can be considered to be dissected into portions with positive and negative derivatives respectively, and those with positive derivatives can be considered collectively to form one band and those with negative derivatives to form the other. This is the fundamental point, the further simplification of treating each separate band as being spherically symmetrical being introduced merely in order to obtain the final results in a manageable form. There are, of course, phenomena for which this model is entirely inadequate; for example, it can give no explanation of anisotropic effects. More complicated models are then necessary, but in every case it is desirable to choose the simplest model having a reasonable physical basis that gives a non-zero result.

THE THERMAL CONDUCTIVITY

8·3. The thermal conductivity is determined by the heat flow produced by a temperature gradient in conditions such that there is no electric current. There must therefore be a slight redistribution of the conduction electrons so as to set up an electric field of the right amount to counteract the drift velocities of the electrons due to the temperature gradient and to reduce the electric current to zero. In an isotropic metal the electric field set up is in the same direction as the temperature gradient, which we take to be along the x-axis, and the Boltzmann equation is then

$$-\frac{2\pi e}{h}\mathscr{E}_x\frac{\partial f}{\partial k_1}+v_1\frac{\partial f}{\partial T}\frac{\partial T}{\partial x}=-\frac{f-f_0}{\tau}. \tag{8·3·1}$$

Since the electric field and temperature gradient are always small and their squares and products can be neglected, the solution of (8·3·1) is obtained by putting $f=f_0$ on the left-hand side. Now $\dfrac{\partial f_0}{\partial T}=T\dfrac{\partial}{\partial T}\left(\dfrac{E-\zeta}{T}\right)\dfrac{\partial f_0}{\partial E}$, and so

$$f=f_0+\tau v_1\frac{\partial f_0}{\partial E}\left(e\mathscr{E}_x+T\frac{\partial}{\partial x}\frac{\zeta}{T}+\frac{E}{T}\frac{\partial T}{\partial x}\right), \tag{8·3·2}$$

and the electric and heat current densities are

$$J=-\frac{\epsilon}{4\pi^3}\int \mathbf{v}f\,d\mathbf{k}=\mathscr{K}_1\left(\epsilon^2\mathscr{E}_x+\epsilon T\frac{\partial}{\partial x}\frac{\zeta}{T}\right)+\mathscr{K}_2\frac{\epsilon}{T}\frac{\partial T}{\partial x}, \tag{8·3·3}$$

$$w=\frac{1}{4\pi^3}\int \mathbf{v}Ef\,d\mathbf{k}=\mathscr{K}_2\left(-\epsilon\mathscr{E}_x-T\frac{\partial}{\partial x}\frac{\zeta}{T}\right)-\mathscr{K}_3\frac{1}{T}\frac{\partial T}{\partial x}, \tag{8·3·4}$$

where
$$\mathscr{K}_n=-\frac{1}{4\pi^3}\int \tau v_1^2 E^{n-1}\frac{\partial f_0}{\partial E}\,d\mathbf{k}=-\frac{1}{12\pi^3}\int \tau v^2 E^{n-1}\frac{\partial f_0}{\partial E}\,d\mathbf{k}, \tag{8·3·5}$$

since, on account of the isotropy, v_1^2 can be replaced by $\tfrac{1}{3}v^2$. Just as in § 8·2, one integration can be carried out in the integral for $\mathscr{K}_n$, the result being, to the second order in kT/ζ,

$$\mathscr{K}_n=-\frac{1}{12\pi^3}\int \frac{\partial f_0}{\partial E}\,dE\int \frac{\tau v^2 E^{n-1}\,dS}{|\,\mathrm{grad}_k E\,|}$$

$$=\frac{1}{12\pi^3}\left[\left[\int_{E=\zeta}\frac{\tau v^2 E^{n-1}\,dS}{|\,\mathrm{grad}_k E\,|}+\tfrac{1}{6}\pi^2 k^2 T^2\left\{\frac{\partial^2}{\partial E^2}\int\frac{\tau v^2 E^{n-1}\,dS}{|\,\mathrm{grad}_k E\,|}\right\}_{E=\zeta}\right]\right.$$

$$=\frac{1}{12\pi^3}\int_{E=\zeta}\frac{\tau v^2 E^{n-1}\,dS}{|\,\mathrm{grad}_k E\,|}+\frac{k^2 T^2}{72\pi}\left[\left[\int_{E=\zeta}\frac{(n-1)(n-2)\tau v^2\zeta^{n-3}\,dS}{|\,\mathrm{grad}_k E\,|}\right.\right.$$

$$\left.\left.+\left\{2(n-1)\zeta^{n-2}\frac{\partial}{\partial E}\int\frac{\tau v^2\,dS}{|\,\mathrm{grad}_k E\,|}+\zeta^{n-1}\frac{\partial^2}{\partial E^2}\int\frac{\tau v^2\,dS}{|\,\mathrm{grad}_k E\,|}\right\}_{E=\zeta}\right]\right]. \tag{8·3·6}$$

The electrical conductivity σ, given by $\epsilon^2 \mathscr{K}_1$ has already been discussed in § 8·2. To determine the thermal conductivity κ we must impose the condition $J = 0$, and use the resulting equation to eliminate $\mathscr{E}_x$. This gives

$$\kappa = \frac{w}{-\partial T / \partial x} = \frac{\mathscr{K}_1 \mathscr{K}_3 - \mathscr{K}_2^2}{\mathscr{K}_1 T}, \qquad (8 \cdot 3 \cdot 7)$$

and on substituting the expressions given by (8·3·6) into the numerator it is found that the only surviving terms are those involving the second term in (8·3·6) (i.e. the term containing $(n-1)(n-2)\zeta^{n-3}$).

The final expression for κ is

$$\kappa = \frac{k^2 T}{36\pi} \int_{E-\zeta} \frac{\tau v^2 dS}{|\operatorname{grad}_k E|}, \qquad (8 \cdot 3 \cdot 8)$$

and, combining this with (8·21·1), the Lorenz number is given by

$$L = \frac{\kappa}{\sigma T} = \frac{\pi^2}{3} \left(\frac{k}{\epsilon}\right)^2. \qquad (8 \cdot 3 \cdot 9)$$

There is no need to give the special forms that κ takes for particular models, since they can all be obtained from the corresponding expressions for σ by using (8·3·9).

8·31. *The Lorenz number.* It is not possible to discuss in detail the separate expressions for σ and κ until the mechanism of conductivity has been considered. At this state it is sufficient to note that the calculations of Chapter IX show that τ is proportional to $E^{\frac{3}{2}}$ and to $1/T$. This means that $1/\sigma$ is proportional to T and that κ is constant, which agree well with the experimental data at room temperatures. The Lorenz number, on the other hand, does not depend upon τ at all or on the energy-level structure. Its value given by (8·3·9) is $2·7 \times 10^{-13}$ e.s.u., in very good agreement with the mean value of $2·72 \times 10^{-13}$ e.s.u. for a number of metals at $18°$ C. (the metals given in Table VIII 1, with the exception of W and Bi, which are abnormal). There is, however, a variation in L from metal to metal and a slight increase with temperature as shown by the figures given in the table, but the variation is not large. (The average for thirteen metals at $0°$ C. is $2·7 \times 10^{-13}$ e.s.u., and the average for fifteen metals at $100°$ C. is $2·73 \times 10^{-13}$ e.s.u.)

The fact that L is independent of the properties of a metal is a consequence of the existence of a time of relaxation (and also of the assumption that the electron gas is degenerate). It has been shown by Kohler (1941) that the Wiedemann-Franz law holds (for anisotropic metals) if the conduction electrons are scattered without change of energy. The existence of a time of relaxation is therefore a sufficient, but not a necessary, condition for the constancy of the Lorenz number. It is found experimentally that as the temperature decreases, L in general falls below the value given by (8·3·9), and therefore the concept of a time of relaxation loses its usefulness at low temperatures. A full discussion of the behaviour at low temperatures is given in §§ 9·72 and 9·84, as well as a treatment of those poorly conducting metals such as bismuth for which L is greater than the normal value.

Table VIII 1. *Lorenz numbers at 0 and 100° C. in* 10^{-13} *e.s.u.*

	0° C.	100° C.		0° C.	100° C.
Mg	—	2·57	Ag	2·57	2·63
Al	2·43	2·47	Cd	2·69	2·70
Ni	—	2·53	Sn	2·80	2·76
Cu	2·48	2·60	W	3·34	3·55
Zn	2·57	2·60	Ir	2·76	2·76
Mo	2·90	3·10	Pt	2·80	2·88
Rh	2·85	2·82	Au	2·61	2·67
Pd	2·57	2·63	Pb	2·74	2·85
			Bi	3·68	3·20

THE THERMOELECTRIC EFFECTS

8·4. Since there are many possible definitions of the thermoelectric effects some description of them is necessary. If two wires of different metals are joined together at both ends and the two junctions are kept at different temperatures, an electromotive force is set up. This is known as the Seebeck effect, and was discovered in 1821. The electromotive force is proportional to the temperature difference provided it is small.

The Peltier effect was discovered in 1834 and is as follows. If two metals are joined together and kept at a constant temperature while a current passes through the junction, heat is generated or absorbed at the junction in addition to the Joule heat. This heat is directly proportional to the current. The Peltier coefficient Π_{12} is defined as the heat emitted per second when unit current flows from conductor 1 to conductor 2. The Peltier heat is known as the reversible heat, and $\Pi_{12} = -\Pi_{21}$.

The Thomson effect was predicted theoretically in 1854 and found experimentally in 1856, and is as follows. When an electric current J passes between two points of a homogeneous wire whose temperature difference is ΔT, an amount of heat $\mu J \Delta T$ is emitted or absorbed in addition to the Joule heat. The Thomson coefficient is taken to be positive if heat is evolved when a positive current passes from a higher to a lower temperature.

To define the thermoelectric force consider two metals forming an open circuit as shown in fig. VIII 1. The thermoelectric force Θ_{12} is defined to be $V_A - V_D$. It is independent of T_0 and is a function of T' and T'' only. If T' is held fixed and $T'' = T$ is varied, then $d\Theta_{12}/dT$ is a function of T only. It is called the thermoelectric force per degree (or the thermoelectric power), and is positive if $V_A - V_D$ is increased when T'' is increased.

8·41. There exist well-known thermodynamic relations between the thermoelectric effects, which can be obtained as follows. Suppose that the points A and D in fig. VIII 1 are joined by a wire of large resistance, so that any current passing round the closed circuit does so slowly enough for the process to be considered as reversible. Let unit charge pass round the circuit in the direction $ABCDA$. By the first law of thermodynamics the total work done plus the heat given out must be zero. The work done is Θ_{12}, while an amount of heat $(\Pi_{12})_{T'} - (\Pi_{12})_{T''}$ is given out at the junctions. The Thomson heat given out in conductor 2 is $-\int_B^C \mu_2 \dfrac{\partial T}{\partial x} dx = -\int_{T'}^{T''} \mu_2 dT$;

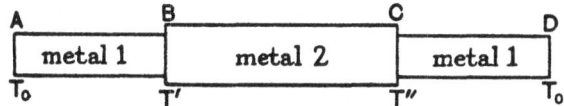

Fig. VIII 1. The thermoelectric circuit.

there is a similar expression for the Thomson heat in the conductor 1. Hence we have

$$\Theta_{12} + (\Pi_{12})_{T'} - (\Pi_{12})_{T''} + \int_{T'}^{T''} (\mu_1 - \mu_2) \, dT = 0. \tag{8·41·1}$$

Further, by the second law of thermodynamics, the total entropy change must be zero in this reversible process, which gives

$$\frac{(\Pi_{12})_{T'}}{T'} - \frac{(\Pi_{12})_{T''}}{T''} + \int_{T'}^{T''} \frac{\mu_1 - \mu_2}{T} dT = 0. \tag{8·41·2}$$

By differentiating (8·41·1) with respect to T'' and putting $T'' = T$, we obtain

$$\frac{d\Theta_{12}}{dT} - \frac{d\Pi_{12}}{dT} + \mu_1 - \mu_2 = 0, \tag{8·41·3}$$

and by differentiating (8·41·2) we find

$$\frac{d}{dT} \frac{\Pi_{12}}{T} = \frac{\mu_1 - \mu_2}{T}. \tag{8·41·4}$$

Eliminating $\mu_1 - \mu_2$ from this and (8·41·3) we have

$$\Pi_{12} = T \, d\Theta_{12}/dT. \tag{8·41·5}$$

If we now substitute this expression for Π_{12} in (8·41·4) we obtain the remaining relation

$$T \, d^2\Theta_{12}/dT^2 = \mu_1 - \mu_2. \tag{8·41·6}$$

Of the three thermoelectric quantities only μ refers to a single metal. Borelius has, however, managed to obtain the absolute thermoelectric force of a single metal in the following way. We define the absolute thermo-electric force per degree $d\Theta/dT$ by

$$\frac{d\Theta}{dT} = \int_0^T \frac{\mu}{T} dT, \qquad (8\cdot41\cdot7)$$

in accordance with (8·41·6). In order to obtain $d\Theta/dT$ it is therefore neces-sary to measure μ for at least one metal down to the absolute zero. When this has been done, $d\Theta/dT$ can be found for other metals either by direct measurement of the relative thermoelectric power or by measuring the Thomson coefficients and integrating equation (8·41·6). The obvious plan of extrapolating the μ, T relation to the absolute zero is useless, since μ behaves in a very complicated way at low temperatures. Instead, use is made of the fact that there is no thermoelectric force between metals in the superconducting states, which is interpreted as meaning that $\mu = 0$ for a superconductor. Hence by measuring the thermoelectric force of a metal against a superconductor we obtain the absolute thermoelectric force for temperatures below the transition temperature of the superconductor, and measurements of the Thomson coefficient enable us to obtain $d\Theta/dT$ at higher temperatures.

8.42. *The thermoelectric force.* Equation (8·3·3) shows that on open circuit in any isotropic metal, homogeneous or otherwise, there is an electric field $\mathscr{E}_x$ given by

$$-\epsilon\mathscr{E}_x = \frac{\partial\zeta}{\partial x} + \mathfrak{S}\frac{\partial T}{\partial x}, \qquad (8\cdot42\cdot1)$$

where

$$\mathfrak{S} = (\mathscr{K}_2 - \zeta\mathscr{K}_1)/\mathscr{K}_1 T. \qquad (8\cdot42\cdot2)$$

The thermoelectric power Θ_{12} is given by

$$\Theta_{12} = -\int_A^D \mathscr{E}_x dx = \frac{1}{\epsilon}[\zeta] + \frac{1}{\epsilon}\int_A^D \mathfrak{S}\frac{dT}{dx} dx = -\frac{1}{\epsilon}\int_{T'}^{T''}(\mathfrak{S}_1 - \mathfrak{S}_2) dT. \qquad (8\cdot42\cdot3)$$

The absolute thermoelectric force of a metal, as defined in the preceding section, is therefore

$$\Theta = -\frac{1}{\epsilon}\int_0^T \mathfrak{S} dT, \qquad (8\cdot42\cdot4)$$

and the absolute thermoelectric force per degree is

$$d\Theta/dT = -\mathfrak{S}/\epsilon. \qquad (8\cdot42\cdot5)$$

The zero-order approximation for $\mathfrak{S}$ is obtained at once from (8·42·2) and (8·3·6). It is

$$\mathfrak{S} = \tfrac{1}{3}\pi^2 k^2 T\left(\frac{\partial}{\partial E}\log\int\frac{\tau v^2 dS}{|\mathrm{grad}_k E|}\right)_{E=\zeta}, \qquad (8\cdot42\cdot6)$$

which, on account of (8·21·2), can be written in the form

$$\mathfrak{S} = \tfrac{1}{3}\pi^2 k^2 T \left(\frac{\partial}{\partial E} \log \sum_s n_s \tau_s v_s^2\right)_{E-\zeta}, \tag{8·42·7}$$

where the summation is taken over the various separate bands. For a single band (8·42·7) takes the simple form

$$\mathfrak{S} = \tfrac{1}{3}\pi^2 k^2 T \left(\frac{1}{n}\frac{dn}{dE} + \frac{1}{\tau v^2}\frac{d(\tau v^2)}{dE}\right)_{E-\zeta}. \tag{8·42·8}$$

For free electrons $n \propto E^{\frac{1}{2}}$ and $\tau \propto E^{\frac{1}{2}}$ (see (2·71·3) and (9·35·3)), so that $\partial \log(n\tau v^2)/\partial E = 3/E$, and (8·42·7) gives

$$\mathfrak{S} = \pi^2 k^2 T/\zeta. \tag{8·42·9}$$

For an inverted band of standard form the only difference is that E must be replaced by $A - E$, where A is the energy of the top of the band. Hence $n\tau v^2 \propto (A - E)^3$, and $\mathfrak{S}$ is then $-\pi^2 k^2 T/\zeta'$, where $\zeta' = A - \zeta$. For two overlapping bands of standard form, the second being inverted, $\mathfrak{S}$ is given by

$$\mathfrak{S} = \pi^2 k^2 T \frac{(n_1 \tau_1 v_1^2/\zeta_1) - (n_2 \tau_2 v_2^2/\zeta_2)}{n_1 \tau_1 v_1^2 + n_2 \tau_2 v_2^2}, \tag{8·42·10}$$

where $\zeta_2 = A - \zeta_1$. This can also be written in the form

$$\mathfrak{S} = \pi^2 k^2 T \frac{n_1 \tau_1/(m_1 \zeta_1) - n_2 \tau_2/(m_2 \zeta_2)}{n_1 \tau_1/m_1 + n_2 \tau_2/m_2}, \tag{8·42·11}$$

or, alternatively,

$$\mathfrak{S} = \frac{\sigma_1 \mathfrak{S}_1 + \sigma_2 \mathfrak{S}_2}{\sigma_1 + \sigma_2}, \tag{8·42·12}$$

where σ_1, σ_2 and $\mathfrak{S}_1$, $\mathfrak{S}_2$ are the values of σ and $\mathfrak{S}$ if only one band is present. This last form for $\mathfrak{S}$ is valid whether the bands are normal or inverted.

8·43. *The Thomson effect.* In an isotropic metal the heat Q produced per second in unit volume is given by

$$Q = J\mathscr{E} - \partial w/\partial x,$$

since $J\mathscr{E}$ is the electrical energy supplied and $\partial w/\partial x$ is the heat flowing out. If we express Q in terms of J and $\partial T/\partial x$ by means of equations (8·3·3) and (8·3·4), we find

$$Q = \frac{J^2}{\sigma} + J\frac{T}{\epsilon}\frac{\partial}{\partial x}\frac{\mathscr{K}_2 - \zeta\mathscr{K}_1}{\mathscr{K}_1 T} + \frac{\partial}{\partial x}\left(\kappa\frac{\partial T}{\partial x}\right), \tag{8·43·1}$$

the first term being the Joule heat, the second the thermoelectric heat, and the third being due to the heat conduction. If we consider a homogeneous conductor in which there is a temperature gradient and a current, the

Thomson coefficient μ is so defined that the term in Q which is linear in J is $-\mu J\,\partial T/\partial x$. We therefore have

$$\mu = -\frac{T}{\epsilon}\frac{\partial \mathfrak{S}}{\partial T},$$

which is equivalent to the relation (8·41·6) obtained by thermodynamic arguments.

8·44. *The Peltier effect.* When the temperature is constant, heat is emitted whenever a current passes through a region in which the composition is variable. Since T is constant, equation (8·43·1) shows that the reversible heat is an exact differential, and it is therefore immaterial whether we suppose the change in composition to take place gradually or suddenly. If we consider two metals 1 and 2, the Peltier coefficient is given by

$$\Pi_{12} = -T(\mathfrak{S}_2 - \mathfrak{S}_1)/\epsilon, \tag{8·44·1}$$

which is in accordance with the thermodynamic relation (8·41·5).

The validity of the thermodynamic arguments concerning the thermoelectric effects has been a matter of controversy since the earliest days, and no entirely convincing argument has ever been advanced for applying thermodynamics to transport phenomena. According to the present theory, the relations between the various thermoelectric phenomena are consequential upon the occurrence of the same integral $\mathscr{K}_2$ in both (8·3·3) and (8·3·4). More generally, they follow, in anisotropic crystals, from the symmetry relations (8·1·14) when $H=0$, and these symmetry relations can be deduced from the fact that the distribution function is the solution of an integral equation with a symmetrical nucleus. On the other hand, when a magnetic field is present the integral equation is no longer symmetrical, and the correct symmetry relations (8·1·14) do not seem to be establishable by purely thermodynamic arguments.

8·45. *Survey of the experimental data.* Some typical results are shown in fig. VIII 2 for the absolute thermoelectric force per degree, and further values are given in Table VIII 2. (The behaviour at very low temperatures is still more complicated and cannot at the moment be reconciled with the theory.) Now since $d\Theta/dT$ is obtained by integrating μ/T (see (8.41.7)), any anomalies in μ at low temperatures persist in $d\Theta/dT$ at high temperatures, and hence it is a fairer test of the theory to compare the calculated and observed values of μ rather than of $d\Theta/dT$. If then we find agreement for μ at high temperatures, any discrepancy in $d\Theta/dT$ is to be ascribed to the breakdown of the theory at low temperatures.

Table VIII 2. *Absolute thermoelectric powers at 0° and 100° C.*
in microvolts per degree

	Ag	Au	Mg	$Zn_,$	$Zn_\perp$	Al	Pb	$Bi_,$	$Bi_\perp$
0° C.	1·3	1·7	−1·3	0·4	2·2	−1·6	−1·2	−110	−54
100° C.	2·0	2·2	−1·5	0·8	3·4	−2·1	−1·4	− 95	−59

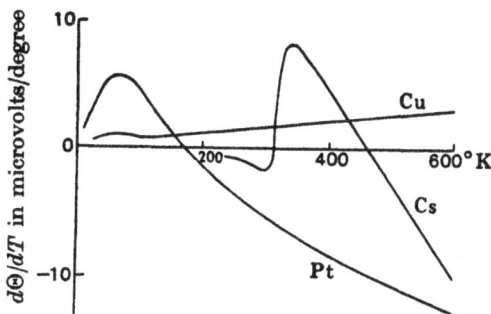

Fig. VIII 2. The absolute thermoelectric force per degree as a
function of the temperature.

We should expect the theory to be reasonably correct for the monovalent
metals, but this is not so. The Thomson coefficients of copper, silver, gold
and lithium are all positive instead of negative at ordinary temperatures.
This can only mean that the calculation of the thermoelectric effects
requires an accurate knowledge of the energy levels and that the assumption
that the electrons are free is not such a good one in this problem as it is, for
example, in the calculation of the Hall effect. We could, of course, obtain
agreement by using formula (8·42·11) and adjusting the constants accor-
dingly, and we could then try to fit the results for the Hall coefficient, etc.
(see (8·521·6)).

The main objection to this procedure is that, for the monovalent metals,
we have no reliable evidence as to what would be a reasonable value for
n_2 except that it must be considerably smaller than n_1. The ratio of n_2 to
n_1 is a measure of the relative proportion of the energy levels with energy ζ
that are in the regions of the energy surface with anomalous curvature, and
owing to our ignorance of the configuration of the energy surfaces it is
impossible to give a reasonable estimate of how much this should be. It is
less difficult to apply the theory to divalent metals where the s-band is
nearly full and the p-band nearly empty, or to the transition metals which
have a nearly full d-band and a nearly empty s-band, but in most clear-cut
cases the effect of one band predominates and the other can be neglected.

In some respects the theory is reasonably satisfactory. The proportionality of the Thomson coefficient to T and of the Peltier coefficient to
T^2 is in good agreement with the experimental results for not too low

temperatures. Further, in some cases the magnitude of μ is given correctly. For example, if we assume that for Pd and Pt there is 0·6 electron per atom in the s-band and, on account of their low mobility, ignore the contribution of the holes in the d-band, the calculated values of μ/T are about -2×10^{-2} μV./(degree)^2 for both metals. The observed values are $-3·4 \times 10^{-2}$ for Pd and $-1·8 \times 10^{-2} \mu$V./(degree)^2 for Pt. The smallness of the Hall coefficients (Table VIII 5, p. 215), however, indicates that the holes in the d-band reduce the effect of the s-electrons, and so the agreement is not so good as it seems.

A further satisfactory piece of evidence is that the Thomson coefficients of the liquid alkali metals, with the exception of lithium, agree quite well with the values calculated on the assumption that the electrons are free (see Table VIII 3). (The experimental results, due to Bidwell, have been reduced to absolute values by Sommerfeld (1934). Bidwell obtained his values for μ by measuring the thermoelectric force relative to platinum, applying a correction to make them relative to lead, and then using (8·41·6). There are various discontinuities in the gradients of his curves for Θ, and these, magnified by the differentiation required to obtain μ, may be the cause of the apparent discrepancy between the observed and calculated values of μ for solid sodium and caesium.) When a metal is in the liquid state the energy-level system ought to be fairly simple, since the zone structure characteristic of the solid is lost, and this confirms us in ascribing the failure of the elementary theory, as applied to most metals, to our inadequate knowledge of the energy-level system and not to the breakdown of the foundations of the theory.

Table VIII 3. *Thomson coefficients μ in microvolts/degree of the alkalis just above and just below the melting-point*

		Li	Na	K	Rb	Cs
μ/T:	Solid	0·022	(0·005)	$-0·036$	$-0·036$	(0·008)
	Liquid	0·030	$-0·048$	$-0·043$	$-0·085$	$-0·076$
	Calculated	$-0·016$	$-0·023$	$-0·036$	$-0·041$	$-0·048$

THE GALVANOMAGNETIC AND THERMOMAGNETIC EFFECTS

8·5. When a magnetic field is present as well as electric and thermal currents the resulting phenomena, which were first observed by Hall in 1879, are extremely complex even in isotropic media. The only model for which the calculations have been carried to completion is that in which there are a number of overlapping bands each of which has spherical symmetry, and we shall therefore carry through the calculations for this case first and give the generalizations later.

The effects can be classified in various ways. The first is into galvano-magnetic effects which arise when the primary current is electrical and into thermomagnetic effects which arise when it is thermal. Another subdivision is into 'transverse effects in a transverse magnetic field', where the primary current and the electric field (or temperature gradient) are perpendicular to one another and to the magnetic field, 'longitudinal effects in a transverse magnetic field' where they are parallel to one another but at right angles to the magnetic field, and 'longitudinal effects in a longitudinal magnetic field', where all the fields are parallel. The second and third groups include the electrical and thermal magneto-resistance effects and the influence of a magnetic field on the thermoelectric power. There are many examples in the first group, the most important being the four defined below which are often referred to without qualification as the transverse galvano- and thermomagnetic effects. They are usually described by certain coefficients which may be defined by considering a metal strip with its surface in the xy-plane, carrying an electric current J_x or a thermal current w_x in the x-direction and subjected to a magnetic field H in the z-direction. The following cases are then of practical importance:

If $J_x \neq 0$ but the transverse current $J_y = 0$, there exists a transverse electric field $\mathscr{E}_y$. The ratio $\mathscr{E}_y/HJ_x$ is known as the Hall coefficient R.

If $J_x \neq 0$ and $w_y = 0$, there exists a transverse temperature gradient $\partial T/\partial y$, and the ratio $(-\partial T/\partial y)/HJ_x$ is known as the Ettingshausen coefficient A_E.

If $\partial T/\partial x \neq 0$ and $J_y = 0$, then $\mathscr{E}_y \neq 0$, and the ratio $\mathscr{E}_y/(-H\,\partial T/\partial x)$ is known as the Ettingshausen-Nernst coefficient B_{EN}. The experimental values are also sometimes expressed in terms of the ratio $\mathscr{E}_y/Hw_x$, which is denoted by A_{EN}. Clearly $A_{EN} = B_{EN}/\kappa$, where κ is the thermal conductivity.

If $\partial T/\partial x \neq 0$ and $w_y = 0$, then $\partial T/\partial y \neq 0$, and the ratio $(\partial T/\partial y)/(H\,\partial T/\partial x)$ is known as the Righi-Leduc coefficient B_{RL}. An alternative coefficient is defined by

$$A_{RL} = (-\partial T/\partial y)/Hw_x = B_{RL}/\kappa.$$

In many cases the thermal conditions in the y-direction are not specified, and it is then necessary to distinguish between 'adiabatic' conditions where $w_y = 0$ and 'isothermal' conditions where $\partial T/\partial y = 0$. In the conditions in which the experiments are normally carried out, the Hall and Ettingshausen-Nernst effects are 'isothermal' in the y-direction, while the other two are necessarily 'adiabatic'. When it is necessary to distinguish between the 'isothermal' and 'adiabatic' values of a coefficient an index i or a is added.

8·51. *Solution of the Boltzmann equation for an isotropic metal.* If we assume that the energy levels are given by $E = \frac{1}{2}\hbar^2 |\mathbf{k}|^2/m$, the Boltzmann equation (8·11·2) can be solved by putting

$$f = f_0 - \mathbf{k} . \mathbf{c}(E) \, \partial f_0 / \partial E. \tag{8·51·1}$$

Then

$$(\mathbf{v} \times \mathbf{H}) . \mathrm{grad}_{\mathbf{k}} f = - (\mathbf{v} \times \mathbf{H}) . \mathbf{c}(E) \, \partial f_0 / \partial E = - \mathbf{v} . \{\mathbf{H} \times \mathbf{c}(E)\} \, \partial f_0 / \partial E,$$

and, since $\mathbf{v} = \hbar \mathbf{k} / m$, (8·11·2) becomes

$$- \epsilon \mathscr{E} . \mathbf{v} + T \mathbf{v} . \mathrm{grad}_{\mathbf{r}} \frac{E - \zeta}{T} + \frac{\epsilon}{\hbar c} \mathbf{v} . \{\mathbf{H} \times \mathbf{c}(E)\} = \frac{m}{\hbar \tau} \mathbf{v} . \mathbf{c}(E),$$

so that

$$(m/\hbar\tau) \, \mathbf{c}(E) - (\epsilon/\hbar c) \, \mathbf{H} \times \mathbf{c}(E) = - \epsilon \mathscr{E} + T \, \mathrm{grad}_{\mathbf{r}} \{(E - \zeta)/T\}. \tag{8·51·2}$$

The general solution of this is easily obtained by forming the scalar and vector products with $\mathbf{H}$. It is given by

$$\left[\left(\frac{m}{\hbar\tau} \right)^2 + \left(\frac{\epsilon H}{\hbar c} \right)^2 \right] \mathbf{c}(E) = \frac{m}{\hbar\tau} \mathbf{P} + \frac{\epsilon}{\hbar c} \mathbf{H} \times \mathbf{P} + \frac{\epsilon^2 \tau}{\hbar m c^2} \mathbf{H} (\mathbf{H}.\mathbf{P}), \tag{8·51·3}$$

where

$$\mathbf{P} = - \epsilon \mathscr{E} + T \, \mathrm{grad}_{\mathbf{r}} \{(E - \zeta)/T\}. \tag{8·51·4}$$

The expression (8·51·3) is a solution of the Boltzmann equation for all values of H. It must, however, be remembered that the derivation of the Boltzmann equation breaks down when $\epsilon H \tau$ is of the same order as mc. It is then necessary to take into account the quantization of the electron orbits due to the magnetic field, and the theory becomes very complicated. A preliminary discussion of the changes introduced by the quantization has been given by Titeica (1935), whose paper should be consulted for details.

When the electric fields and temperature gradients are in the xy-plane and when $\mathbf{H} = (0, 0, H)$, we have $\mathbf{c} = (c_1, c_2, 0)$ and (8·51·2) becomes

$$\left. \begin{aligned} \frac{m}{\hbar\tau} c_1 + \frac{\epsilon H}{\hbar c} c_2 &= - \epsilon \mathscr{E}_x - T \frac{\partial}{\partial x} \frac{\zeta}{T} - \frac{E}{T} \frac{\partial T}{\partial x}, \\ - \frac{\epsilon H}{\hbar c} c_1 + \frac{m}{\hbar\tau} c_2 &= - \epsilon \mathscr{E}_y - T \frac{\partial}{\partial y} \frac{\zeta}{T} - \frac{E}{T} \frac{\partial T}{\partial y}. \end{aligned} \right\} \tag{8·51·5}$$

The electric current density is $(J_x, J_y, 0)$, where

$$J_x = - \frac{\epsilon}{4\pi^3} \int v_1 f \, d\mathbf{k} = \frac{\epsilon\hbar}{4\pi^3 m} \int k_1^2 c_1 \frac{\partial f_0}{\partial E} \, d\mathbf{k} = \frac{(2m)^{\frac{3}{2}} \epsilon}{3\pi^2 \hbar^4} \int_0^\infty E^{\frac{3}{2}} c_1 \frac{\partial f_0}{\partial E} \, dE \tag{8·51·6}$$

and

$$J_y = \frac{(2m)^{\frac{3}{2}} \epsilon}{3\pi^2 \hbar^4} \int_0^\infty E^{\frac{3}{2}} c_2 \frac{\partial f_0}{\partial E} \, dE. \tag{8·51·7}$$

Similarly, the thermal current density is $(w_x, w_y, 0)$, where

$$w_x = - \frac{(2m)^{\frac{3}{2}}}{3\pi^2 \hbar^4} \int_0^\infty E^{\frac{3}{2}} c_1 \frac{\partial f_0}{\partial E} \, dE, \quad w_y = - \frac{(2m)^{\frac{3}{2}}}{3\pi^2 \hbar^4} \int_0^\infty E^{\frac{3}{2}} c_2 \frac{\partial f_0}{\partial E} \, dE. \tag{8·51·8}$$

8·511. A solution similar to that given above holds for an inverted band of standard form, but various changes of sign are required, since, as we know from general principles and as appears from the explicit expressions derived in the next few sections, electrons in an inverted band can formally be replaced by positive holes (Peierls, 1929). There are many ways of introducing the required changes in sign, but the most satisfactory way is to use the 'hole distribution function' $f_0^+(E')$, where $E' = A - E$, $\zeta' = A - \zeta$ (the apex of the band having energy A) and

$$f_0^+(E') = 1 - f_0(E) = \frac{1}{e^{(E'-\zeta')/kT} + 1}. \tag{8·511·1}$$

If we are dealing with either a single normal band or a single inverted band the formulae can be written in identical forms, but since we usually wish to consider models containing two bands and since it is impossible to take the energy zero at the apex of both bands simultaneously, the formulae for one of the bands are then complicated by the explicit appearance of a constant giving the energy overlap of the bands.

If for the inverted band we write $E = A - \frac{1}{2}\hbar^2 |\mathbf{k}|^2/m$ and $\mathbf{v} = -\hbar\mathbf{k}/m$, we can put

$$f = f_0 + \mathbf{k} \cdot \mathbf{c}(E) \, \partial f_0/\partial E, \tag{8·511·2}$$

and the equation for $\mathbf{c}(E)$ is the same as (8·51·2), but with $\mathbf{H}$ replaced by $-\mathbf{H}$. Equivalently we may write

$$f = f_0 + \mathbf{k} \cdot \mathbf{c}^+(E') \, \partial f_0^+/\partial E', \tag{8·511·3}$$

where $\mathbf{c}^+ = (c_1^+, c_2^+, 0)$ and

$$\left.\begin{aligned}
\frac{m}{\hbar\tau} c_1^+ - \frac{\epsilon H}{\hbar c} c_2^+ &= -\epsilon \mathscr{E}_x + T \frac{\partial}{\partial x} \frac{\zeta'}{T} + \frac{E'}{T} \frac{\partial T}{\partial x}, \\
\frac{\epsilon H}{\hbar c} c_1^+ + \frac{m}{\hbar\tau} c_2^+ &= -\epsilon \mathscr{E}_y + T \frac{\partial}{\partial y} \frac{\zeta'}{T} + \frac{E'}{T} \frac{\partial T}{\partial y}.
\end{aligned}\right\} \tag{8·511·4}$$

The electric current density is $(J_x, J_y, 0)$, where

$$J_x = -\frac{\epsilon}{4\pi^3} \int v_1 f \, d\mathbf{k} = \frac{\epsilon \hbar}{4\pi^3 m} \int k_1^2 c_1^+ \frac{\partial f_0^+}{\partial E'} \, d\mathbf{k} = \frac{(2m)^{\frac{3}{2}} \epsilon}{3\pi^2 \hbar^4} \int_0^\infty E'^{\frac{3}{2}} c_1^+ \frac{\partial f_0^+}{\partial E'} \, dE' \tag{8·511·5}$$

and

$$J_y = \frac{(2m)^{\frac{3}{2}} \epsilon}{3\pi^2 \hbar^4} \int_0^\infty E'^{\frac{3}{2}} c_2^+ \frac{\partial f_0^+}{\partial E'} \, dE'. \tag{8·511·6}$$

The thermal current density is given by $(w_x, w_y, 0)$, where

$$w_x = \frac{1}{4\pi^3} \int v_1 E f \, d\mathbf{k} = -\frac{(2m)^{\frac{3}{2}}}{3\pi^2 \hbar^4} \int_0^\infty E E'^{\frac{3}{2}} c_1^+ \frac{\partial f_0^+}{\partial E'} \, dE'$$

$$= -\frac{A}{\epsilon} J_x + \frac{(2m)^{\frac{3}{2}}}{3\pi^2 \hbar^4} \int_0^\infty E'^{\frac{5}{2}} c_1^+ \frac{\partial f_0^+}{\partial E'} \, dE'. \tag{8·511·7}$$

Similarly
$$w_y = -\frac{A}{\epsilon} J_y + \frac{(2m)^{\frac{3}{2}}}{3\pi^2\hbar^4} \int_0^\infty E'^{\frac{3}{2}} c_2^+ \frac{\partial f_0^+}{\partial E'} dE'. \tag{8·511·8}$$

Further, the solutions of (8·51·5) and (8·511·4) are given by

$$c_1 = \frac{(m/\hbar\tau) P_x - \alpha P_y}{\alpha^2 + (m/\hbar\tau)^2}, \quad c_2 = \frac{\alpha P_x + (m/\hbar\tau) P_y}{\alpha^2 + (m/\hbar\tau)^2}, \quad \alpha = \frac{\epsilon H}{\hbar c}, \tag{8·511·9}$$

where
$$\mathbf{P} = -\epsilon\mathscr{E} - T \operatorname{grad}(\zeta/T) - (E \operatorname{grad} T)/T, \tag{8·511·10}$$

and
$$c_1^+ = \frac{(m/\hbar\tau) P_x^+ + \alpha P_y^+}{\alpha^2 + (m/\hbar\tau)^2}, \quad c_2^+ = \frac{-\alpha P_x^+ + (m/\hbar\tau) P_y^+}{\alpha^2 + (m/\hbar\tau)^2}, \tag{8·511·11}$$

where
$$\mathbf{P}^+ = -\epsilon\mathscr{E} + T \operatorname{grad}(\zeta'/T) + (E' \operatorname{grad} T)/T. \tag{8·511·12}$$

THE ISOTHERMAL EFFECTS

8·52. When thermal gradients are absent, we have $\mathbf{P} = \mathbf{P}^+ = -\epsilon\mathscr{E}$. Hence, by (8·51·6), (8·51·7) and (8·511·9), the current density is given by

$$J_x = \frac{\epsilon^2}{3\pi^2\hbar^4}(i_1\mathscr{E}_x - i_4\mathscr{E}_y), \quad J_y = \frac{\epsilon^2}{3\pi^2\hbar^4}(i_4\mathscr{E}_x + i_1\mathscr{E}_y) \tag{8·52·1}$$

for a normal band, where

$$i_1 = -(2m)^{\frac{3}{2}}\frac{m}{\hbar}\int_0^\infty \frac{E^{\frac{3}{2}}}{\tau\{\alpha^2 + (m/\hbar\tau)^2\}}\frac{\partial f_0}{\partial E}dE, \quad i_4 = -(2m)^{\frac{3}{2}}\alpha\int_0^\infty \frac{E^{\frac{3}{2}}}{\alpha^2 + (m/\hbar\tau)^2}\frac{\partial f_0}{\partial E}dE. \tag{8·52·2}$$

If we have two overlapping bands of standard form, the first being normal with effective mass m_1 and the second being inverted with effective mass m_2, we have

$$\left.\begin{aligned} J_x &= \frac{\epsilon^2}{3\pi^2\hbar^4}\{(i_1^{(1)} + i_1^{(2)})\mathscr{E}_x - (i_4^{(1)} - i_4^{(2)})\mathscr{E}_y\}, \\ J_y &= \frac{\epsilon^2}{3\pi^2\hbar^4}\{(i_4^{(1)} - i_4^{(2)})\mathscr{E}_x + (i_1^{(1)} + i_1^{(2)})\mathscr{E}_y\}, \end{aligned}\right\} \tag{8·52·3}$$

where $i_1^{(1)}$ and $i_4^{(1)}$ are given by (8·52·2) with m, τ replaced by m_1, τ_1, while $i_1^{(2)}$ and $i_4^{(2)}$ are given by (8·52·2) with m, τ replaced by m_2, τ_2, E replaced by E' and f_0 replaced by f_0^+.

8·521. *The Hall effect.* It will be seen from the equations of the preceding section that a transverse electric field must be set up in order to counteract the deflecting effect of the magnetic field and reduce the transverse electric current to zero. The Hall coefficient R is defined by

$$R = \mathscr{E}_y/(HJ_x), \tag{8·521·1}$$

and when the currents are given by (8·52·3) it is

$$R = -\frac{3\pi^2\hbar^4}{H\epsilon^2}\frac{i_4^{(1)} - i_4^{(2)}}{(i_1^{(1)} + i_1^{(2)})^2 + (i_4^{(1)} - i_4^{(2)})^2}. \tag{8·521·2}$$

For small magnetic fields we can neglect the α^2's in the denominators of the integrals for the i's. We then have

$$R = -\frac{3\pi^2\hbar^4}{H\epsilon^2}\frac{i_4^{(1)}-i_4^{(2)}}{(i_1^{(1)}+i_1^{(2)})^2}, \tag{8·521·3}$$

where

$$\left.\begin{aligned}
i_1^{(1)} &= -(2m_1)^{\frac{1}{2}}\frac{\hbar}{m_1}\int_0^\infty \tau E^{\frac{3}{2}}\frac{\partial f_0}{\partial E}dE = (2m_1\zeta_1)^{\frac{1}{2}}\hbar\tau_1/m_1,\\
i_4^{(1)} &= -(2m_1)^{\frac{1}{2}}\alpha\frac{\hbar^2}{m_1^2}\int_0^\infty \tau^2 E^{\frac{3}{2}}\frac{\partial f_0}{\partial E}dE = \alpha(2m_1\zeta_1)^{\frac{1}{2}}(\hbar\tau_1/m_1)^2,
\end{aligned}\right\} \tag{8·521·4}$$

neglecting terms of order $(kT/\zeta_1)^2$, and where $i_1^{(2)}$, $i_4^{(2)}$ are obtained from $i_1^{(1)}$, $i_4^{(1)}$ by replacing m_1, τ_1, ζ_1 by m_2, τ_2, ζ_2. (Note that ζ_1 gives the position of the Fermi energy level relative to the bottom of band 1, while ζ_2 gives its position relative to the top of band 2.) If we now substitute for ζ_1 and ζ_2 from the relations

$$n_1 = \frac{(2m_1\zeta_1)^{\frac{3}{2}}}{3\pi^2\hbar^3}, \quad n_2 = \frac{(2m_2\zeta_2)^{\frac{3}{2}}}{3\pi^2\hbar^3}, \tag{8·521·5}$$

n_1 and n_2 being the number of electrons and holes in the bands 1 and 2 respectively, we find

$$R = -\frac{1}{\epsilon c}\frac{n_1\tau_1^2/m_1^2 - n_2\tau_2^2/m_2^2}{(n_1\tau_1/m_1 + n_2\tau_2/m_2)^2} = -\frac{1}{\epsilon c}\frac{\sigma_1^2/n_1 - \sigma_2^2/n_2}{(\sigma_1+\sigma_2)^2}. \tag{8·521·6}$$

If only one band is present, $\quad R = \pm 1/(n\epsilon c)$, $\tag{8·521·7}$

the lower sign referring to the case in which there are n free electrons in a normal band and the upper to the case when there are n vacant levels in an inverted band. We can therefore write (8·521·6) in the form

$$R = \frac{R_1\sigma_1^2 + R_2\sigma_2^2}{(\sigma_1+\sigma_2)^2}, \tag{8·521·8}$$

and this formula is valid whether the bands are normal or inverted.

The generalization of the above calculations to arbitrary values of H is straightforward, the result being

$$R = -\frac{1}{\epsilon c}\frac{\dfrac{\sigma_{01}^2}{n_1} - \dfrac{\sigma_{02}^2}{n_2} + \left(\dfrac{H}{\epsilon c}\right)^2\dfrac{n_1-n_2}{n_1^2 n_2^2}\sigma_{01}^2\sigma_{02}^2}{(\sigma_{01}+\sigma_{02})^2 + \left(\dfrac{H}{\epsilon c}\right)^2\dfrac{(n_1-n_2)^2}{n_1^2 n_2^2}\sigma_{01}^2\sigma_{02}^2}, \tag{8·521·9}$$

σ_{01}, σ_{02} being the partial conductivities due to the two bands, when $H = 0$.

8·522. Survey of the experimental data. When the current is carried by electrons in one band only, the Hall coefficient gives at once the effective number of electrons or holes. Further, since by (8·21·4) and (8·521·7) we have

$$R\sigma = \pm \epsilon\tau(\zeta)/mc,$$

the value of τ/m can be obtained. Unfortunately, it is only for the mono-valent elements that we can expect the hypothesis to be true; for other metals σ and R are given by formulae which reduce, when the metal is treated as isotropic, to (8·21·5) and (8·521·6). There are too many arbitrary parameters in the latter formulae for them to be determined merely by measurements of R and σ, but some of these parameters can be determined in other ways. For example, when the number of valency electrons is just sufficient to be able to fill the lower band completely (an example is a divalent cubic metal), the number of holes in the lower band must be equal to the number of electrons in the upper band. Hence $n_1 = n_2$, and the number of parameters is reduced by one.

There have been many attempts to find all the parameters by combining various measurements. The most recent of these is by Ariyama (1938), whose calculation for the divalent metals Be, Mg, Zn and Cd are outlined in § 8·523. His results are not unreasonable, but it is impossible, from the formulae at present known, to obtain correctly all the properties of any metal including, say, the specific heat, the magnetic susceptibility, the electrical conductivity, the Hall coefficient, the change of resistance in a magnetic field, the Thomson coefficient and the refractive index. It is in fact only in exceptional cases such as the alkalis that we can reproduce correctly the magnitudes of more than two or three of these effects, but it is also true to say that by choosing the parameters properly we can repro-duce the magnitudes of any particular combination of say three quantities. In general, we may say that if a parameter can be determined unambig-uously by the measurement of one quantity then its value is likely to be correct. If it depends upon the combination of two or more measurements, the accuracy of the determination is very considerably reduced.

Table VIII 4. *The temperature variation of the Hall coefficient in 10^{-25} Gaussian units*

$T°$ K. ...	290	20·3	14·5
Cu	−5·5	− 7·3	− 7·3
Ag	−8·9	−11·3	−11·1
Au	−8·1	−10·9	−10·9
Pd	−7·6	−15·4	−15·5
Cd	6·1	22·4	24·8

The temperature variation of the Hall effect is in general small (it should in fact be zero according to equation (8·521·6)), but such evidence as exists suggests that the Hall coefficients increase in absolute magnitude as the temperature is lowered (see Table VIII 4; Meissner, 1935, pp. 338 ff.). This behaviour could be explained by the two-band model if it is assumed that σ_1 and σ_2 have somewhat different temperature variations, but this could only explain a moderate increase in R, and such large increases as are found for cadmium must (if genuine) require an entirely different interpretation.

For the monovalent metals we can reasonably assume that the valency electrons are all in one band, and we can then use the formula $R = -1/(nec)$ to deduce n, and τ can then be determined from the values of $-R\sigma = e\tau/mc$. The results are given in Table VIII 5.

Table VIII 5. *Hall coefficients and conductivities of the monovalent metals at* 0° C. *in Gaussian units*

	$n_a \times 10^{-22}$	$R \times 10^{25}$	$n \times 10^{-22}$ calculated	$\sigma \times 10^{-16}$	$-R\sigma \times 10^7$
Li	4·8	−19	3·7	10·5	2·0
Na	2·6	−28	2·5	20·8	5·8
K	1·4	−47	1·5	13·3	6·2
Rb	1·1	—	—	7·7	—
Cs	0·9	−87	0·8	5·0	4·3
Cu	8·5	− 5·5	12	57	3·2
Ag	5·9	− 8·9	8	60	5·4
Au	5·9	− 8·1	9	44	3·5

It is possible to define an effective number of electrons by writing $n_{eff.} = 1/(|R|ec)$, but, if the electrons are distributed between two bands which are of comparable importance, the values of $n_{eff.}$ have no physical significance. For, small values of $|R|$ would imply large values of $n_{eff.}$, whereas in general they mean that the numbers of electrons and holes are nearly equal and are of normal order of magnitude. For example, $n_{eff.}$ for zinc would be given as 9×10^{22}, while calculations to be outlined in § 8·524 show that the number of electrons and holes is of the order of $0·6 \times 10^{22}$ per c.c. Similarly, $n_{eff.}$ for bismuth would be $1·2 \times 10^{19}$, whereas the number of electrons and holes is approximately 7×10^{18} per c.c.

The field variation of the Hall coefficient is negligible except for metals of the bismuth group, and even for these the experimental evidence is so conflicting that it cannot reasonably be analysed at the moment.

8·523. *The magneto-resistance effect.* The transverse Hall electric field eliminates on the average the transverse motion of the electrons, but there are nevertheless second-order changes in the paths of the electrons due to the action of the magnetic field, and these result in an increase in the electrical resistance, which, since it is due to the mean square deviations of the electrons, must be proportional to H^2 for small fields. If, however, all the electrons which contribute to the electric current have the same velocity, the Hall field exactly cancels the effect of the magnetic field, and the magneto-resistance effect vanishes. This occurs if the electrons are treated as being perfectly free, and so more complicated models must be considered in order to find a non-zero effect (Peierls, 1931).

To find the effect of a magnetic field on the electrical conductivity we revert to equations (8·52·3) and impose the condition $J_y = 0$, as in the discussion of the Hall effect. Then, on eliminating $\mathscr{E}_y$, we find

$$\sigma = \frac{J_x}{\mathscr{E}_x} = \frac{\epsilon^2}{3\pi^2\hbar^4} \frac{(i_1^{(1)} + i_1^{(2)})^2 + (i_4^{(1)} - i_4^{(2)})^2}{i_1^{(1)} + i_1^{(2)}}. \tag{8·523·1}$$

It is sufficient to evaluate this to the zero order in kT/ζ but without assuming H to be small. The result is

$$\sigma = \epsilon^2 \frac{\left(\dfrac{n_1\tau_1}{m_1} + \dfrac{n_2\tau_2}{m_2}\right)^2 + \left(\dfrac{\epsilon H}{c}\right)^2 (n_1 - n_2)^2 \dfrac{\tau_1^2\tau_2^2}{m_1^2 m_2^2}}{\dfrac{n_1\tau_1}{m_1} + \dfrac{n_2\tau_2}{m_2} + \left(\dfrac{\epsilon H}{c}\right)^2 \left(\dfrac{n_1\tau_2}{m_2} + \dfrac{n_2\tau_1}{m_1}\right)\dfrac{\tau_1\tau_2}{m_1 m_2}}. \tag{8·523·2}$$

Now in the absence of a magnetic field the conductivity is

$$\sigma_0 = \epsilon^2\left(\frac{n_1\tau_1}{m_1} + \frac{n_2\tau_2}{m_2}\right) = \sigma_{01} + \sigma_{02},$$

and so the relative change in resistivity is

$$\frac{\rho - \rho_0}{\rho_0} = \frac{\sigma_0 - \sigma}{\sigma} = \frac{\left(\dfrac{\epsilon H}{c}\right)^2 \dfrac{n_1\tau_1}{m_1}\dfrac{n_2\tau_2}{m_2}\left(\dfrac{\tau_1}{m_1} + \dfrac{\tau_2}{m_2}\right)^2 \left(\dfrac{n_1\tau_1}{m_1} + \dfrac{n_2\tau_2}{m_2}\right)^{-2}}{1 + \left(\dfrac{\epsilon H}{c}\right)^2 (n_1 - n_2)^2 \dfrac{\tau_1^2}{m_1^2}\dfrac{\tau_2^2}{m_2^2}\left(\dfrac{n_1\tau_1}{m_1} + \dfrac{n_2\tau_2}{m_2}\right)^{-2}} \tag{8·523·3}$$

$$= \frac{\left(\dfrac{H}{\epsilon c}\right)^2 \left(\dfrac{\sigma_{01}}{n_1} + \dfrac{\sigma_{02}}{n_2}\right)^2 \dfrac{\sigma_{01}\sigma_{02}}{(\sigma_{01} + \sigma_{02})^2}}{1 + \left(\dfrac{H}{\epsilon c}\right)^2 (n_1 - n_2)^2 \dfrac{\sigma_{01}^2\sigma_{02}^2}{n_1^2 n_2^2}\dfrac{1}{(\sigma_{01} + \sigma_{02})^2}}, \tag{8·523·4}$$

where σ_{01} and σ_{02} are the partial conductivities due to the electrons in bands 1 and 2 with zero magnetic field (Jones, 1936; Sondheimer and Wilson, 1947).

At normal temperatures the change in resistance is small, but it increases considerably as the temperature is lowered, in accordance with the predictions of equation (8·523·4). Interest therefore centres round the low-temperature results, but there are some experimental data at normal temperatures, though they have not a high degree of accuracy. If, for small fields, we write $\Delta\rho/\rho = BH^2$, and for simplicity put $n_1 = n_2 = n$, $\sigma_1 = \sigma_2 = \frac{1}{2}\sigma$, then

$$B = \sigma_0^2/(2n\epsilon c)^2, \tag{8·523·5}$$

which is sufficiently accurate to give the order of magnitude of the effect. The values of B for zinc and cadmium are of the order of 10^{-12}, while σ_0 is of the order of 10^{17} e.s.u., and $2nec$ is of the order of 10^{23}. The relation (8·523·5) is therefore correct as regards order of magnitude.

The above value of B has only been derived by considering a special model. Calculations given later (in § 8·552 for anisotropic metals and in § 8·64 for semi-conductors) show that in all cases worked out so far $\Delta\rho/\rho$ is of the order of $H^2\sigma_0^2/(n_a ec)^2$ for small fields, where n_a is the number of atoms per unit volume. The peculiarities of special models, therefore, only affect the multiplying constant and the behaviour in strong fields. While the two-band isotropic model has many obvious limitations, it has the advantage that the final result can be obtained explicitly and its consequences rigorously deduced. It is, however, inadequate for discussing essentially anisotropic effects such as the longitudinal change in resistance.

8·524. Ariyama (1938) has tried to improve the agreement between the theoretical and experimental results by using the full formulae (8·521·6) and (8·523·3) to determine the effective masses in the two bands for the metals Be, Mg, Zn and Cd. He makes the simplifying assumptions $n_1 = n_2 = n$ and $\tau_1 = \tau_2 = \tau$. These cannot be entirely correct because the metals are hexagonal and the first Brillouin zone contains slightly fewer than two electrons per atom, but they are sufficiently accurate for illustrative purposes. The only remaining parameters are n, τ, m_1 and m_2, and we can determine these from the measured values of

$$\sigma = ne^2\tau\frac{m_1+m_2}{m_1 m_2}, \quad R = \frac{1}{nec}\frac{m_1-m_2}{m_1+m_2}, \quad B = \frac{e^2\tau^2}{m_1 m_2 c^2}$$

and one further quantity. Ariyama assumed that $\sigma/(M\Theta)^2$, where M is the atomic weight, for these divalent metals would be the same as for the adjacent monovalent metals (see § 9·5). The values he obtained are shown in Table VIII 6. In view of all the simplifying assumptions made, the values deduced are not unreasonable. (It must also be borne in mind that much of the experimental data on the Hall and other effects is highly unsatisfactory. Many recent compilations are in fact taken uncritically from Campbell's book (1923), which in turn quotes results obtained in the 1880's. In the absence of accurate measurements of recent date, the experimental values can only be considered as representing the order of magnitude of R.). The more interesting measurements at low temperatures are discussed in § 10·51.

8·525. The formulae for two bands can be expressed in terms of the parameters for the separate bands in various ways, all of which are equivalent to (8·521·2) and (8·523·1). The following is perhaps the simplest to

Table VIII 6. *The parameters of divalent metals deduced*
from the magneto-resistance effects

	Be	Mg	Zn	Cd
$\sigma \times 10^{-16}$	16·2	22·5	16·3	13·2
$R \times 10^{25}$	27	-10	11	6
$B \times 10^{12}$	100	20	7	9
$n_a \times 10^{-22}$	12·5	4·3	6·7	4·7
$n_1, n_2 \times 10^{-22}$	0·5	1·6	0·6	1·4
m_1/m	1·5	5	0·4	1·4
m_2/m	2·2	3	0·5	1·8

comprehend in the general case of arbitrary H and any degree of degeneracy (Chambers, 1952 a). Let σ_1 and R_1 be the electrical conductivity and Hall coefficient for band 1 alone for arbitrary values of H. Then

$$\sigma_1 = \frac{\epsilon^2}{3\pi^2\hbar^4} i_1^{(1)}\left(1 + \frac{i_4^{(1)2}}{i_1^{(1)2}}\right), \quad HR_1\sigma_1 = -\frac{i_4^{(1)}}{i_1^{(1)}}, \qquad (8·525·1)$$

which give

$$\frac{\epsilon^2}{3\pi^2\hbar^4} i_1^{(1)} = \frac{\sigma_1}{1 + (HR_1\sigma_1)^2}, \quad \frac{\epsilon^2}{3\pi^2\hbar^4} i_4^{(1)} = -\frac{HR_1\sigma_1^2}{1 + (HR_1\sigma_1)^2}. \qquad (8·525·2)$$

Substituting these and the corresponding expressions for $i_1^{(2)}$, $i_4^{(2)}$ into (8·521·2) and (8·523·1) we find, after some simplification,

$$\sigma = \frac{(\sigma_1 + \sigma_2)^2 + H^2\sigma_1^2\sigma_2^2(R_1 + R_2)^2}{\sigma_1 + \sigma_2 + H^2\sigma_1\sigma_2(\sigma_1 R_1^2 + \sigma_2 R_2^2)} \qquad (8·525·3)$$

and

$$R = \frac{\sigma_1^2 R_1 + \sigma_2^2 R_2 + H^2\sigma_1^2\sigma_2^2 R_1 R_2(R_1 + R_2)}{(\sigma_1 + \sigma_2)^2 + H^2\sigma_1^2\sigma_2^2(R_1 + R_2)^2}, \qquad (8·525·4)$$

of which equations (8·523·2) and (8·521·9) are particular cases.

THE THERMOMAGNETIC EFFECTS

8·53. The treatment of the general case in which there is a temperature gradient as well as an electric field follows the same lines as the discussion given above, but the results are more cumbersome. After a straightforward calculation starting from equations (8·51·6)–(8·51·8) and (8·511·5)–(8·511·12) and utilizing the fact that $\zeta_1 + \zeta_2 = A$, we find

$$J_x = \frac{\epsilon}{3\pi^2\hbar^4}\left[i_1^{(1)}\left(\epsilon\mathscr{E}_x + T\frac{\partial}{\partial x}\frac{\zeta_1}{T}\right) + i_1^{(2)}\left(\epsilon\mathscr{E}_x - T\frac{\partial}{\partial x}\frac{\zeta_2}{T}\right) + (i_2^{(1)} - i_2^{(2)})\frac{1}{T}\frac{\partial T}{\partial x} \right.$$
$$\left. - i_4^{(1)}\left(\epsilon\mathscr{E}_y + T\frac{\partial}{\partial y}\frac{\zeta_1}{T}\right) + i_4^{(2)}\left(\epsilon\mathscr{E}_y - T\frac{\partial}{\partial y}\frac{\zeta_2}{T}\right) - (i_5^{(1)} + i_5^{(2)})\frac{1}{T}\frac{\partial T}{\partial y}\right], \quad (8·53·1)$$

$$J_y = \frac{\epsilon}{3\pi^2\hbar^4}\left[i_4^{(1)}\left(\epsilon\mathscr{E}_x + T\frac{\partial}{\partial x}\frac{\zeta_1}{T}\right) - i_4^{(2)}\left(\epsilon\mathscr{E}_x - T\frac{\partial}{\partial x}\frac{\zeta_2}{T}\right) + (i_5^{(1)} + i_5^{(2)})\frac{1}{T}\frac{\partial T}{\partial x}\right.$$

$$\left. + i_1^{(1)}\left(\epsilon\mathscr{E}_y + T\frac{\partial}{\partial y}\frac{\zeta_1}{T}\right) + i_1^{(2)}\left(\epsilon\mathscr{E}_y - T\frac{\partial}{\partial y}\frac{\zeta_2}{T}\right) + (i_2^{(1)} - i_2^{(2)})\frac{1}{T}\frac{\partial T}{\partial y}\right], \quad (8\cdot53\cdot2)$$

$$w_x = -\frac{1}{\epsilon}(\zeta_1 + \zeta_2)J_x^{(2)}$$

$$-\frac{1}{3\pi^2\hbar^4}\left[i_2^{(1)}\left(\epsilon\mathscr{E}_x + T\frac{\partial}{\partial x}\frac{\zeta_1}{T}\right) - i_2^{(2)}\left(\epsilon\mathscr{E}_x - T\frac{\partial}{\partial x}\frac{\zeta_2}{T}\right) + (i_3^{(1)} + i_3^{(2)})\frac{1}{T}\frac{\partial T}{\partial x}\right.$$

$$\left. - i_5^{(1)}\left(\epsilon\mathscr{E}_y + T\frac{\partial}{\partial y}\frac{\zeta_1}{T}\right) - i_5^{(2)}\left(\epsilon\mathscr{E}_y - T\frac{\partial}{\partial y}\frac{\zeta_2}{T}\right) - (i_6^{(1)} - i_6^{(2)})\frac{1}{T}\frac{\partial T}{\partial y}\right], \quad (8\cdot53\cdot3)$$

$$w_y = -\frac{1}{\epsilon}(\zeta_1 + \zeta_2)J_y^{(2)}$$

$$-\frac{1}{3\pi^2\hbar^4}\left[i_5^{(1)}\left(\epsilon\mathscr{E}_x + T\frac{\partial}{\partial x}\frac{\zeta_1}{T}\right) + i_5^{(2)}\left(\epsilon\mathscr{E}_x - T\frac{\partial}{\partial x}\frac{\zeta_2}{T}\right) + (i_6^{(1)} - i_6^{(2)})\frac{1}{T}\frac{\partial T}{\partial x}\right.$$

$$\left. + i_2^{(1)}\left(\epsilon\mathscr{E}_y + T\frac{\partial}{\partial y}\frac{\zeta_1}{T}\right) - i_2^{(2)}\left(\epsilon\mathscr{E}_y - T\frac{\partial}{\partial y}\frac{\zeta_2}{T}\right) + (i_3^{(1)} + i_3^{(2)})\frac{1}{T}\frac{\partial T}{\partial y}\right], \quad (8\cdot53\cdot4)$$

where

$$i_n^{(1)} = -(2m_1)^{\frac{3}{2}}\frac{m_1}{\hbar}\int_0^\infty \frac{E^{n+\frac{1}{2}}}{\tau_1\{\alpha^2 + (m_1/\hbar\tau_1)^2\}}\frac{\partial f_0}{\partial E}dE \quad (n=1,2,3),$$

$$i_n^{(1)} = -(2m_1)^{\frac{3}{2}}\alpha\int_0^\infty \frac{E^{n-\frac{1}{4}}}{\alpha^2 + (m_1/\hbar\tau_1)^2}\frac{\partial f_0}{\partial E}dE \quad (n=4,5,6),$$

$$\left.\right\} \quad (8\cdot53\cdot5)$$

and where the $i_n^{(2)}$'s are obtained by replacing m_1, τ_1, ζ_1 by m_2, τ_2, ζ_2 and E, f_0 by E', f_0^+ (Sondheimer and Wilson, 1947). $J_x^{(2)}$ and $J_y^{(2)}$ are the contributions to the electric current density components from band 2.

The equations (8·53·1)–(8·53·4) exhibit the symmetry relations required to justify, for a general external field, the procedure of treating vacant levels in an inverted band as being equivalent to positively charged particles. (All terms containing even powers of ϵ have the same signs for both bands, while all terms containing odd powers have opposite signs for the two bands. Note that i_n ($n=4,5,6$) is an odd function of ϵ.)

The choice of the energy zero at the bottom of band 1 leads to a slight lack of symmetry in the contributions of the two bands to the heat current. The arbitrariness in the energy zero entails an arbitrariness in the first terms of the expressions (8·53·3) and (8·53·4), but these terms do not affect any of the results.

The integrals i_n can be evaluated to any desired accuracy, but the results are excessively complicated unless we confine our attention to a highly degenerate gas and expand the integrals in powers of kT/ζ_1 and kT/ζ_2, stopping at the first term that gives a non-vanishing result. If we write

$$i_n^{(1)} = \sum_m (kT/\zeta_1)^{2m} (i_n^{(1)})_m, \quad i_n^{(2)} = \sum_m (kT/\zeta_2)^{2m} (i_n^{(2)})_m, \qquad (8\cdot53\cdot6)$$

then

$$(i_2^{(r)})_0 = \zeta_r(i_1^{(r)})_0, \quad (i_3^{(r)})_0 = \zeta_r^2(i_1^{(r)})_0, \quad (i_5^{(r)})_0 = \zeta_r(i_4^{(r)})_0, \quad (i_6^{(r)})_0 = \zeta_r^2(i_4^{(r)})_0 \quad (r = 1, 2),$$
$$(8\cdot53\cdot7)$$

and (Sondheimer, 1948) it is found that only six combinations of the i_n's occur, namely

$$\begin{aligned} V_r &= (i_1^{(r)})_0, \quad W_r = (i_4^{(r)})_0, \\ \zeta_r^2 X_r &= \zeta_r^2(i_1^{(r)})_1 - 2\zeta_r(i_2^{(r)})_1 + (i_3^{(r)})_1, \quad \zeta_r^2 Y_r = \zeta_r^2(i_4^{(r)})_1 - 2\zeta_r(i_5^{(r)})_1 + (i_6^{(r)})_1, \\ \zeta_r^2 P_r &= (i_2^{(r)})_1 - \zeta_r(i_1^{(r)})_1, \quad \zeta_r^2 Q_r = (i_5^{(r)})_1 - \zeta_r(i_4^{(r)})_1. \end{aligned} \right\} \quad (8\cdot53\cdot8)$$

Now, omitting the suffix r, we have

$$V = \frac{m}{\hbar\tau} \frac{(2m\zeta)^{\frac{3}{2}}}{\alpha^2 + (m/\hbar\tau)^2}, \quad W = \alpha \frac{(2m\zeta)^{\frac{3}{2}}}{\alpha^2 + (m/\hbar\tau)^2}, \qquad (8\cdot53\cdot9)$$

and the other quantities can be readily expressed in terms of V and W, the results being

$$X = \tfrac{1}{3}\pi^2 V, \quad Y = \tfrac{1}{3}\pi^2 W, \quad P = \tfrac{1}{3}\pi^2 \partial V/\partial\zeta, \quad Q = \tfrac{1}{3}\pi^2 \partial W/\partial\zeta. \quad (8\cdot53\cdot10)$$

We now apply the above formulae to discuss the more important special cases.

8.531. *The longitudinal effects.* The most important of the longitudinal effects is the change in the thermal conductivity.

The conditions under which κ, the thermal conductivity, is measured do not seem to have been standardized. No electric current flows, so that $J_x = J_y = 0$ and $\partial T/\partial x \neq 0$, but the thermal conditions in the y direction are a matter of some speculation. In general, it would seem that the experimental conditions are such that, approximately, $w_y = 0$, since the specimen is usually in the form of a long strip whose ends are kept at a constant difference of temperature while no attempt is made to control the temperature of the edges, which are only cooled by radiation. It will therefore be assumed that the correct condition is $w_y = 0$. Then

$$\kappa = \frac{w_x}{-\partial T/\partial x} = \frac{k^2}{3\pi^2\hbar^4} T \frac{(X_1 + X_2)^2 + (Y_1 - Y_2)^2}{X_1 + X_2}. \qquad (8\cdot531\cdot1)$$

Now $X = \tfrac{1}{3}\pi^2 V$, $Y = \tfrac{1}{3}\pi^2 W$, and

$$\sigma = \frac{J_x}{\mathscr{E}_x} = \frac{\epsilon^2}{3\pi^2\hbar^4} \frac{(V_1 + V_2)^2 + (W_1 - W_2)^2}{V_1 + V_2}, \qquad (8\cdot531\cdot2)$$

so that

$$\kappa/\sigma T = \tfrac{1}{3}\pi^2(k/\epsilon)^2 \qquad (8\cdot531\cdot3)$$

for all values of H. This result can only be expected to hold at high temperatures (where the change in resistance is small), and the main interest therefore lies in discussing what happens at low temperatures where a time of relaxation cannot be defined. This problem is dealt with in § 10·514. In the meantime, it may be noted that, on account of (8·523·4) and (8·531·3), κ can be written as

$$\frac{\kappa_0 - \kappa}{\kappa} = \frac{\left(\dfrac{3H\epsilon}{\pi^2 k^2 cT}\right)^2 \left(\dfrac{\kappa_{01}}{n_1} + \dfrac{\kappa_{02}}{n_2}\right)^2 \dfrac{\kappa_{01}\kappa_{02}}{(\kappa_{01}+\kappa_{02})^2}}{1 + \left(\dfrac{3H\epsilon}{\pi^2 k^2 cT}\right)^2 \dfrac{(n_1 - n_2)^2}{n_1^2 n_2^2} \dfrac{\kappa_{01}^2 \kappa_{02}^2}{(\kappa_{01}+\kappa_{02})^2}}. \tag{8·531·4}$$

The calculation of the effect of a magnetic field on the thermo-electric power is straightforward but lengthy. Reference can be made to Sondheimer's paper for the details.

8·532. The transverse effects. In view of the complexity of the results we shall only give the explicit expressions for the various coefficients for the case of a single band in the limit $H = 0$. The calculations are then simple and we find

$$A_E = \frac{3\pi^2 \hbar^4 T}{\epsilon H} \frac{QV - PW}{(V^2 + W^2) X} = \frac{T}{nc\tau} \frac{\partial \tau}{\partial \zeta}, \tag{8·532·1}$$

$$B_{EN}^i = \frac{k^2 T}{\epsilon H} \frac{QV - PW}{V^2 + W^2} = \frac{\pi^2 k^2 T}{3mc} \frac{\partial \tau}{\partial \zeta}, \tag{8·532·2}$$

$$B_{RL} = -Y/(HX) = -\epsilon\tau/mc. \tag{8·532·3}$$

If we introduce instead of B_{EN} the expression $A_{EN} = B_{EN}/\kappa$, the following relations exist between the coefficients:

$$B_{RL} = R\sigma, \tag{8·532·4}$$

$$A_{EN}^i = A_E/T, \tag{8·532·5}$$

and

$$\frac{R A_{EN}^i}{A_E A_{RL}} = \frac{\pi^2}{3} \left(\frac{k}{\epsilon}\right)^2, \tag{8·532·6}$$

where $A_{RL} = B_{RL}/\kappa$. These also apply to a single inverted band. It should be noted that R and B_{RL} are negative or positive according as the conduction is predominantly due to electrons or positive holes. However, B_{EN} and A_E are even functions of ϵ, and so conduction by positive holes does not give rise to a change in sign. There are, however, many metals, including the monovalent metals Cu, Ag, Au, which have negative Ettingshausen and Ettingshausen-Nernst coefficients at high temperatures. This is clearly a more complicated effect than the occurrence of positive Hall and Righi-Leduc coefficients, and some more general model than those so far considered would be necessary for its elucidation. The type of model required is indicated by equations (8·532·1) and (8·532·2), which show that

A_E and B^i_{EN} are proportional to the energy derivative of the time of relaxation, and are therefore negative provided that τ is a decreasing function of the energy at the surface of the Fermi distribution. No model which has this property at high temperatures has so far been proposed; in addition to giving negative values for B^i_{EN} and A_E, it would presumably explain the positive thermoelectric powers of Cu, Ag and Au.

8·533. The generalizations of the above formulae have been given by Sondheimer (1948) for the case of two overlapping bands for all temperatures and fields. Reference should be made to his paper for details (see also, Chambers, 1952a). The only quantities for which simple relations exist are κ, B_{RL} and the 'isothermal' thermoelectric power $\mathfrak{S}^i$. There is an exact correspondence between κ and B_{RL} on the one hand and σ and $R\sigma$ on the other, so that from equations (8·525·3) and (8·525·4) we have

$$\kappa = \frac{(\kappa_1+\kappa_2)^2 + H^2(\kappa_1 B^{(2)}_{RL} + \kappa_2 B^{(1)}_{RL})^2}{\kappa_1 + \kappa_2 + H^2(\kappa_1 B^{(2)}_{RL}{}^2 + \kappa_2 B^{(1)}_{RL}{}^2)} \tag{8·533·1}$$

and

$$B_{RL} = \frac{\kappa_1 B^{(1)}_{RL} + \kappa_2 B^{(2)}_{RL} + H^2 B^{(1)}_{RL} B^{(2)}_{RL} (\kappa_1 B^{(2)}_{RL} + \kappa_2 B^{(1)}_{RL})}{\kappa_1 + \kappa_2 + H^2(\kappa_1 B^{(2)}_{RL}{}^2 + \kappa_2 B^{(1)}_{RL}{}^2)}. \tag{8·533·2}$$

If we have isothermal conditions in the y-direction so that $\partial T/\partial y = 0$, the thermoelectric power is determined by

$$\mathfrak{S}^i = \frac{i^{(1)}_2 - \zeta_1 i^{(1)}_1 - (i^{(2)}_2 - \zeta_1 i^{(2)}_1)}{(i^{(1)}_1 + i^{(2)}_1) T}. \tag{8·533·3}$$

Hence, if $\mathfrak{S}^i_1$ is the value of $\mathfrak{S}^i$ for band 1 alone,

$$i^{(1)}_2 - \zeta_1 i^{(1)}_1 = i^{(1)}_1 T \mathfrak{S}^i_1 = \frac{3\pi^2 \hbar^4}{e^2} \frac{T\sigma_1 \mathfrak{S}^i_1}{1 + (HR_1\sigma_1)^2}, \tag{8·533·4}$$

by (8·525·2), and

$$\mathfrak{S}^i = \frac{\sigma_1 \mathfrak{S}^i_1 + \sigma_2 \mathfrak{S}^i_2 + H^2 \sigma_1 \sigma_2 (\sigma_2 R^2_2 \mathfrak{S}^i_1 + \sigma_1 R^2_1 \mathfrak{S}^i_2)}{\sigma_1 + \sigma_2 + H^2 \sigma_1 \sigma_2 (\sigma_1 R^2_1 + \sigma_2 R^2_2)}. \tag{8·533·5}$$

The quantity which is normally determined experimentally is the 'adiabatic' thermoelectric power $\mathfrak{S}^a$, obtained by imposing the condition $w_y = 0$. The equations for $\mathfrak{S}^a$ are very complicated unless $H = 0$, in which case $\mathfrak{S}^a = \mathfrak{S}^i$.

8·54. *Survey of the experimental data.* The experimental data are both meagre and contradictory, there being large discrepancies between the values reported by different workers. In many cases this is probably due to the presence of ferromagnetic impurities. The data used here have been taken from the articles by Borelius (1935) and Meissner (1935). The experimental results for the Hall coefficient have already been discussed and it therefore only remains to investigate the three relations (8·532·4), (8·532·5) and (8·532·6).

Values of $A_E/(A_{EN}T)$ are given in Table VIII 7. According to equation (8·532·5) this quantity should be unity for all metals. It will be seen that, taking into consideration the experimental error, the agreement is not unreasonable. In Table VIII 7 values of $B_{RL}/(R\sigma)$ are given which, according to (8·532·4), should be unity for all metals. It will be seen that the relation holds as regards order of magnitude.

Table VIII 7. *Values of $A_E/(A_{EN}T)$*

| | $A_E/(A_{EN}T)$ from | | | | $A_E/(A_{EN}T)$ from | |
	Borelius, p. 431	Meissner, p. 383			Borelius, p. 431	Meissner, p. 383
Cu	1·05	0·76		Pd	1·23	0·82
Ag	1·22	1·34		Al	1·43	1·73
Au	0·67	0·67		Bi	0·94	2·88
Zn	1·37	4·25		As	0·98	—
Cd	0·75	7·6		Sb	1·25	0·66

Table VIII 8. *Values of $B_{RL}/(R\sigma)$*

| | $B_{RL}/(R\sigma_0)$ from | |
	Borelius, p. 431	Meissner, p. 383
Cu	0·78	0·65
Ag	0·87	0·46
Au	1·03	0·89
Zn	0·85	0·70
Cd	0·74	1·24
Ir	0·69	—
Pd	0·59	0·62
Pt	1·9	—
Al	0·69	0·49
Bi	2·4	(0·035)
As	—	0·26
Sb	0·39	0·36
W	—	0·69
Mo	—	0·74

Finally, values of $RA_{EN}/(A_E A_{RL} L_n)$, where $L_n = \frac{1}{3}\pi^2(k/\epsilon)^2$, are given in Table VIII 9, and it will be seen that the values are of the order of unity, as predicted.

Table VIII 9. *Values of $RA_{EN}/(A_E A_{RL} L_n)$*

| | $RA_{EN}/(A_E A_{RL} L_n)$ from | | | | $RA_{EN}/(A_E A_{RL} L_n)$ from | |
	Borelius, p. 431	Meissner, p. 383			Borelius, p. 431	Meissner, p. 383
Cu	1·09	1·86		Pd	1·56	1·17
Ag	0·84	1·56		Al	0·98	0·90
Au	1·45	1·52		Bi	—	1·15
Zn	0·78	3·2		As	—	5·7
Cd	1·69	1·06		Sb	—	4·1

GALVANOMAGNETIC EFFECTS IN ANISOTROPIC CONDUCTORS

8·55. It is possible to solve the Boltzmann equation in the general case as follows. If we take the magnetic field along the z-axis and write

$$f = f_0 - \Phi \, \partial f_0 / \partial E,$$

then (8·11·2) becomes

$$\frac{\epsilon H}{\hbar^2 c}\left(\frac{\partial E}{\partial k_2}\frac{\partial \Phi}{\partial k_1} - \frac{\partial E}{\partial k_1}\frac{\partial \Phi}{\partial k_2}\right) - \frac{\Phi}{\tau} = -\frac{1}{\hbar}\mathbf{P}.\operatorname{grad}_k E, \qquad (8·55·1)$$

where

$$\mathbf{P} = -\epsilon \mathscr{E} + T \operatorname{grad}_r \{(E - \zeta)/T\}. \qquad (8·55·2)$$

This is a linear first-order equation for Φ, which can be solved by standard methods. The solution is

$$\Phi = \frac{\hbar c}{\epsilon H}\exp\left(\frac{\hbar^2 c}{\epsilon H}\int\frac{dk_1}{\tau\,\partial E/\partial k_2}\right)\int\left[\frac{\tau\mathbf{P}.\operatorname{grad}_k E}{-\tau\,\partial E/\partial k_2}\exp\left(-\frac{\hbar^2 c}{\epsilon H}\int\frac{dk_1}{\tau\,\partial E/\partial k_2}\right)\right]dk_1, \qquad (8·55·3)$$

where the integrations are taken over the curve given by $E(\mathbf{k}) = \text{constant}$, $k_3 = \text{constant}$, that is, over the curve whose gradient is

$$\frac{dk_2}{dk_1} = -\frac{\partial E/\partial k_1}{\partial E/\partial k_2}. \qquad (8·55·4)$$

The solution (8·55·3) is apparently unsymmetrical in k_1 and k_2, but it can be transformed into a symmetrical form by integrating by parts and using (8·55·4). A single integration by parts gives

$$\Phi = \frac{\tau}{\hbar}\mathbf{P}.\operatorname{grad}_k E$$

$$+ \frac{1}{\hbar}\exp\left(\frac{\hbar^2 c}{\epsilon H}\int\frac{dk_1}{\tau\,\partial E/\partial k_2}\right)\int\left[\frac{\tau\Omega(\tau\mathbf{P}.\operatorname{grad}_k E)}{\tau\,\partial E/\partial k_2}\exp\left(\frac{-\hbar^2 c}{\epsilon H}\int\frac{dk_1}{\tau\,\partial E/\partial k_2}\right)\right]dk_1, \qquad (8·55·5)$$

where

$$\Omega = \frac{\partial E}{\partial k_1}\frac{\partial}{\partial k_2} - \frac{\partial E}{\partial k_2}\frac{\partial}{\partial k_1}, \qquad (8·55·6)$$

and the process can be repeated indefinitely, each of the integrated terms being expressible in terms of Ω.

In principle all the galvanomagnetic and thermomagnetic effects can be obtained from the expression (8·55·3), which can be expanded as an ascending series in H by integrating successively by parts, or as a descending series in $1/H$ by expanding the exponentials. The expressions obtained are, however, purely formal and cannot in general be evaluated explicitly. The only case for which the calculations can be carried out exactly is when the energy is a quadratic function of $\mathbf{k}$. The solution is then more easily obtained from the expression (8·51·3) which is the exact solution of the

Boltzmann equation when the energy surfaces are spheres. For, if the energy is a quadratic function of **k**, a linear transformation in **k** space will reduce the energy surfaces to spheres, so that the case of a general quadratic function can be deduced from the spherical case. Detailed calculations for spheroidal energy surfaces have been carried out by Jones (1936), whose paper may be consulted if the complete formulae are required. The results have the same form as those given in the preceding sections except that the σ^2's occurring in the various formulae have to be replaced by $\sigma_\parallel \sigma_\perp$, where $\sigma_\parallel$ and $\sigma_\perp$ are the conductivities parallel and perpendicular to the axis of symmetry of the crystal.

8·551. The general form of (8·55·1) for an isothermal metal is

$$\frac{1}{\tau}\Phi + \frac{\epsilon}{\hbar}\mathscr{E}.\operatorname{grad} E + \frac{\epsilon}{\hbar^2 c}\mathbf{H}.\Omega\Phi = 0, \tag{8·551·1}$$

where Ω is the operator
$$\Omega = \operatorname{grad} E \times \operatorname{grad}_{\mathbf{k}}. \tag{8·551·2}$$

The solution of (8·551·1) in ascending powers of H, obtained by iteration (Jones and Zener, 1934); is

$$\Phi = -\frac{\epsilon}{\hbar}\left[\tau\mathscr{E}.\operatorname{grad} E - \frac{\epsilon}{\hbar^2 c}\tau\mathbf{H}.\Omega(\tau\mathscr{E}.\operatorname{grad} E)\right.$$
$$\left. + \frac{\epsilon^2}{\hbar^4 c^2}\tau\mathbf{H}.\Omega\{\tau\mathbf{H}.\Omega(\tau\mathscr{E}.\operatorname{grad} E)\} + \dots\right], \tag{8·551·3}$$

and the electric current is

$$\mathbf{J} = \frac{\epsilon}{4\pi^3\hbar}\int\operatorname{grad} E\,\Phi\frac{\partial f_0}{\partial E}\,d\mathbf{k}. \tag{8·551·4}$$

If we neglect terms of order H^3 and higher, then in an isotropic metal the current is of the form

$$\mathbf{J} = \sigma_0\mathscr{E} + \lambda\mathscr{E} \times \mathbf{H} + \mu\mathscr{E}H^2 + \nu\mathbf{H}(\mathscr{E}.\mathbf{H}), \tag{8·551·5}$$

as can be seen from considerations of symmetry or by evaluating (8·551·4). For a cubic metal there is an additional term given by

$$J_x = \xi\mathscr{E}_x H_x^2, \quad J_y = \xi\mathscr{E}_y H_y^2, \quad J_z = \xi\mathscr{E}_z H_z^2, \tag{8·551·6}$$

while for metals with only one axis of symmetry there are further terms. For simplicity we consider cubic metals, and take the coordinate axes along the crystal axes. If the current is $(J_x, 0, 0)$ and the magnetic field is $(H_x, 0, H_z)$, then the electric field is $(\mathscr{E}_x, \mathscr{E}_y, 0)$, where

$$J_x = \{\sigma_0 + \mu H^2 + (\nu + \xi)H_x^2\}\mathscr{E}_x + \lambda\mathscr{E}_y H_z, \quad 0 = -\lambda\mathscr{E}_x H_z + (\sigma_0 + \mu H^2)\mathscr{E}_y.$$

Hence the Hall coefficient and conductivity are given by

$$R = \frac{\mathscr{E}_y}{HJ_x} = \frac{\lambda}{\sigma_0^2}\sin\theta, \tag{8·551·7}$$

and

$$\sigma = \frac{J_x}{\mathscr{E}_x} = \sigma_0\left[1 + \left\{\left(\frac{\mu}{\sigma_0} + \frac{\lambda^2}{\sigma_0^2}\right)\sin^2\theta + \frac{\mu+\nu+\xi}{\sigma_0}\cos^2\theta\right\}H^2\right], \tag{8·551·8}$$

where

$$\mathbf{H} = (\cos\theta, 0, \sin\theta)H.$$

The conductivity constants are easily obtained by substituting (8·551·3) into (8·551·4) and comparing the result, when the simplifications due to the cubic symmetry are taken into account, with (8·551·5) and (8·551·6) (Seitz, 1950). We find

$$\sigma_0 = -\frac{\epsilon^2}{4\pi^3\hbar^2}\int\tau\left(\frac{\partial E}{\partial k_1}\right)^2\frac{\partial f_0}{\partial E}\,d\mathbf{k}, \quad \lambda = \frac{\epsilon^3}{4\pi^3\hbar^4 c}\int\tau\frac{\partial E}{\partial k_1}\,\Omega_3\left(\tau\frac{\partial E}{\partial k_2}\right)\frac{\partial f_0}{\partial E}\,d\mathbf{k},$$

$$\mu = -\frac{\epsilon^4}{4\pi^3\hbar^6 c^2}\int\tau\frac{\partial E}{\partial k_1}\,\Omega_3\left\{\tau\Omega_3\left(\tau\frac{\partial E}{\partial k_1}\right)\right\}\frac{\partial f_0}{\partial E}\,d\mathbf{k},$$

$$\nu = -\frac{\epsilon^4}{4\pi^3\hbar^6 c^2}\int\tau\frac{\partial E}{\partial k_1}\left[\Omega_1\left\{\tau\Omega_2\left(\tau\frac{\partial E}{\partial k_2}\right)\right\} + \Omega_2\left\{\tau\Omega_1\left(\tau\frac{\partial E}{\partial k_1}\right)\right\}\right]\frac{\partial f_0}{\partial E}\,d\mathbf{k},$$

$$\mu+\nu+\xi = -\frac{\epsilon^4}{4\pi^3\hbar^6 c^2}\int\tau\frac{\partial E}{\partial k_1}\,\Omega_1\left\{\tau\Omega_1\left(\tau\frac{\partial E}{\partial k_1}\right)\right\}\frac{\partial f_0}{\partial E}\,d\mathbf{k}. \tag{8·551·9}$$

By integrating by parts, the expressions for μ and $\mu+\nu+\xi$ can be transformed into

$$\mu = \frac{\epsilon^4}{4\pi^3\hbar^6 c^2}\int\tau\left\{\Omega_3\left(\tau\frac{\partial E}{\partial k_1}\right)\right\}^2\frac{\partial f_0}{\partial E}\,d\mathbf{k}, \quad \mu+\nu+\xi = \frac{\epsilon^4}{4\pi^3\hbar^6 c^2}\int\tau\left\{\Omega_1\left(\tau\frac{\partial E}{\partial k_1}\right)\right\}^2\frac{\partial f_0}{\partial E}\,d\mathbf{k}, \tag{8·551·10}$$

which shows that they are both negative since $\partial f_0/\partial E$ is negative. In addition, by Schwarz's inequality, $\lambda^2 \leqslant -\mu\sigma_0$, so that σ is less than σ_0, and the resistance is always increased by the presence of a magnetic field.

8·552. For perfectly free electrons, the integrals are given by

$$\sigma_0 = -\frac{2(2m)^{\frac{1}{2}}\epsilon^2}{3\pi^2\hbar^3}\int_0^\infty\tau E^{\frac{3}{2}}\frac{\partial f_0}{\partial E}\,dE, \quad \lambda = \frac{4\epsilon^3}{3\pi^2\hbar^3(2m)^{\frac{1}{2}}c}\int_0^\infty\tau^2 E^{\frac{3}{2}}\frac{\partial f_0}{\partial E}\,dE,$$

$$\mu = -\nu = \frac{4\epsilon^4}{3\pi^2\hbar^3 m(2m)^{\frac{1}{2}}c^2}\int_0^\infty\tau^3 E^{\frac{3}{2}}\frac{\partial f_0}{\partial E}\,dE, \quad \xi = 0. \tag{8·552·1}$$

and, when the electrons are completely degenerate, $\lambda^2 = -\mu\sigma_0$, $\mu+\nu+\xi = 0$, so that the conductivity is independent of the magnetic field. If, however, E and τ deviate slightly from spherical symmetry, the conductivity constants will be given by the above expressions but with slightly different numerical multipliers. We shall then have $\Delta\rho/\rho = BH^2$, where B depends upon the orientation of H relative to the electric current and where

$$B_\perp = a\epsilon^2\tau(\zeta)^2/m^2c^2, \quad B_\parallel = b\epsilon^2\tau(\zeta)^2/m^2c^2, \tag{8.552.2}$$

a and b, instead of being identically zero, being pure numbers of the order of unity. The values of a and b have been calculated by Davis (1939) for the case in which the departure of E and τ from spherical symmetry is given by the spherical harmonic of order four with cubic symmetry. In the present state of the theory, however, the cubic parts of E and τ cannot be calculated from first principles, and it is best to treat a and b as adjustable parameters. We may write (8·552·2) as

$$B_\perp = a\sigma_0^2/(nec)^2, \quad B_\parallel = b\sigma_0^2/(nec)^2, \tag{8.552.3}$$

so that the present calculations give results of exactly the same form as those of the two-band model discussed in § 8·523, the only difference being that there the coefficients a and b are determined by the model (with $b = 0$), whereas here they are adjustable parameters which cannot be calculated. The discussion in §§ 8·523 and 8·524 (and § 10·5) shows that (8·552·3) gives good agreement with the observations regarding the magnitude and temperature dependence of $B_\perp$ if a is of order unity, and in the present state of the theory we have insufficient knowledge of the energy levels to be able to separate conclusively the isotropic and anisotropic contributions to $B_\perp$, especially as they are probably of the same order of magnitude and have the same temperature variation. On the other hand, $B_\parallel$ is of the same order as, but rather less than, $B_\perp$, and since any isotropic model gives $B_\parallel = 0$ it is necessary to consider anisotropic effects in order to arrive at a non-zero value of $B_\parallel$. The two-band model is, however, the only one which can at present be used if we wish to discuss saturation effects in large fields, so that the theory is still far from complete.

THE ELECTRICAL PROPERTIES OF BISMUTH

8·56. The five valency electrons of bismuth are capable of filling a Brillouin zone exactly, but owing to the comparative smallness of the energy discontinuities at the zone boundaries some of the electrons overlap into the next zone and leave a number of holes in the almost filled band. The number of free electrons is, however, so small that it can be altered appreciably by the addition of suitable impurities. Experiments on the electrical resistance of bismuth to which known amounts of impurities were added have been carried out by Thompson (1936), whose main results are as follows.

The resistance of perfectly pure bismuth increases with the temperature in the normal manner. When small amounts of lead are added, humps appear in the resistance curve, and when there is a sufficient amount of lead present the resistance curve possessses a maximum, the position of which moves to higher temperatures as the lead content increases. The general behaviour is shown in fig. VIII 3. The interpretation of these results is as follows. Since lead has fewer valency electrons than bismuth, the replacement of an atom of bismuth by an atom of lead in the crystal reduces the number of electrons in the upper Brillouin zone. If sufficient lead is added there will be no electrons in the upper Brillouin zone, but there will be some vacant levels in the lower zone. The resistance of such an alloy behaves normally at low temperatures—the resistance increases with the temperature since the free path decreases—but, when the temperature is such that the thermal energy is of the same order as the energy required to raise an electron to the upper zone, the resistance must decrease, since the decrease in the free path will be more than offset by the increase in the number of electrons excited, and of the number of holes

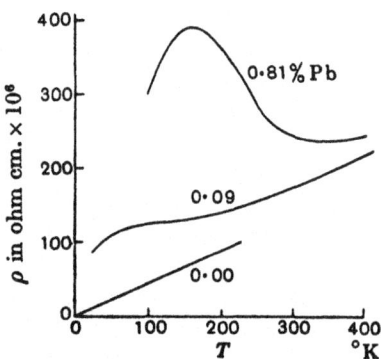

Fig. VIII 3. The specific resistance of bismuth-lead alloys parallel to the principal axis.

produced, as the temperature is raised. When the number of holes is increased by the addition of more lead, the energy required to excite an electron is increased, and thus the excitation of the electrons only becomes apparent at a higher temperature, which is what is observed.

The resistance parallel to the principal axis is influenced much more by impurities than is the resistance perpendicular to the axis. For example, $0 \cdot 1 \%$ of lead is sufficient to produce a maximum in $\rho_{\shortparallel}$, while 1% is required to produce a maximum in $\rho_{\perp}$. Since the average resistance is $\frac{1}{3}\rho_{\shortparallel} + \frac{2}{3}\rho_{\perp}$, about $0 \cdot 7 \%$ of lead would be required to produce a maximum in the resistance of polycrystalline bismuth, though the exact figure depends on the magnitudes of $\rho_{\shortparallel}$ and $\rho_{\perp}$. If we assume that the maximum first appears when there are no electrons in the upper zone, we deduce that the number of electrons per atom in the upper zone of pure bismuth is of the order of 7×10^{-3}. This estimate is, however, probably too large for several reasons. In the first place we have disregarded the anisotropy of bismuth and considered the average resistance. Secondly, it is fairly certain that the addition of lead reduces the number of electrons in the upper zone and increases the number of holes at the same time, whereas we have assumed that the number of holes remains constant until there are no electrons left in the upper zone. Further, tin is about three times as efficacious as lead in affecting the resistance. In view of all these factors, a figure of about 10^{-3} electron per atom is a reasonable estimate for polycrystalline bismuth.

Impurities such as selenium, which have more than five valency electrons per atom, have quite a different effect on the resistance. In these alloys the resistance behaves normally as regards temperature variation, while the first trace of impurity decreases the resistance. This is what we should expect, since the addition of electrons cannot produce the state of affairs which must exist if the resistance as a function of the temperature is to have a maximum, namely that there should be no electrons in the upper zone. When a small amount of selenium is added, the extra electrons must go into the upper zone in order for the conductivity to be increased, while when more selenium is added the holes again begin to be filled up and the conductivity decreases.

8·561. The galvanomagnetic effects in bismuth are exceptionally large and are easily measurable at room temperatures. They have therefore attracted a great deal of attention from both the experimental and the theoretical sides. The large magnitude of the effects is almost entirely due to the number of free electrons and positive holes being small.

Jones (1936) has made detailed calculations of the Hall and magneto-resistance effects based on the solution of the Boltzmann equation discussed in § 8·55, it being assumed that the energy surfaces are spheroids. For an account of the anisotropic effects the reader is referred to Jones's paper, but for a semi-quantitative survey it is sufficient to use the formulae for isotropic metals and to relate them to the data for polycrystalline bismuth.

Since the Brillouin zone for bismuth can just accommodate five electrons per atom we must have $n_1 = n_2$. In this case τ_1/m_1 and τ_2/m_2 can be found from the expressions

$$R\sigma = -\frac{\epsilon}{c}\left(\frac{\tau_1}{m_1} - \frac{\tau_2}{m_2}\right), \quad B = \frac{\epsilon^2}{c^2}\frac{\tau_1\tau_2}{m_1 m_2}, \qquad (8\cdot561\cdot1)$$

where B is the coefficient in the relation $\Delta\rho/\rho = BH^2$. The experimental values at $0°$ C. for polycrystalline bismuth are $R\sigma = -5\cdot4 \times 10^{-5}$ e.s.u. and $B = 1\cdot3 \times 10^{-9}$. Then the equations (8·561·1) give $\tau_1/m_1 = 4\cdot5 \times 10^{15}$ and $\tau_2/m_2 = 1\cdot1 \times 10^{15}$. Further, since $\sigma = 8\cdot3 \times 10^{15}$ e.s.u., (8·21·5) gives $n = 7 \times 10^{18}$. Thus the present calculations give the number of conduction electrons as $2\cdot5 \times 10^{-4}$ per atom, whereas the behaviour of the alloys of bismuth indicates that the number is considerably larger than this.

We cannot determine the effective masses without some independent evidence about the time of relaxation. If we assume that τ_1 and τ_2 are both about the same as τ for rubidium (we choose rubidium since the Debye Θ is of the same order of magnitude for rubidium and bismuth), we find $m_1/m = 6.7 \times 10^{-3}$ and $m_2/m = 2.7 \times 10^{-2}$. These values of n, m_1 and m_2 are of the same order of magnitude as those required to explain the normal diamagnetism of bismuth.

The corresponding data and results for arsenic and antimony are as follows. Arsenic,

$$R\sigma = +1.3 \times 10^{-6} \, \text{e.s.u.}; \quad B = 1.6 \times 10^{-11}; \quad \tau_1/m_1 = 2 \times 10^{14}, \ \tau_2/m_2 = 3 \times 10^{14}.$$

Antimony,

$$R\sigma = +5 \times 10^{-6} \, \text{e.s.u.}; \quad B = 9 \times 10^{-11}; \quad \tau_1/m_1 = 4 \times 10^{14}, \ \tau_2/m_2 = 8 \times 10^{14}.$$

The very small effective masses required to explain the galvanomagnetic effects and the still smaller values required to account for the diamagnetic susceptibility lead to very large curvatures of the energy surfaces over a considerable energy range. There is some theoretical justification, based upon the perturbation calculations of § 2·51, for believing that large curvatures can occur near an energy discontinuity, but no numerical calculations have been made. It should be noted that the different methods of determining the various parameters give widely differing results, and that it is impossible to reconcile all the results in the present state of the theory. This is perhaps best shown by the following comparison.

The values of the parameters given above for bismuth fit the observed values of the galvanomagnetic effects and the normal diamagnetism reasonably well. They give surprisingly large values for the free paths. (The free path l for a single band is defined to be $l = \tau \bar{v}$, where $m^* \bar{v} = \frac{1}{2} h (3n/\pi)^{\frac{1}{3}}$, and l_1 and l_2 come out to be $3 \times 10^{-5} \, \text{cm}$. and $7 \times 10^{-6} \, \text{cm}$. respectively.) But other evidence suggests that l_1 and l_2 are considerably larger than these values. In the first place, the theory of the mechanism of conductivity shows that only the longer lattice vibrations can scatter the electrons in a semi-metal (§ 9·351) and therefore that τ_1 and τ_2 for bismuth should be much greater than τ for rubidium. Although the other predictions of equation (9·351·1), on which this result is based, do not appear to be verified, nevertheless large free paths seem to be demanded by other phenomena. The magnitudes of the conductivity of thin bismuth wires (Eucken and Förster, 1934; Justi, Kohler and Lautz, 1951) and of the anomalous skin effect in bismuth (the penetration of an electromagnetic field into a metal when the free path is longer than the classical skin depth, see § 8·74) can both be explained if it is assumed that l_1 and l_2 are of the order of 10^{-4} cm. and that n is of the order of 3×10^{17} per c.c. (Chambers and Pippard, 1952). This gives the order of magnitude of the parameters as $n/n_a = 10^{-5}$, $\tau_1/m_1 = \tau_2/m_2 = 5 \times 10^{16}$. This value of n/n_a agrees with that deduced from the de Haas-van Alphen effect, but it gives much too large values for the Hall coefficient and the magneto-resistance effect.

CONDUCTION IN SEMI-CONDUCTORS

8·6. All the general formulae so far given apply equally to semi-conductors, the only difference being that the approximations used in the final evaluation of the integrals for a degenerate gas no longer apply to electrons with a Maxwellian distribution. We now give the most important results for semi-conductors, omitting all reference to purely thermal effects since these are always masked by the thermal properties of the lattice, the effect of which predominates over that of the few free electrons. In order to proceed any distance it is necessary to know how the time of relaxation depends upon the velocity, and we shall usually assume that the free path $l = \tau v$ is independent of the velocity (see § 9·36). The theory can easily be extended to deal with any given energy variation of the free path.

It is possible to carry out the calculations from the start in terms of the Maxwell function or to consider them as particular cases of the general formulae given here. For most purposes either method has little advantage over the other, but for the calculation of the thermoelectric effects it is simpler to start from the Fermi function since ζ occurs explicitly in the final result, and, when two conductors are in equilibrium with one another, ζ is the same for both. For problems involving two conductors it is not sufficient to know the variable part of ζ, and the absolute value of ζ is not deducible directly from the constants in the Maxwell function. It is convenient to use the velocity $\mathbf{v}$ instead of the wave vector $\mathbf{k}$. The equilibrium distribution function is then

$$n\left(\frac{m}{2\pi kT}\right)^{\frac{3}{2}} e^{-\frac{1}{2}mv^2/kT}\, du\, dv\, dw, \tag{8·6·1}$$

and this must be the same as

$$\frac{1}{4\pi^3} \frac{dk_1 dk_2 dk_3}{e^{(E-\zeta)/kT} + 1} = 2\left(\frac{m}{h}\right)^3 \frac{du\, dv\, dw}{e^{(E-\zeta)/kT} + 1}.$$

Hence
$$e^{\zeta/kT} = \tfrac{1}{2} n h^3 (2\pi m kT)^{-\frac{3}{2}}. \tag{8·6·2}$$

8·61. When the only fields are an electric field and temperature gradient along the x direction, the current densities are given by (8·3·3) and (8·3·4)

with
$$\mathscr{K}_s = \tfrac{4}{3}\pi \frac{nl}{kT}\left(\frac{m}{2\pi kT}\right)^{\frac{3}{2}} \int_0^\infty E^{s-1} e^{-\frac{1}{2}mv^2/kT}\, v^3\, dv = \frac{4}{3}\frac{nl}{(2\pi m)^{\frac{1}{2}}} s!(kT)^{s-\frac{3}{2}}.$$
$$\tag{8·61·1}$$

The electrical conductivity is therefore

$$\sigma = \epsilon^2 \mathscr{K}_1 = \frac{4}{3} \frac{n\epsilon^2 l}{(2\pi m k T)^{\frac{1}{2}}}, \tag{8·61·2}$$

and the thermal conductivity is

$$\kappa = \frac{\mathscr{K}_1 \mathscr{K}_3 - \mathscr{K}_2^2}{\mathscr{K}_1 T} = \tfrac{4}{3} n l \left(\frac{2k^3 T}{\pi m}\right)^{\frac{1}{2}},$$

as already found in § 1·5.

The thermoelectric effects are determined by $\mathfrak{S}$, where

$$\frac{d\Theta}{dT} = -\frac{1}{\epsilon} \mathfrak{S} = -\frac{1}{\epsilon} \left(\frac{\mathscr{K}_2}{\mathscr{K}_1 T} - \frac{\zeta}{T}\right). \tag{8·61·3}$$

If ζ and $\mathscr{K}_s$ are given by (8·6·2) and (8·61·1), $\mathfrak{S}$ is

$$\mathfrak{S} = 2k - \zeta/T = 2k - k \log n + \tfrac{3}{2} k \log T + a, \tag{8·61·4}$$

where

$$a = -k \log \{\tfrac{1}{2} h^3 (2\pi m k)^{-\frac{3}{2}}\}. \tag{8·61·5}$$

Now unless $e^{\zeta/kT}$ is much less than unity, classical statistics do not hold. In semi-conductors, therefore, $-\zeta/T$ is positive and of the order of k at least, so that $\mathfrak{S}$ will in general lie between $3k$ and $10k$ (the upper limit depending upon how small a conductivity is observable). This gives a value of some tenths of a millivolt per degree (or even of a millivolt or so) for the thermoelectric power of a semi-conductor, whereas the thermoelectric powers of metals are measured in microvolts per degree.

The above result can easily be generalized to the case in which τ is proportional to v^p. The result is that

$$\mathfrak{S} = (\tfrac{1}{2} p + \tfrac{5}{2} - \xi) k \quad (\xi = \zeta/kT, \ \xi \sim -\infty), \tag{8·61·6}$$

which can be compared with the value for a degenerate gas of

$$\mathfrak{S} = (\tfrac{1}{6} p + \tfrac{1}{3}) \pi^2 k/\xi \quad (\xi \sim \infty). \tag{8·61·7}$$

The way in which $\mathfrak{S}/k$ passes over from (8·61·6) to (8·61·7) as ξ ranges from $-\infty$ to ∞ is shown by the curves in fig. VIII 4 for the extreme values of p.

A particularly important special case is that of a thermoelectric couple formed by two specimens of the same substance but with different impurity contents. We then have

$$\frac{d\Theta_{12}}{dT} = -\frac{1}{\epsilon}(\mathfrak{S}_1 - \mathfrak{S}_2) = \frac{1}{\epsilon}\frac{\zeta_1 - \zeta_2}{T} = \frac{k}{\epsilon} \log \frac{n_1}{n_2}. \tag{8·61·8}$$

For deficit semi-conductors in which the current is carried by positive holes, the sign of ϵ must be reversed in equations (8·61·3) and (8·61·8).

8·611. If we wish to generalize the preceding results to include cases in which there are both electrons and holes present we may start from equations (8·53·1)–(8·53·4) or apply the general formula (8·533·5).

To calculate the thermoelectric power from first principles it is only necessary to consider equation (8·53·1) which, in the absence of a magnetic field, can be written as

$$J_x = \frac{\epsilon}{3\pi^2\hbar^4}\left[(i_1^{(1)}+i_1^{(2)})\left(\epsilon\mathscr{E}_x + \frac{\partial\zeta_1}{\partial x}\right) + \{i_2^{(1)} - \zeta_1 i_1^{(1)} - (i_2^{(2)} - \zeta_2 i_1^{(2)})\}\frac{1}{T}\frac{\partial T}{\partial x}\right], \quad (8\cdot611\cdot1)$$

since $\zeta_1 + \zeta_2$ is constant, being equal to the separation $-\Delta E$ of the bands. (Since $E = 0$ corresponds to the bottom of band 1, ζ and ζ_1 are the same, while

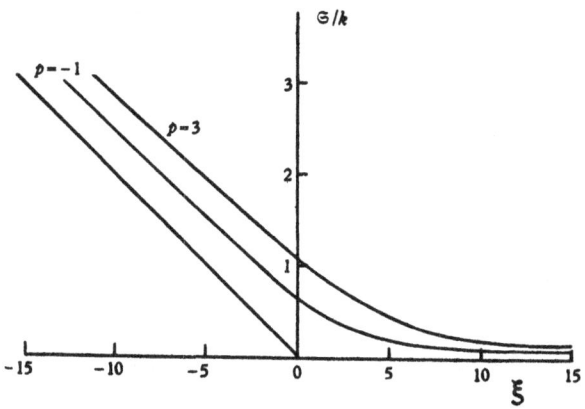

Fig. VIII 4. $\mathscr{G}/k$ as a function of ζ for two different values of the scattering coefficient.

$-\zeta_2$ gives the height of the energy level ζ above the top of band 2.) Now for electrons in a nearly empty band $f_0(E) = e^{-(E-\zeta_1)/kT}$ approximately, and for holes in a nearly empty band $f_0(E) = 1 - e^{(E-\zeta_1)/kT} = 1 - f_0^+(E')$ approximately, i.e. $f_0^+(E') = e^{-(E'-\zeta_2)/kT}$ approximately, where $E' = -\Delta E - E$, $\zeta_2 = -\Delta E - \zeta_1$. We therefore have

$$i_s^{(1)} = \frac{(2m_1)^{\frac{3}{2}}}{kT}\frac{\hbar}{m_1}\int_0^\infty \tau_1 E^{s+\frac{1}{2}} e^{-(E-\zeta_1)/kT}\,dE = s!\,2\hbar(kT)^s l_1 m_1 e^{\zeta_1/kT}$$

$$(8\cdot611\cdot2$$

and
$$i_s^{(2)} = s!\,2\hbar(kT)^s l_2 m_2 e^{\zeta_2/kT}. \quad (8\cdot611\cdot3)$$

On open circuit we have $J_x = 0$, and the electric field set up is given, as usual, by

$$-\epsilon\mathscr{E}_x = \frac{\partial\zeta_1}{\partial x} + \mathscr{G}\frac{\partial T}{\partial x},$$

where
$$\mathscr{G} = \frac{i_2^{(1)} - i_2^{(2)}}{(i_1^{(1)} + i_1^{(2)})T} - \frac{\zeta_1 i_1^{(1)} - \zeta_2 i_1^{(2)}}{(i_1^{(1)} + i_1^{(2)})T}. \quad (8\cdot611\cdot4)$$

If we now insert the expressions (8·611·2) and (8·611·3) for $i_s^{(1)}$ and $i_s^{(2)}$ and use n_1 and n_2 instead of $e^{\zeta_1/kT}$ and $e^{\zeta_2/kT}$ (equation (8·6·2)), we obtain the result

$$\mathfrak{S} = \frac{n_1 l_1/m_1^{\frac{3}{2}} - n_2 l_2/m_2^{\frac{3}{2}}}{n_1 l_1/m_1^{\frac{3}{2}} + n_2 l_2/m_2^{\frac{3}{2}}} 2k - \frac{\zeta_1 n_1 l_1/m_1^{\frac{3}{2}} - \zeta_2 n_2 l_2/m_2^{\frac{3}{2}}}{n_1 l_1/m_1^{\frac{3}{2}} + n_2 l_2/m_2^{\frac{3}{2}}} \frac{1}{T}. \qquad (8·611·5)$$

It should be noted that n_1 and n_2, the numbers of free electrons and holes per unit volume, are not independent quantities, but are determined by the temperature, the number of impurities and the band structure according to the principles detailed in § 5·2. If either n_2 or n_1 is zero, $\mathfrak{S}$ reduces to $2k - \zeta_1/T$ or $-(2k - \zeta_2/T)$ respectively, but in the general case the temperature variation of $\mathfrak{S}$ is complicated and it is possible for $\mathfrak{S}$ to change sign. The generalized formula for the electrical conductivity is

$$\sigma = \frac{4}{3} \frac{\epsilon^2}{(2\pi kT)^{\frac{1}{2}}} \left(\frac{n_1 l_1}{m_1^{\frac{1}{2}}} + \frac{n_2 l_2}{m_2^{\frac{1}{2}}} \right), \qquad (8·611·6)$$

as is seen by substituting for $i_1^{(1)}$ and $i_1^{(2)}$ in the expression

$$\sigma = \epsilon^2 (i_1^{(1)} + i_1^{(2)})/(3\pi^2 \hbar^4).$$

8·62. The galvanomagnetic effects are relatively simple for semi-conductors and are of considerable importance in determining the nature of the conductivity. Since both electrons and 'positive holes' can occur when the conduction is intrinsic we shall derive the expression for the Hall coefficient for the case in which carriers of both types are present. For arbitrary magnetic fields R is given by (8·521·2), i.e. by

$$R = -\frac{3\pi^2 \hbar^4}{H\epsilon^2} \frac{i_4^{(1)} - i_4^{(2)}}{(i_1^{(1)} + i_1^{(2)})^2 + (i_4^{(1)} - i_4^{(2)})^2}, \qquad (8·62·1)$$

where, by (8·52·2) and (8·6·2),

$$i_1 = \frac{\sqrt{2}(\pi m)^{\frac{3}{2}}}{(kT)^{\frac{5}{2}}} nl\hbar^4 \int_0^\infty e^{-\frac{1}{2}mv^2/kT} \frac{v^5}{v^2 + \beta^2 l^2} dv, \qquad (8·62·2)$$

$$i_4 = \frac{\sqrt{2}(\pi m)^{\frac{3}{2}}}{(kT)^{\frac{5}{2}}} nl^2\hbar^4 \beta \int_0^\infty e^{-\frac{1}{2}mv^2/kT} \frac{v^4}{v^2 + \beta^2 l^2} dv \qquad (8·62·3)$$

and $\beta = \epsilon H/mc$. For small values of H we have

$$i_1 = \frac{(2\pi)^{\frac{3}{2}} nl\hbar^4}{(mkT)^{\frac{1}{2}}}, \quad i_4 = \frac{\pi^2 nl^2\hbar^4 \beta}{kT}, \qquad (8·62·4)$$

and so

$$R = -\frac{3\pi}{8\epsilon c} \frac{n_1 l_1^2/m_1 - n_2 l_2^2/m_2}{(n_1 l_1/m_1^{\frac{1}{2}} + n_2 l_2/m_2^{\frac{1}{2}})^2}, \qquad (8·62·5)$$

which could be derived as a special case of (8·525·4). If either n_1 or n_2 is zero, this reduces to the familiar form

$$R = \pm 3\pi/(8n\epsilon c). \qquad (8·62·6)$$

8·63. In semi-conductors, unlike metals, there is a change in resistance due to a magnetic field when the electrons occupy one band only. This is because the electrons cannot be considered to have the same velocity even in the zero approximation, and hence the Hall field does not compensate entirely the transverse motion of the individual electrons. For simplicity, therefore, we shall only consider free electrons in one energy band (Harding, 1933). According to (8·523·1) we then have

$$\sigma = \frac{e^2}{3\pi^2\hbar^4} \frac{i_1^2 + i_4^2}{i_1}.$$ (8·63·1)

To evaluate this to the first order in H^2 we must obtain the second approximation to i_1, which we do by writing

$$\frac{v^5}{v^2 + \beta^2 l^2} = v^3 - \beta^2 l^2 v + O(\beta^4 l^4).$$

Then
$$i_1 = \frac{(2\pi)^{\frac{1}{2}} n l \hbar^4}{(mkT)^{\frac{1}{2}}} \left(1 - \frac{m\beta^2 l^2}{2kT}\right),$$ (8·63·2)

and so
$$\sigma = \sigma_0 \left\{1 - \left(\frac{1}{2} - \frac{\pi}{8}\right) \frac{m\beta^2 l^2}{kT}\right\} = \sigma_0 \left\{1 - 0·38 \left(\frac{H\sigma_0}{nec}\right)^2\right\}.$$ (8·63·3)

For strong magnetic fields the integrals must be evaluated exactly. They can all be reduced to one of the following types:

$$\int_0^\infty v^n e^{-\frac{1}{2}mv^2/kT} dv = \frac{1}{2}\Gamma(\frac{1}{2}n + \frac{1}{2})(2kT/m)^{\frac{1}{2}n + \frac{1}{2}},$$

$$\int_0^\infty \frac{v}{v^2 + \beta^2 l^2} e^{-\frac{1}{2}mv^2/kT} dv = \frac{1}{2}e^{w^2}\int_{w^2}^\infty \frac{e^{-t}}{t} dt = \frac{1}{2}e^{w^2} E_1(w^2),$$

and
$$I = \int_0^\infty \frac{e^{-\frac{1}{2}mv^2/kT}}{v^2 + \beta^2 l^2} dv,$$

where
$$w^2 = \frac{1}{2}m\beta^2 l^2/kT \quad \text{and} \quad E_1(x) = \int_x^\infty \frac{e^{-t}}{t} dt.$$

To obtain the integral I in a convenient form we find the differential equation satisfied by $I(x)$, where

$$I(x) = \int_0^\infty \frac{e^{-xv^2}}{v^2 + \beta^2 l^2} dv.$$

It is
$$\frac{dI}{dx} - \beta^2 l^2 I = -\frac{1}{2}\left(\frac{\pi}{x}\right)^{\frac{1}{2}},$$

of which the solution is
$$I(x) = \frac{\sqrt{\pi}}{\beta l} e^{\beta^2 l^2 x} \int_{\beta l x^{\frac{1}{2}}}^{\gamma} e^{-t^2} dt.$$

We can take βl to be positive, and since $I(\infty) = 0$ the upper limit γ must be ∞, so that

$$I\left(\frac{m}{2kT}\right) = \frac{\sqrt{\pi}}{\beta l} e^{w^2} \int_w^\infty e^{-t^2} dt = \frac{\pi}{2\beta l} e^{w^2} F(w),$$

where $1 - F(w)$ is the error function.

Introducing these integrals into the expressions for i_1 and i_4, we find, after a little manipulation,

$$\frac{\sigma_0 - \sigma}{\sigma} = \frac{1 - w^2 + w^4 e^{w^2} E_1(w^2)}{\{1 - w^2 + w^4 e^{w^2} E_1(w^2)\}^2 + \pi w^2 \{\frac{1}{2} - w^2 + \sqrt{\pi} w^3 e^{w^2} F(w)\}^2} - 1 \tag{8·63·4}$$

and

$$\frac{R}{R_0} = \frac{1 - 2w^2 + 2\sqrt{\pi} w^3 e^{w^2} F(w)}{1 - w^2 + w^4 e^{w^2} E_1(w^2)} \frac{\sigma_0}{\sigma}, \tag{8·63·5}$$

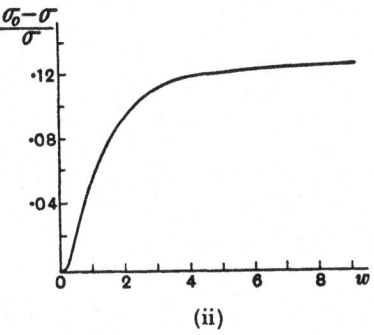

Fig. VIII 5. The change in electrical resistance in a magnetic field and the Hall coefficient of a semi-conductor in which the free path is independent of the energy. $w = \frac{3}{4}\sqrt{\pi} H\sigma_0/(nec)$.

where R_0 is the Hall coefficient for vanishing magnetic field. In Fig. VIII 5, $\Delta\sigma/\sigma$ and R/R_0 are plotted against $w = \frac{3}{4}\sqrt{\pi} H\sigma_0/(nec)$, and both curves show the phenomenon of saturation. The limiting values can be found by inserting the asymptotic expansions of $E_1(w^2)$ and $F(w)$, and are

$$\lim_{H\to\infty} \frac{\sigma_0 - \sigma}{\sigma} = \frac{32}{9\pi} - 1 = 0\cdot132, \quad \lim_{H\to\infty} \frac{R}{R_0} = \frac{8}{3\pi} = 0\cdot849. \tag{8·63·6}$$

They can also be obtained more easily directly, by evaluating i_1 and i_4 for the limit of $H = \infty$.

8·631. The occurrence of a non-zero change in resistance in the preceding calculations is due to the electron gas being non-degenerate. If, therefore, we consider the more general case of an arbitrary electron concentration characterized by the parameter $\xi = \zeta/kT$, $(\sigma_0 - \sigma)/\sigma$ must decrease from the value (8·63·4) to zero as ξ ranges from $-\infty$ to ∞. Similarly, R/R_0 must

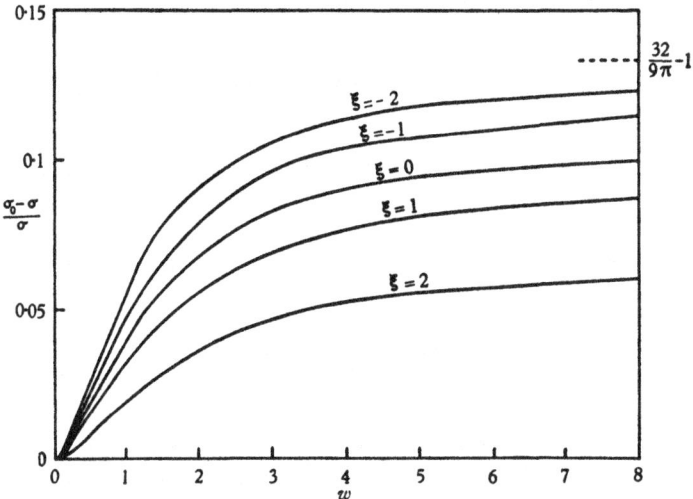

Fig. VIII 6. The change in resistance in a magnetic field for different degrees of degeneracy (scattering index $p = -1$).

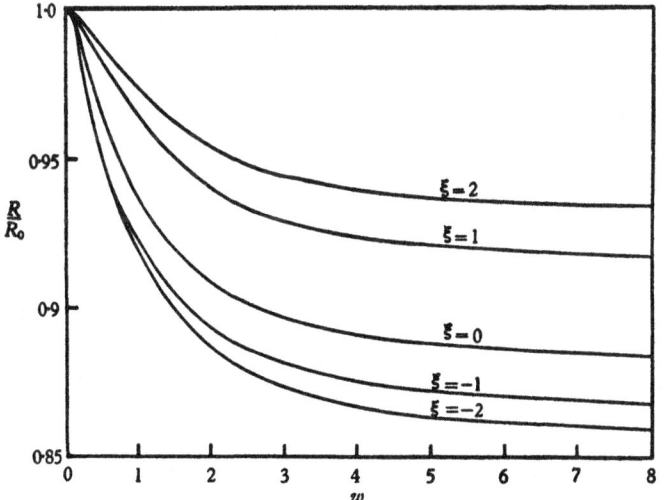

Fig. VIII 7. The Hall coefficient for different degrees of degeneracy (scattering index $p = -1$).

increase from the value (8·63·5) to unity. The general expressions are easily evaluated numerically and are shown in figs. VIII 6 and VIII 7 for five values of ξ.

As a further generalization we can consider the case in which τ is proportional to v^p. The integrals are easily evaluated in the limiting cases

$H = 0$ and $H = \infty$, the results being

$$R_0 = -\frac{3\sqrt{\pi}\,\Gamma(\tfrac{5}{2}+p)}{4[\Gamma(\tfrac{5}{2}+\tfrac{1}{2}p)]^2}\frac{1}{nec} \quad (\xi = \zeta/kT \sim -\infty), \qquad (8\cdot631\cdot1)$$

$$\frac{R}{R_0} = 1 - \left[\frac{\Gamma(\tfrac{5}{2}+2p)}{\Gamma(\tfrac{5}{2}+p)} + \frac{[\Gamma(\tfrac{5}{2}+p)]^2}{[\Gamma(\tfrac{5}{2}+\tfrac{1}{2}p)]^2} - 2\frac{\Gamma(\tfrac{5}{2}+\tfrac{3}{2}p)}{\Gamma(\tfrac{5}{2}+\tfrac{1}{2}p)}\right]w^2 + \dots \quad (\xi \sim -\infty),$$
$$(8\cdot631\cdot2)$$

$$R_{H=\infty} = -1/(nec) \quad (\text{all } \xi), \qquad (8\cdot631\cdot3)$$

$$\frac{\sigma_0}{\sigma} = 1 + \left[\frac{\Gamma(\tfrac{5}{2}+\tfrac{3}{2}p)}{\Gamma(\tfrac{5}{2}+\tfrac{1}{2}p)} - \frac{[\Gamma(\tfrac{5}{2}+p)]^2}{[\Gamma(\tfrac{5}{2}+\tfrac{1}{2}p)]^2}\right]w^2 + \dots \quad (\xi \sim -\infty), \qquad (8\cdot631\cdot4)$$

$$\frac{\sigma_0}{\sigma_{H=\infty}} = \frac{\Gamma(\tfrac{5}{2}+\tfrac{1}{2}p)\,\Gamma(\tfrac{5}{2}-\tfrac{1}{2}p)}{[\Gamma(\tfrac{5}{2})]^2} \quad (\xi \sim -\infty), \qquad (8\cdot631\cdot5)$$

where $$w = \frac{3\sqrt{\pi}}{4\Gamma(\tfrac{5}{2}+\tfrac{1}{2}p)}\frac{H\sigma_0}{nec} = -\frac{\Gamma(\tfrac{5}{2}+\tfrac{1}{2}p)}{\Gamma(\tfrac{5}{2}+p)}H\sigma_0 R_0. \qquad (8\cdot631\cdot6)$$

As an alternative to (8·631·4) we may write

$$\frac{\sigma_0}{\sigma} = 1 + \left[\frac{[\Gamma(\tfrac{5}{2}+\tfrac{1}{2}p)]^2\,\Gamma(\tfrac{5}{2}+\tfrac{3}{2}p)}{[\Gamma(\tfrac{5}{2}+p)]^2\,\Gamma(\tfrac{5}{2}+\tfrac{1}{2}p)} - 1\right]H^2\sigma_0^2 R_0^2 + \dots, \qquad (8\cdot631\cdot7)$$

which is probably the simplest form, since σ_0 and R_0 are directly measurable quantities. The corresponding results for a completely degenerate gas are $R_H = -1/(nec)$ and $\sigma_0/\sigma_H = 1$ for all H. For intermediate values of ζ and H the results must be obtained in numerical form, some typical curves being given in figs. VIII 8, 9, 10 and 11.

8·64. It is found that the magneto-resistance effect in germanium can be much larger than that given by (8·63·4). There are various generalizations of the theory that can be invoked to obtain an increased magneto-resistance effect.

In the first place we can consider a two-band model, and in this case σ is given by (8·525·3) just as for metals, but $\sigma_1, \sigma_2, R_1, R_2$ are no longer independent of H. The resulting formula for σ gives a magneto-resistance effect which is essentially the same as that for metals in which the electrons occupy two bands. It can hardly be of importance for extrinsic semi-conductors, but would have to be taken into account when dealing with intrinsic semi-conductors, since electrons and holes are then both present.

The second generalization to be considered arises from the energy variation of the free path. The simplest theory is based upon the free path being independent of the energy, but a consideration of the various possible scattering mechanisms (§ 9·361) shows that this is not necessarily true.

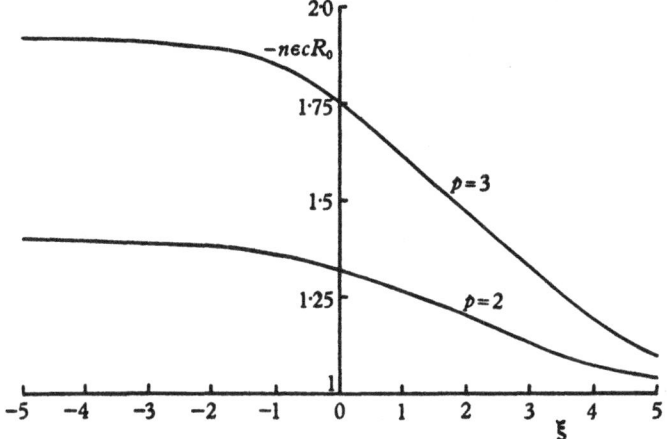

Fig. VIII 8. The Hall coefficient R_0 for $H = 0$ as a function of ξ for different values of the scattering index.

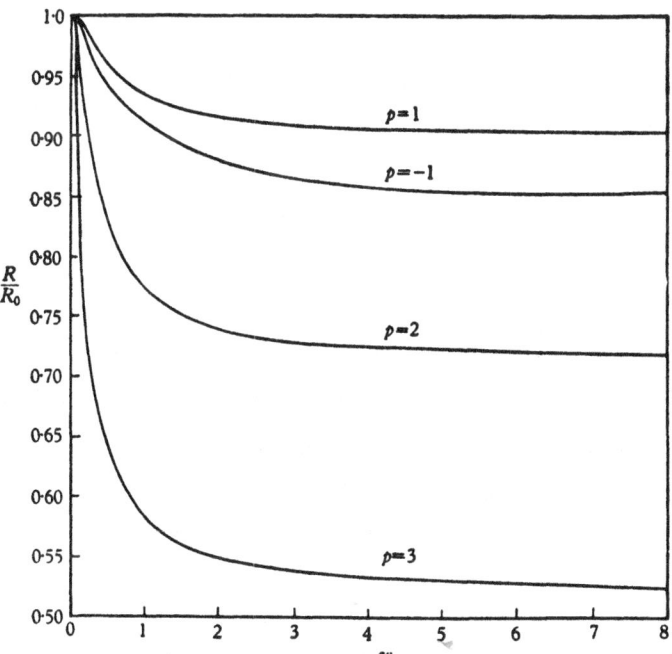

Fig. VIII 9. The Hall coefficient R for a non-degenerate electron gas as a function of H for different values of the scattering index.

At low temperatures, in particular, where the most important scatterers are the ionized atoms from which the free electrons are derived, the free path varies as E^2. In general, the energy variation of l is complicated, and we therefore put $\tau = \mu v^p$ and consider p as an adjustable parameter, chosen

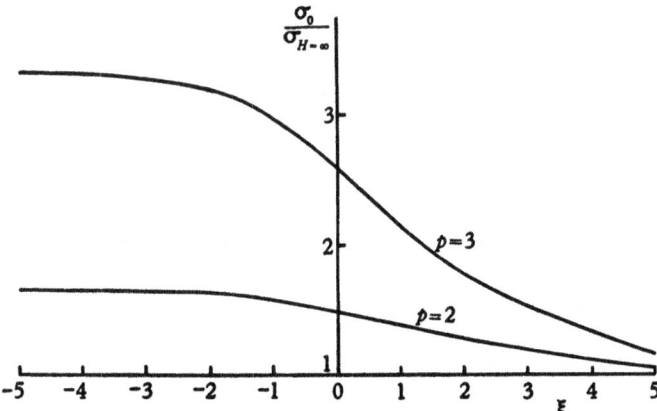

Fig. VIII 10. $\lim_{H \to \infty} \sigma_0/\sigma$ as a function of ξ for different values of the scattering index.

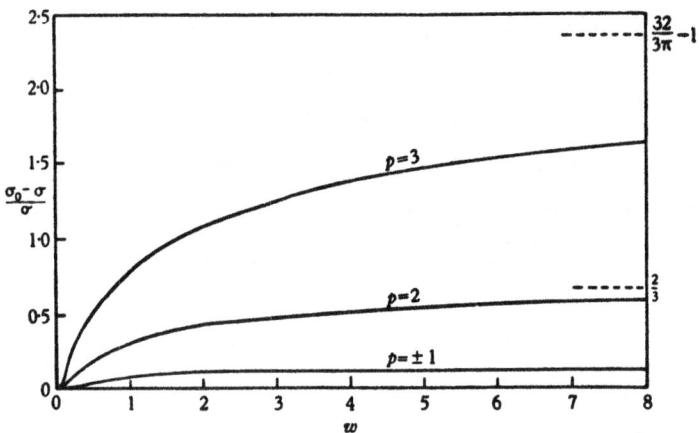

Fig. VIII 11. $(\sigma_0 - \sigma)/\sigma$ as a function of H for a non-degenerate electron gas for different values of the scattering index.

at any particular temperature to give the best representation of τ. The limiting values of p are -1 (scattering by the acoustical lattice vibrations) and 3 (coulomb scattering by ionized atoms). We can then apply the theory given in § 8·631, and it will be seen that the magneto-resistance effect can be extremely large if p is in the neighbourhood of 3.

The third generalization that must be considered is the inclusion of anisotropic effects. For this purpose we can use the solution of the Boltzmann equation given in § 8·551, the only difference being that the electrons cannot be treated as completely degenerate. If we treat τ and E as spherically symmetrical we regain the formulae of the preceding sections, but if we consider them to have only cubic symmetry, we shall obtain

$$\sigma_{\perp} = \sigma_0 \{1 - a(H\sigma_0/n\epsilon c)^2 + ...\} \tag{8·64·1}$$

instead of (8·63·3), and

$$\sigma_{\shortparallel} = \sigma_0 \{1 - b(H\sigma_0/n\epsilon c)^2 + ...\} \tag{8·64·2}$$

instead of $\sigma_{\shortparallel} = \sigma_0$, where a and b are pure numbers which can be treated as adjustable parameters.

Seitz (1950) has calculated a and b when there is a slight departure from spherical symmetry, and his results have been used by Estermann and Foner (1950) and by Pearson and Suhl (1951) to analyse their results on the magneto-resistance of germanium. In their experiments and in those of Dunlap (1950), specimens of various purities were used, some of which were 'positively' and others 'negatively' conducting. The conductivities ranged from about 10^{11} to 10^{14} e.s.u., the Hall coefficients from 10^{-16} to 10^{-19} Gaussian units and the temperatures from 20 to 300° K. It was found that for sufficiently small fields the conductivity was reasonably well given by (8·64·1) and (8·64·2) with a between 1 and 2, and b sometimes of the same order as a but sometimes only one-tenth of a. The values of a are somewhat larger than those which are possible for an isotropic medium (0·55 for $p = 3$), but the evidence is not conclusive since the H^2 region in the resistance-magnetic field curves is comparatively small and the resistance shows signs of saturation (at fields of the order of 10^4 gauss at temperatures lower than 100° K.). The saturation value of the resistance appears to be in the neighbourhood of $2\rho_0$, which could be explained by choosing $p = 3$, as shown by the curves of figs. VIII 10 and 11.

On the other hand, the existence of a large longitudinal effect seems to be decisively in favour of the anisotropy of the crystal being the dominant factor. Pearson and Suhl found that while the longitudinal effect was considerably smaller than the transverse one for positively conducting germanium, the opposite was true for negatively conducting germanium when the current was in the (1, 0, 0) direction. At 77° K. they found $\Delta\rho_{\shortparallel}/\rho_0 = 2\cdot2$ and $\Delta\rho_{\perp}/\rho_0 = 1\cdot1$ for $H = 100,000$ gauss. The magneto-resistance constants determined by Pearson and Suhl are given in Table VIII 10 (for the definitions see (8·551·5) and (8·551·6)). The constants have been chosen so as to give the best general picture, since it is impossible to fit all the observations by any one set of constants.

Table VIII 10. *Magneto-resistance constants of germanium*

	Positively conducting		Negatively conducting	
	300° K.	77° K.	300° K.	77° K.
$\sigma_0 \times 10^{-11}$	23	40	0·8	5
$R \times 10^{18}$	9	5	− 5	− 5·4
$\lambda \times 10^{-6}$	51	77	− 30	− 15
$\mu \times 10^{-2}$	−41	−2760	−1·8	−585
$\nu \times 10^{-2}$	37	2470	1·7	513
$\xi \times 10^{-2}$	0	0	−1·0	−460

For perfectly free electrons and a free path which is independent of the energy

$$\lambda = R\sigma_0^2, \quad \mu = -\nu = -4R^2\sigma_0^3/\pi, \quad \xi = 0.$$

It will be seen that these relations, with the exception of $\xi = 0$ for the negatively conducting material, are approximately obeyed, and therefore that very large magneto-resistance effects can be caused by comparatively small departures of the conductivity constants from their values for free electrons.

Dunlap found a considerable variation of the Hall coefficient with the magnetic field, a good negatively conducting specimen having its Hall coefficient reduced by 8 % by a field of 12,000 gauss, while that of a high resistance positively conducting specimen was reduced by 22 %. This variation could readily be explained by the formulae for an isotropic medium if the scattering index p is reasonably large (see fig. VIII 9).

CONDUCTION IN THIN FILMS

8·7. In general the conductivity of a metal is independent of the dimensions of the specimen, but this must cease to be true for films or wires whose thicknesses or diameters are so small as to be comparable with the free path of the electrons. The conductivity must then be less than that of the bulk metal, since some of the free paths will be reduced by termination at the surface of the metal.

An elementary theory of the phenomenon was given by J. J. Thomson (1901) based on Drude's theory of conduction and on the additional assumption that the scattering at the surface is random. Consider a plane film of thickness t and let the free path in the bulk metal be l_0. Then the free path of an electron starting from a point at distance z from the surface and moving in a direction making an angle θ with the z-axis is $l(\theta)$, where $l = l_0$ only for $\theta_1 \leqslant \theta \leqslant \theta_2$, with θ_1 and θ_2 being given by

$$\cos\theta_1 = (t-z)/l_0, \quad \cos\theta_2 = -z/l_0. \tag{8·7·1}$$

For other directions

$$l(\theta) = (t - z)/\cos\theta \quad (0 \leqslant \theta \leqslant \theta_1),$$
$$\text{and} \qquad l(\theta) = -z/\cos\theta \quad (\theta_2 \leqslant \theta \leqslant \pi). \tag{8·7·2}$$

The mean free path $\bar{l}$ is therefore the average of $l(\theta)$ over all values of z and θ, i.e.

$$\bar{l} = \frac{1}{2t} \int_0^t dz \int_0^\pi l(\theta) \sin\theta \, d\theta$$

$$= \frac{1}{2t} \int_0^t dz \left[\int_{(t-z)/l_0}^1 \frac{t-z}{\cos\theta} d\cos\theta + \int_{-z/l_0}^{(t-z)/l_0} l_0 d\cos\theta - \int_{-1}^{-z/l_0} \frac{z}{\cos\theta} d\cos\theta \right]$$

$$= \tfrac{3}{4}t + \tfrac{1}{2}t \log l_0/t. \tag{8·7·3}$$

Hence

$$\frac{\sigma}{\sigma_0} = \frac{\bar{l}}{l_0} = \frac{3t}{4l_0} + \frac{t}{2l_0} \log \frac{l_0}{t}. \tag{8·7·4}$$

8·71. The theory given above suffers from the defect that it ignores free paths which start from the surface, and these become increasingly important as the thickness is reduced. A more elaborate theory (Fuchs, 1938; see also Chambers, 1950; Sondheimer, 1952) can be obtained as follows by solving the Boltzmann equation for the distribution function $f(\mathbf{v}, \mathbf{r})$, which must depend explicitly on the coordinate z. If there is an electric field $\mathscr{E}$ parallel to the x-axis, the Boltzmann equation is

$$-\frac{e\mathscr{E}}{m} \frac{\partial f}{\partial v_1} + v_3 \frac{\partial f}{\partial z} = -\frac{f - f_0}{\tau}, \tag{8·71·1}$$

and if we put

$$f = f_0(\mathbf{v}) + f_1(\mathbf{v}, z) \tag{8·71·2}$$

and neglect squares of $\mathscr{E}$, we have

$$v_3 \frac{\partial f_1}{\partial z} + \frac{1}{\tau} f_1 = \frac{e\mathscr{E}}{m} \frac{\partial f_0}{\partial v_1}. \tag{8·71·3}$$

The general solution of this is

$$f_1 = \frac{e\mathscr{E}\tau}{m} \frac{\partial f_0}{\partial v_1} \{1 + \phi(\mathbf{v}) e^{-z \cdot \tau v_3}\}, \tag{8·71·4}$$

where the arbitrary function $\phi(\mathbf{v})$ is determined by the boundary conditions at the surface of the film. It has two different analytic forms ϕ^+ and ϕ^- according as v_3 is positive or negative.

We shall assume that a proportion p of the electrons are reflected specularly at the surface and that a proportion $1 - p$ are reflected randomly. Denote the function (8·71·4) by f_1^+ when $v_3 > 0$, and let the corresponding distribution function for $v_3 < 0$ be denoted by f_1^-. Then by symmetry

$$f_1^-(v_1, v_2, v_3, z) = f_1^+(v_1, v_2, -v_3, t - z) \quad (v_3 < 0), \tag{8·71·5}$$

and so $\quad f_1^-(\mathbf{v}, z) = \dfrac{e\mathscr{E}\tau}{m}\dfrac{\partial f_0}{\partial v_1}\{1 + \phi^+(v_1, v_2, -v_3)\,e^{(t-z)/\tau v_3}\}\quad (v_3 < 0),$

i.e. $\qquad\qquad \phi^-(v_1, v_2, v_3) = \phi^+(v_1, v_2, -v_3)\,e^{t/\tau v_3}\quad (v_3 < 0).$ $\qquad$ (8·71·6)

The electrons arriving at the surface $z = 0$ have the distribution function $f_0 + f_1^-(\mathbf{v}, 0)$. The proportion of those that are scattered specularly give a distribution function $p\{f_0 + f_1^-(v_1, v_2, -v_3, 0)\}$ for the electrons leaving the surface, while those that are scattered randomly give a contribution g which is independent of the direction of $\mathbf{v}$. Hence

$$f_0 + f_1^+(v_1, v_2, v_3, 0) = p\{f_0 + f_1^-(v_1, v_2, -v_3, 0)\} + g,$$

and on substitution from (8·71·4) and (8·71·6) we find that

$$g = f_0(1-p) + \frac{e\mathscr{E}\tau}{m}\frac{\partial f_0}{\partial v_1}[1 - p + \phi^+(\mathbf{v})\,(1 - p\,e^{-t/\tau v_3})].$$

This expression for g can only be independent of the direction of $\mathbf{v}$ if the second term vanishes, which gives

$$\phi^+(\mathbf{v}) = -(1-p)/(1 - p\,e^{-t/\tau v_3})\quad (v_3 > 0), \qquad (8·71·7)$$

and, by (8·71·6),

$$\phi^-(\mathbf{v}) = -e^{t/\tau v_3}(1-p)/(1 - p\,e^{t/\tau v_3})\quad (v_3 < 0). \qquad (8·71·8)$$

The current density $j(z)$ is $-2e\left(\dfrac{m}{h}\right)^3\int v_1 f\,d\mathbf{v}$, and the average of this with respect to z is

$$J = \frac{1}{t}\int_0^t j(z)\,dz = -2\frac{e^2\mathscr{E}}{mt}\left(\frac{m}{h}\right)^3\int_0^t dz\int \tau v_1\{1 + \phi(\mathbf{v})\,e^{-z/\tau v_3}\}\frac{\partial f_0}{\partial v_1}\,d\mathbf{v},$$

which can be evaluated by introducing polar coordinates in the velocity space, the polar axis being along the z-axis so that

$$v_3 = v\cos\theta, \qquad v_1^2 + v_2^2 = v^2\sin^2\theta.$$

Then $\quad J = -\dfrac{2\pi e^2\mathscr{E}}{mt}\left(\dfrac{m}{h}\right)^3\displaystyle\int_0^t dz\int_0^\infty \tau v^3\frac{\partial f_0}{\partial v}dv\int_0^\pi (1 + \phi\,e^{-z/\tau v\cos\theta})\sin^3\theta\,d\theta.$

Also $\qquad\qquad \sigma_0 = -\dfrac{8}{3}\dfrac{\pi e^2}{m}\left(\dfrac{m}{h}\right)^3\displaystyle\int_0^\infty \tau v^3\frac{\partial f_0}{\partial v}dv,$

and, since f_0 is the Fermi function, only the value $\bar{v}$ for $E = \zeta$ occurs in the integral so that we may write $\tau\bar{v} = l_0$. Then

$$\frac{\sigma}{\sigma_0} = 1 - \frac{3}{8\lambda}(1-p) + \frac{3}{4\lambda}\int_0^\pi \sin^3\theta\,|\cos\theta\,|\frac{(1-p)^2 e^{-\lambda/|\cos\theta|}}{1 - p\,e^{-\lambda/|\cos\theta|}}\,d\theta, \qquad (8·71·9)$$

where $\lambda = t/l_0$.

The integral can be expressed in terms of exponential integrals as follows. Put $\xi = \lambda/|\cos\theta|$. Then

$$\int_0^\pi \frac{\sin^3\theta\,|\cos\theta|\,e^{-\lambda/|\cos\theta|}}{1 - p\,e^{-\lambda/|\cos\theta|}}\,d\theta = 2\int_\lambda^\infty \frac{e^{-\xi}}{1 - p\,e^{-\xi}}\left(\frac{\lambda^2}{\xi^3} - \frac{\lambda^4}{\xi^5}\right)d\xi$$

$$= 2\sum_{s=1}^\infty p^{s-1}\int_\lambda^\infty e^{-s\xi}\left(\frac{\lambda^2}{\xi^3} - \frac{\lambda^4}{\xi^5}\right)d\xi.$$

This can be integrated by parts to be expressible in terms of exponentials and of $E_1(s\lambda)$, where

$$E_1(x) = \int_x^\infty \frac{e^{-\xi}}{\xi}\,d\xi. \tag{8·71·10}$$

The complete expression for σ is then

$$\frac{\sigma}{\sigma_0} = 1 - \frac{3(1-p)}{8\lambda} + \frac{3(1-p)^2}{4\lambda}\sum_{s=1}^\infty p^{s-1}[(\tfrac{1}{2} - \tfrac{5}{8}s\lambda - \tfrac{1}{12}s^2\lambda^2 + \tfrac{1}{12}s^3\lambda^3)\,e^{-s\lambda}$$
$$+ (s^2\lambda^2 - \tfrac{1}{12}s^4\lambda^4)\,E_1(s\lambda)]. \tag{8·71·11}$$

For thick films the first approximation to σ/σ_0 is

$$\frac{\sigma}{\sigma_0} = 1 - \frac{3(1-p)}{8\lambda}. \tag{8·71·12}$$

For very thin films the leading term can be obtained by using the approximation $E_1(x) = \log(1/x)$ for very small x, and summing the resulting series:

$$\sum_{s=1}^\infty s^2 p^{s-1} = \frac{1+p}{(1-p)^3}.$$

This gives

$$\frac{\sigma}{\sigma_0} \sim \frac{3}{4}\frac{1+p}{1-p}\lambda \log\frac{1}{\lambda}, \tag{8·71·13}$$

but this approximation has only a very limited range of validity and in general it is necessary to calculate σ/σ_0 from (8·71·9) or (8·71·11).

8·72. An extension of the preceding theory to square cross-section wires has been given by MacDonald and Sarginson (1950). Their result is that, for large values of λ and for $p = 0$,

$$\sigma/\sigma_0 = 1 - 3/(4\lambda),$$

while for very thin wires

$$\sigma/\sigma_0 = 1 \cdot 115\lambda.$$

For thick wires the decrease in conductivity is just double that for a film of the same thickness, while for thin wires the conductivity decreases much more rapidly than for films.

Corresponding formulae have been obtained by Dingle (1950) for circular wires. His formulae reduce to

$$\frac{\sigma}{\sigma_0} = 1 - \frac{3}{4\lambda}(1-p)$$

for thick wires, and

$$\frac{\sigma}{\sigma_0} = \frac{1+p}{1-p}\lambda$$

for very thin wires.

8·73. The experimental results (Appleyard and Lovell, 1937) for the conductivity of thin films of caesium at 70° K. are shown in fig. VIII 12,

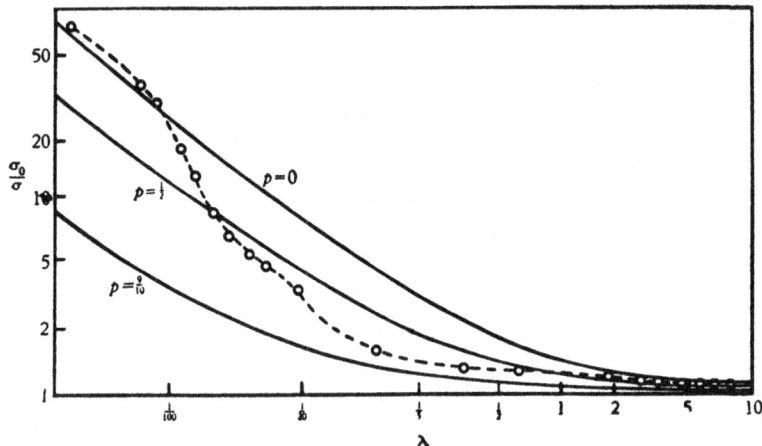

Fig. VIII 12. The ratio of the resistivity of a thin film of caesium to the bulk resistivity as a function of the thickness.

together with three theoretical curves for $p = 0$, $\frac{1}{2}$ and $\frac{9}{10}$ and $l_0 = 1450$ A., and it is clear that the observations do not fit any one curve. The results for very thin films (thicknesses less than 12 A., i.e. two atomic layers) are of doubtful significance, but the results for thicker films indicate that random scattering predominates when the film is thin, but that the scattering becomes increasingly specular as the thickness increases.

Experiments on thin wires have been carried out by MacDonald and Sarginson (1950). Their results have similar characteristics to those of Appleyard and Lovell, and no single value of p will fit their experimental data.

The experimental difficulties are diminished if the measurements are carried out at low temperatures where the free path is of the order of 10^{-2} cm., since relatively thick films can be used. Andrew (1949) measured the resistances of tin foils at 3·8° K., with thicknesses ranging from 3×10^{-4}

to 0·2 cm., and found that the results fitted the theoretical formula excellently, provided that $p = 0$. He obtained the values $l_0 = 9·5 \times 10^{-3}$ cm. and $\sigma_0/l_0 = 4·5 \times 10^{22}$ e.s.u., corresponding to a value of 0·43 for n/n_a, the number of conduction electrons per atom. Andrew obtained similar results for mercury wires at 2·5° K., the values of the parameters being $p = 0$, $l_0 = 5·6 \times 10^{-3}$ cm., $\sigma_0/l_0 = 2·5 \times 10^{22}$ e.s.u. and $n/n_a = 0·15$.

8·74. *The anomalous skin effect.* The penetration of electromagnetic fields into conductors can be discussed by the method of § 8·71 if the electric field as well as the current density is treated as a function of position. Let the metal occupy the space $z > 0$, and let there be an electric field incident normally on the surface. The field is parallel to the x-axis and is taken to be the real part of $\mathscr{E}(z) e^{2\pi i \nu t}$. (We omit time factors, carry through the calculation with complex fields and currents and take real parts at the end.) The distribution function is found by using exactly the same arguments as in § 8·71 except that the metal is now infinite and that equation (8·71·3) must be integrated treating $\mathscr{E}$ as a function of z. If we define $\mathscr{E}(z)$ for $z < 0$ as being equal to $\mathscr{E}(-z)$ we find that

$$f_1 = \frac{\epsilon}{mv_3}\frac{\partial f_0}{\partial v_1} e^{-z/\tau v_3} \left[p \int_{-\infty}^{z} \mathscr{E}(s)\, e^{s/\tau v_3}\, ds + (1-p) \int_{0}^{z} \mathscr{E}(s)\, e^{s/\tau v_3}\, ds \right] \quad (v_3 > 0),$$

$$f_1 = -\frac{\epsilon}{mv_3}\frac{\partial f_0}{\partial v_1} e^{-z/\tau v_3} \int_{s}^{\infty} \mathscr{E}(s)\, e^{s/\tau v_3}\, ds \qquad\qquad (v_3 < 0),$$

$$(8·74·1)$$

and that the current density is

$$j(z) = \frac{3\sigma_0}{4l_0}\left[p \int_{-\infty}^{\infty} k\!\left(\frac{|z-s|}{l_0}\right) \mathscr{E}(s)\, ds + (1-p) \int_{0}^{\infty} k\!\left(\frac{|z-s|}{l_0}\right) \mathscr{E}(s)\, ds \right],$$

$$(8·74·2)$$

where
$$k(x) = E_1(x) - x^2 E_3(x), \quad E_n(x) = \int_{x}^{\infty} \frac{e^{-\zeta}}{\zeta^n}\, d\zeta. \qquad (8·74·3)$$

The relation between $\mathscr{E}(z)$ and $j(z)$ is given by Maxwell's equations. If the frequency is low enough for the displacement current to be neglected, we have

$$d^2\mathscr{E}/dz^2 = 8\pi^2 i \nu j(z)/c^2, \qquad (8·74·4)$$

as the required relation between the complex electric field and the complex current density, and the combination of (8·74·2) and (8·74·4) gives an integral equation for $\mathscr{E}(z)$.

If $l_0 \to 0$, then $j(z) \to \sigma_0 \mathscr{E}(z)$, and the equations have the solution

$$\mathscr{E}(z) = e^{-(1+i)z/\delta}, \qquad (8·74·5)$$

where
$$\delta = c/(4\pi^2 \nu \sigma_0)^{\frac{1}{2}}. \qquad (8·74·6)$$

This exponential decrease in the electric field constitutes the phenomenon of the ordinary skin effect, and δ is the 'classical skin depth'. If, however, l_0/δ cannot be considered as infinitely small, there are deviations from the classical theory, a phenomenon known as the anomalous skin effect (London, 1940; Pippard, 1947).

The integral equation for $\mathscr{E}(z)$ is of a standard type which can be solved by Fourier integrals (Titchmarsh, 1937, chapter 11). The solution, which has been given by Reuter and Sondheimer (1948), is relatively simple when $p = 1$ but is difficult when $p = 0$. The reader is referred to their paper for details of the elegant but involved calculations (see also Sondheimer, 1952) and for extensions of the theory to frequencies where relaxation effects are important (i.e. where $2\pi\nu\tau$ cannot be considered small compared with unity).

At radio frequencies the measurements give the surface resistivity R rather than the total resistance of the metal. The surface impedance Z is defined as the ratio of the (complex) electric field at the surface of the metal to the (complex) total current per unit surface area, that is,

$$Z = R + iX = \mathscr{E}(0) \bigg/ \int_0^\infty j(z)\,dz = -8\pi^2 i\nu c^{-2}\,\mathscr{E}(0)/\mathscr{E}'(0), \qquad (8\cdot74\cdot7)$$

by $(8\cdot74\cdot4)$. In the classical limit, $(8\cdot74\cdot5)$ gives

$$R = X = 2\pi(\nu/\sigma_0 c^2)^{\frac{1}{2}}. \qquad (8\cdot74\cdot8)$$

In the extreme case where $l_0 \gg \delta$, it can be shown that

$$\left.\begin{array}{ll} Z = \tfrac{8}{9}\pi(4\sqrt{3}\nu^2 l_0/\sigma_0 c^4)^{\frac{1}{3}}(1 + \sqrt{3}i) & (p = 1), \\ Z = \pi(4\sqrt{3}\nu^2 l_0/\sigma_0 c^4)^{\frac{1}{3}}(1 + \sqrt{3}i) & (p = 0). \end{array}\right\} \qquad (8\cdot74\cdot9)$$

Measurements of R can therefore be used to determine σ_0/l_0 and p, provided that conditions deviate measurably from those in which the classical theory holds. Since

$$\sigma_0/l_0 = ne^2/(m^*\bar{v}), \qquad (8\cdot74\cdot10)$$

where

$$m^*\bar{v} = \tfrac{1}{2}h(3n/\pi)^{\frac{1}{3}}, \qquad (8\cdot74\cdot11)$$

the values of σ_0/l_0 deduced should be independent of the temperature and should give n directly.

The measurements have to be carried out at low temperatures where l_0 is large. Chambers (1952 b) used a frequency of 1200 Mc/s and temperatures in the range 2° to 90° K. His results were in much better agreement with $p = 0$ than with $p = 1$, and it therefore seems probable that diffuse scattering is predominant both in thin films and in the anomalous skin effect.

Some values of σ_0/l_0 obtained by Chambers are given in Table VIII 11, together with the values of n/n_a, the number of free electrons per atom, deduced from (8·74·10) and (8·74·11). (The corresponding results for bismuth have already been discussed in § 8·561.) The estimates of n/n_a are of the right order of magnitude, though except for copper they differ significantly from those found in other ways. Better results might be obtained for the multivalent metals by considering the electrons to occupy two bands.

Table VIII 11. *Estimates of σ_0/l_0 in e.s.u. and n/n_a deduced from the anomalous skin effect*

	$(\sigma_0/l_0) \times 10^{-11}$	n/n_a
Cu	13·9	1·0
Ag	8·3	0·68
Au	7·6	0·6
Sn	8·6	1·1
Pb	8·5	1·24
Al	18·4	2·2

REFERENCES

Andrew, E. R. (1949). The size variation of resistivity for mercury and tin. *Proc. Phys. Soc. A*, **62**, 77.

Appleyard, E. T. S. and Lovell, A. C. B. (1937). The electrical conductivity of thin metallic films. *Proc. Roy. Soc. A*, **158**, 718.

Ariyama, K. (1938). The electronic states of divalent metals. *Sci. Pap. Inst. Phys. Chem. Res., Tokyo*, **34**, 344.

Borelius, G. (1935). Grundlagen des metallischen Zustandes. Physikalische Eigenschaften. *Handbuch der Metallphysik*, **1** (Leipzig).

Campbell, L. L. (1923). *Galvanomagnetic and thermomagnetic effects* (London).

Chambers, R. G. (1950). The conductivity of thin wires in a magnetic field. *Proc. Roy. Soc. A*, **202**, 378.

Chambers, R. G. (1952a). The two band effect in conduction. *Proc. Phys. Soc. A*, **65**, 903.

Chambers, R. G. (1952b). The anomalous skin effect. *Proc. Roy. Soc. A*, **215**, 481.

Chambers, R. G. and Pippard, A. B. (1952). The mean free path of conduction electrons in bismuth. *Proc. Phys. Soc. A*, **65**, 955.

Davis, L. (1939). Change of resistance in a magnetic field. *Phys. Rev.* **56**, 93.

Dingle, R. B. (1950). The electrical conductivity of thin wires. *Proc. Roy. Soc. A*, **201**, 545.

Dunlap, W. C. (1950). Some properties of high resistivity p-type germanium. *Phys. Rev.* **79**, 286.

Estermann, I. and Foner, A. (1950). Magnetoresistance of germanium samples between 20° and 300° K. *Phys. Rev.* **79**, 365.

Eucken, A. and Förster, F. (1934). The mean free path of electrons in bismuth. *Nach. Ges. Wiss. Göttingen*, **1**, 43.

Fuchs, K. (1938). The conductivity of thin metallic films. *Proc. Camb. Phil. Soc.* **34**, 100.

Harding, J. W. (1933). The change of resistance of a semi-conductor in a magnetic field. *Proc. Roy. Soc. A*, **140**, 205.

Jones, H. (1936). The theory of the galvanomagnetic effects in bismuth. *Proc. Roy. Soc. A*, **155**, 653.

Jones, H. and Zener, C. (1934). The theory of the change of resistance in a magnetic field. *Proc. Roy. Soc.* A, **145**, 268.

Justi, E., Kohler, M. and Lautz, G. (1951). The thermoelectric power of metal films. *Z. Naturforsch.* **6a**, 456.

Kohler, M. (1941). Electrical and thermal phenomena in a magnetic field. *Ann. Phys., Lpz.* (5), **40**, 601.

London, H. (1940). The high frequency resistance of superconducting tin. *Proc. Roy. Soc.* A, **176**, 522.

MacDonald, D. K. C. and Sarginson, K. (1950). Size effect variation of the electrical conductivity of metals. *Proc. Roy. Soc.* A, **203**, 223.

Meissner, W. (1935). Galvano- und thermomagnetische Effekte. *Handbuch der Experimentalphysik*, **11** (Leipzig).

Nordheim, L. (1931). The electron theory of metals. *Ann. Phys., Lpz.* (5), **9**, 607.

Pearson, G. L. and Suhl, H. (1951). The magneto-resistance effect in germanium. *Phys. Rev.* **83**, 768.

Peierls, R. (1929). The theory of the galvanomagnetic effects. *Z. Phys.* **53**, 255.

Peierls, R. (1931). The theory of the magnetic change of resistance. *Ann. Phys. Lpz.* (5), **10**, 97.

Pippard, A. B. (1947). The surface impedance of superconductors and normal metals at high frequencies. II. The anomalous skin effect in normal metals. *Proc. Roy. Soc.* A, **191**, 385.

Reuter, G. E. H. and Sondheimer, E. H. (1948). The theory of the anomalous skin effect in metals. *Proc. Roy. Soc.* A, **195**, 336.

Seitz, F. (1950). The resistance of a cubic semi-conductor in a magnetic field. *Phys. Rev.* **79**, 372.

Sommerfeld, A. (1934). The electron theory of metals. *Naturwissenschaften*, **22**, 49.

Sondheimer, E. H. (1948). The theory of the galvanomagnetic and thermomagnetic effects in metals. *Proc. Roy. Soc.* A, **193**, 484.

Sondheimer, E. H. (1952). The mean free path of electrons in metals. *Advances in Physics*, **1**, 1.

Sondheimer, E. H. and Wilson, A. H. (1947). The theory of the magneto-resistance effects in metals. *Proc. Roy. Soc.* A, **190**, 435.

Thompson, N. (1936). The electrical resistance of bismuth alloys. *Proc. Roy. Soc.* A, **155**, 111.

Thomson, J. J. (1901). The theory of electrical conduction through thin metallic films. *Proc. Camb. Phil. Soc.* **11**, 119.

Titchmarsh, E. C. (1937). *The theory of Fourier integrals* (Oxford).

Titeica, S. (1935). The magneto-resistance effect in metals. *Ann. Phys., Lpz.* (5), **22**, 129.

Chapter IX

THE MECHANISM OF CONDUCTIVITY

INTRODUCTION

9·1. Since an electron moves freely through a perfect lattice without disturbance, the conductivity only becomes finite when we take into account the irregularities in the lattice structure. The most important factor in producing irregularities is usually the thermal motion of the metallic ions, but impurities, strains, displaced atoms, etc., play a part which is especially important at low temperatures. The thermal vibrations of the lattice produce a resistance which is normally much larger than that due to the impurities, but it diminishes rapidly as the temperature is lowered so that at sufficiently low temperatures the resistance becomes independent of the temperature, this 'residual resistance' being due to the impurities, etc., and therefore varying from specimen to specimen. If the residual resistance is subtracted from the total resistance a quantity is obtained called the ideal resistance which is assumed to be characteristic of the pure metal (see § 10·4) and due to the thermal vibrations alone.

We first determine the collision operators for scattering by the thermal vibrations and by the impurities, showing that in certain cases the operators are such that a time of relaxation can be unambiguously defined. We then consider those properties of the collision operators which cannot be dealt with by the methods of the preceding chapter.

THE LATTICE VIBRATIONS

9·2. An outline of the general theory of the vibrations of a crystal lattice has already been given in § 6·15. It is sufficient to consider a Bravais lattice in which all the atoms have the same mass and to expand the potential energy as a quadratic function of the displacement R_g of the lattice point g:

$$V = \tfrac{1}{2} \sum_{g,h} \sum_{k,l} V_{gh}^{kl} R_{g,k} R_{h,l}. \qquad (9\cdot2\cdot1)$$

We introduce the normal coordinates α_{qj} for a cyclical crystal containing $(2G+1)^3$ atoms by putting

$$R_g = \frac{1}{2^{\frac{1}{2}}(2G+1)^{\frac{3}{2}}} \sum_q{}' \sum_{j=1}^{3} e_{qj}(\alpha_{qj} e^{iq \cdot r_g} + \alpha_{qj}^* e^{-iq \cdot r_g}), \qquad (9\cdot2\cdot2)$$

where Σ' denotes a summation over half the phase parallelepiped and where the frequencies ν_q and the polarization vectors e_{qj} of the normal vibrations are such as to satisfy the equations of motion

$$M\,d^2R_{g,k}/dt^2 = -\partial V/\partial R_{g,k},$$

i.e.

$$4\pi^2 M v_q^2 e_{qj,k} = \sum_h \sum_l V_{gh}^{kl} e_{qj,l}\, e^{iq\cdot h_a}. \tag{9·2·3}$$

For given q there are three values of ν_q which are the roots of the secular determinant obtained by eliminating the e_q's from (9·2·3). The vectors e_{qj} are then the unit vectors defining the principal axes of the (three-dimensional) quadratic form associated with the three equations (9·2·3). We therefore have

$$e_{qi}\cdot e_{qj} = \delta_{ij}. \tag{9·2·4}$$

If the vectors defining the unit cell are a_i, the values of the propagation vector q are given by

$$q\cdot a_i = 2\pi(0,\ \pm 1,\ \pm 2,\ \dots,\ \pm G)/(2G+1),$$

i.e. they lie in the parallelepiped $-\pi \leqslant q\cdot a_i \leqslant \pi$, and in order to obtain the correct number of normal modes it is necessary, as in §6·15, to restrict q to one-half of the phase parallelepiped. There are then two standing waves to each value of q. The introduction of the normal coordinates reduces the kinetic and potential energies to the sum of squares.

The calculation of the kinetic and potential energies proceeds exactly as in §6·15. The kinetic energy is

$$T = \tfrac{1}{2}M\Sigma(dR_g/dt)^2$$
$$= \tfrac{1}{4}(2G+1)^{-3}M \sum_{qq'}\nolimits' \sum_{jj'} \sum_g e_{qj}\cdot e_{q'j'}[\dot{\alpha}_{qj}\dot{\alpha}_{q'j'}\exp\{i(q+q')\cdot g_a\}$$
$$+ \dot{\alpha}_{qj}\dot{\alpha}_{q'j'}^*\exp\{i(q-q')\cdot g_a\} + \dot{\alpha}_{qj}^*\dot{\alpha}_{q'j'}\exp\{i(-q+q')\cdot g_a\}$$
$$+ \dot{\alpha}_{qj}^*\dot{\alpha}_{q'j'}^*\exp\{-i(q+q')\cdot g_a\}].$$

The summations with respect to g give zero unless $q=q'$, and since $e_{qj}\cdot e_{qj'} = \delta_{jj'}$, we have

$$T = \tfrac{1}{2}M \sum_{q,j}\nolimits' \left\{\left(\frac{dA_{qj}}{dt}\right)^2 + \left(\frac{dB_{qj}}{dt}\right)^2\right\},$$

where A and B are the real and imaginary parts of α respectively. The expression for the potential energy can be similarly transformed, using the equations (9·2·3) to eliminate the potential coefficients. The only terms which contribute anything are those for which $q=q'$, and we obtain

$$V = \tfrac{1}{4}(2G+1)^{-3} \sum_q\nolimits' \sum_{jj'} \sum_{gh} \sum_{kl} V_{gh}^{kl} e_{qj,k}\, e_{qj',l}[\alpha_{qj}\alpha_{qj'}^*\exp\{iq\cdot(g_a-h_a)\}$$
$$+ \alpha_{qj}^*\alpha_{qj'}\exp\{-iq\cdot(g_a-h_a)\}]$$
$$= \pi^2 M(2G+1)^{-3} \sum_q\nolimits' \sum_{jj'} \sum_g \sum_k e_{qj,k}\, e_{qj',k}\, v_{qj}^2(\alpha_{qj}\alpha_{qj'}^* + \alpha_{qj}^*\alpha_{qj'})$$
$$= 2\pi^2 M \sum_{q,j}\nolimits' v_{qj}^2(A_{qj}^2 + B_{qj}^2),$$

and the total energy is

$$\mathscr{H} = \tfrac{1}{2}M \sum_{\mathbf{q},j}' \left\{ \left(\frac{dA_{\mathbf{q}j}}{dt}\right)^2 + \left(\frac{dB_{\mathbf{q}j}}{dt}\right)^2 + 4\pi^2 \nu_{\mathbf{q}j}^2 (A_{\mathbf{q}j}^2 + B_{\mathbf{q}j}^2) \right\} \qquad (9\cdot2\cdot5)$$

The $3(2G+1)^3$ real quantities $A_{\mathbf{q}j}$, $B_{\mathbf{q}j}$ are the normal coordinates. Each normal mode of vibration represents a standing wave, and by combining two modes with the same frequency we can obtain progressive waves. To each value of $\mathbf{q}$ and j there are two normal modes, so that there is a degeneracy present, and the normal coordinates are not unique. In conduction theory it is convenient to have normal coordinates which correspond to progressive waves, since we are interested in transport phenomena. Such coordinates can be obtained by means of a contact transformation or by using polar coordinates in the A, B plane. We shall adopt the first method (Peierls, 1929), and introduce new coordinates ξ, η and momenta p_ξ, p_η defined by

$$A = \frac{1}{2^{\frac{1}{2}}}\left(\xi + \frac{p_\eta}{2\pi M \nu}\right), \quad B = \frac{1}{2^{\frac{1}{2}}}\left(\eta + \frac{p_\xi}{2\pi M \nu}\right), \left.\begin{array}{c}\\\\\end{array}\right\}$$
$$p_A = 2^{-\frac{1}{2}}(p_\xi - 2\pi M \nu \eta), \quad p_B = 2^{-\frac{1}{2}}(p_\eta - 2\pi M \nu \xi), \qquad (9\cdot2\cdot6)$$

the suffixes $\mathbf{q}, j$ being omitted. In these variables the Hamiltonian is

$$\mathscr{H} = \tfrac{1}{2} \sum_{\mathbf{q},j}' \left\{ \frac{1}{M}(p_\xi^2 + p_\eta^2) + 4\pi^2 M \nu_{\mathbf{q}j}^2 (\xi_{\mathbf{q}j}^2 + \eta_{\mathbf{q}j}^2) \right\}, \qquad (9\cdot2\cdot7)$$

and a normal mode is given by

$$\xi_{\mathbf{q}j} = \xi(0) \cos(2\pi\nu_{\mathbf{q}j}t + \beta), \quad p_\xi = -2\pi M \nu_{\mathbf{q}j} \sin(2\pi\nu_{\mathbf{q}j}t + \beta),$$

so that

$$A_{\mathbf{q}j} = 2^{-\frac{1}{2}}\xi(0)\cos(2\pi\nu_{\mathbf{q}j}t + \beta), \quad B_{\mathbf{q}j} = -2^{-\frac{1}{2}}\xi(0)\sin(2\pi\nu_{\mathbf{q}j}t + \beta).$$

In this normal mode

$$\mathbf{R}_{\mathbf{t}} = (2G+1)^{-\frac{3}{2}}\,\xi(0)\,\mathbf{e}_{\mathbf{q}j}\cos(2\pi\nu_{\mathbf{q}j}t - \mathbf{q}\cdot\mathbf{g}_a + \beta),$$

which is a travelling wave. The corresponding η mode represents a wave in the opposite direction. We can make the equations more symmetrical by means of another contact transformation

$$2\pi M \nu_{\mathbf{q}j} \eta_{\mathbf{q}j} = -p(\xi_{-\mathbf{q}j}), \quad p(\eta_{\mathbf{q}j}) = 2\pi M \nu_{\mathbf{q}j} \xi_{-\mathbf{q}j}. \qquad (9\cdot2\cdot8)$$

The Hamiltonian can now be written as

$$\mathscr{H} = \tfrac{1}{2} \sum_{\mathbf{q},j} \left\{ \frac{1}{M} p(\xi_{\mathbf{q}j})^2 + 4\pi^2 M \nu_{\mathbf{q}j}^2 \xi_{\mathbf{q}j}^2 \right\}, \qquad (9\cdot2\cdot9)$$

the summation being over the whole of the phase cell, and the relation between the displacement and these variables is

$$\mathbf{R}_{\mathbf{t}} = (2G+1)^{-\frac{3}{2}} \sum_{\mathbf{q},j} \mathbf{e}_{\mathbf{q}j} \left\{ \xi_{\mathbf{q}j} \cos\mathbf{q}\cdot\mathbf{g}_a - \frac{p(\xi_{\mathbf{q}j})}{2\pi M \nu_{\mathbf{q}j}} \sin\mathbf{q}\cdot\mathbf{g}_a \right\}. \qquad (9\cdot2\cdot10)$$

THE COUPLING BETWEEN THE ELECTRONS AND THE LATTICE

9·3. The problem of calculating the effect of the interaction between the conduction electrons and the lattice vibrations is exceedingly complex, and in order to make any progress at all the theory must be considerably over-simplified. In the first place we must assume that, in the zero order, the electronic wave function is a simple product and rely upon the use of the Fermi distribution function to compensate partly for the neglect of the antisymmetrical properties of the correct wave function. In the second place the self-consistent field is affected by the lattice vibrations, and logically the self-consistent field ought to be different for every different lattice state. We shall, however, neglect these refinements and assume that the motions of the electrons and the nuclei are independent to the first order, and that the coupling between them is caused by the alteration in the potential energy due to the displacement of the lattice. If $\mathbf{R}_g$ denotes the displacement of the lattice point g from its equilibrium position, the perturbing potential energy is $\Delta V = V(\mathbf{r}, \mathbf{R}) - V(\mathbf{r}, 0)$. If U_g is the potential energy of an electron due to the ion g and if we assume that the ions are displaced without being deformed we have

$$\Delta V = \sum_g U(\mathbf{r} - \mathbf{g}_a - \mathbf{R}_g) - \sum_g U(\mathbf{r} - \mathbf{g}_a) = - \sum_g \mathbf{R}_g . \operatorname{grad} U_g. \qquad (9\cdot3\cdot1)$$

This expression for ΔV is clearly not entirely correct, since the ions must be deformed to a certain extent. Calculations based upon a less restrictive assumption have been made by several authors, in particular by Bardeen (1937), but in the present state of the theory these are refinements which are of doubtful value since they only affect the numerical values of parameters which cannot in any case be calculated accurately. It should, however, be mentioned that it is possible that the crudeness of the approximations made in dealing with the interactions between the conduction electrons and the lattice vibrations is the cause of the failure of the present theory to include superconductivity. While it is possible that some new physical principle is required to explain the phenomenon of superconductivity, it may well be that the difficulties are essentially mathematical rather than physical, and that, just as a searching analysis of the theory of the equation of state of a gas reveals the possibility of the existence of a liquid phase, a more exact mathematical treatment of the interaction problem would lead to an explanation of superconductivity. For this and other reasons, which will become apparent as the various theoretical predictions are compared with experiment, a much improved and more general theory of the interaction between the electrons and the lattice is required.

In the zero approximation the total wave function is the product of an electronic wave function and a lattice wave function. Since we assume that each electron moves independently of the others we can concentrate our attention upon one particular electron, and discuss its possible transitions. To calculate the matrix elements of ΔV with respect to the unperturbed electronic wave functions, we have to evaluate integrals of the type

$$\int \psi_{\mathbf{k}'}^{*} \sum_{g} \exp \left(\pm i\mathbf{q} \cdot \mathbf{g}_a \right) \psi_{\mathbf{k}} \mathbf{e}_{qj} \cdot \operatorname{grad} U_g d\tau. \tag{9·3·2}$$

This expression arises from the normal mode $\mathbf{q}$, by substituting (9·2·10) into (9·3·1). Writing (9·3·2) in the form

$$(2G+1)^{-3} \int e^{i(\mathbf{k}-\mathbf{k}'\pm\mathbf{q})\cdot\mathbf{r}} [\sum_{g} \mathbf{e}_{qj} \cdot \operatorname{grad} U_g \exp\{\mp i\mathbf{q} \cdot (\mathbf{r} - \mathbf{g}_a)\} u_{\mathbf{k}'}^{*} u_{\mathbf{k}}] d\tau,$$

$u_{\mathbf{k}}$ being normalized in the unit cell, whereas $\psi_{\mathbf{k}}$ is normalized in the crystal of $(2G+1)^3$ cells, we see that the integral vanishes unless the exponential factor has the same periodicity as the expression in the square bracket, which is periodic in the lattice constant. Therefore (9·3·2) is different from zero only if

$$\mathbf{k} - \mathbf{k}' \pm \mathbf{q} = 2\pi \mathbf{g}_b, \tag{9·3·3}$$

where $\mathbf{g}_b$ is an arbitrary vector of the reciprocal lattice. For the moment we shall assume that $\mathbf{g}_b = 0$, postponing a discussion of this simplification until § 9·9. When the condition (9·3·3) is satisfied, each cell contributes exactly the same amount, and (9·3·2) becomes

$$\int \mathbf{e}_{qj} \cdot \operatorname{grad} U_g \exp\{\mp i\mathbf{q} \cdot (\mathbf{r} - \mathbf{g}_a)\} u_{\mathbf{k}'}^{*} u_{\mathbf{k}} d\tau_0,$$

i.e. $$-\int \mathbf{e}_{qj} \cdot \operatorname{grad} [u_{\mathbf{k}'}^{*} u_{\mathbf{k}} \exp\{\mp i\mathbf{q} \cdot (\mathbf{r} - \mathbf{g}_a)\}] U_g d\tau_0, \tag{9·3·4}$$

since the integrand is periodic. We can obtain an approximate expression for this last integral if we assume that U_g is of short range, so that the exponential factor does not vary much over the region where U_g is appreciable. There are two terms to consider, the first of which contains the gradient of the exponential factor, and is

$$\pm i\mathbf{e}_{qj} \cdot \mathbf{q} \int U_g u_{\mathbf{k}'}^{*} u_{\mathbf{k}} \exp\{\mp i\mathbf{q} \cdot (\mathbf{r} - \mathbf{g}_a)\} d\tau_0.$$

Inside the atomic cores $u_{\mathbf{k}}$ and $u_{\mathbf{k}'}$ are nearly the same, especially if $|\mathbf{k} - \mathbf{k}'|$ is small, and we can therefore obtain an approximate value of the integral by taking the exponential factor as constant and putting $\mathbf{k} = \mathbf{k}'$. The integral is then

$$\pm i\mathbf{e}_{qj} \cdot \mathbf{q} \int U_g |u_{\mathbf{k}}|^2 d\tau_0. \tag{9·3·5}$$

The second term arises from the gradient of $u_{\mathbf{k}} u_{\mathbf{k}'}^*$. We may again take the exponential as constant, obtaining

$$-\int \mathbf{e}_{qj}.\operatorname{grad}(u_{\mathbf{k}'}^* u_{\mathbf{k}})\, U_{\mathbf{g}}\, d\tau_0.$$

Instead of $U_{\mathbf{g}}$ we may write $V = V(\mathbf{r}, 0)$, since the potential in one cell is supposed to be almost entirely due to one atom. To evaluate the integral we proceed as follows. The Schrödinger equation for $u_{\mathbf{k}}$ is

$$\frac{h^2}{8\pi^2 m}(\nabla^2 + 2i\mathbf{k}.\operatorname{grad} - \mathbf{k}^2)\, u_{\mathbf{k}} + (E_{\mathbf{k}} - V)\, u_{\mathbf{k}} = 0.$$

Multiply this by $\partial u_{\mathbf{k}}^*/\partial s$, where s is the direction of $\mathbf{e}_{qj}$, and add the complex conjugate of the corresponding equation obtained by interchanging $\mathbf{k}$ and $\mathbf{k}'$, obtaining

$$-\int V \frac{\partial}{\partial s}(u_{\mathbf{k}'}^* u_{\mathbf{k}})\, d\tau_0$$

$$= -\frac{ih^2}{4\pi^2 m}\int \frac{\partial u_{\mathbf{k}'}^*}{\partial s}(\mathbf{k} - \mathbf{k}').\operatorname{grad} u_{\mathbf{k}}\, d\tau_0 - \left\{E_{\mathbf{k}} - E_{\mathbf{k}'} - \frac{h^2}{8\pi^2 m}(\mathbf{k}^2 - \mathbf{k}'^2)\right\}\int u_{\mathbf{k}} \frac{\partial u_{\mathbf{k}'}^*}{\partial s}\, d\tau_0.$$

$$(9\cdot3\cdot6)$$

Certain terms have been integrated by parts, all the surface integrals vanishing on account of the periodicity of the integrands. The second term on the right-hand side is in general smaller than the first, and for nearly free electrons it is very much smaller, so that we shall neglect it entirely. Also the functions $u_{\mathbf{k}}$ and $u_{\mathbf{k}'}$ are practically constant outside the atomic cores, and are spherically symmetrical and nearly equal inside. Assuming therefore that $u_{\mathbf{k}}$ is spherically symmetrical, we can write $(9\cdot3\cdot6)$ as

$$-\int V \frac{\partial}{\partial s}(u_{\mathbf{k}'}^* u_{\mathbf{k}})\, d\tau_0 = \pm \frac{ih^2}{4\pi^2 m}\int \frac{\partial u_{\mathbf{k}'}^*}{\partial s}\mathbf{q}.\operatorname{grad} u_{\mathbf{k}}\, d\tau_0 = \pm \frac{ih^2}{4\pi^2 m}\mathbf{q}.\mathbf{e}_{qj}\int \left|\frac{\partial u_{\mathbf{k}}}{\partial s}\right|^2 d\tau_0.$$

$$(9\cdot3\cdot7)$$

From $(9\cdot3\cdot5)$ and $(9\cdot3\cdot7)$ we see that only the longitudinal vibrations have any effect, which means that the coupling between the electrons and the lattice vibrations is due mainly to changes in the density of the solid, there being no volume change associated with the transverse vibrations. In a more exact theory the transverse vibrations would play a part, but we should then be compelled to take into account the dispersion of the lattice waves and the fact that for large values of $|\mathbf{q}|$ the waves cannot be considered as exactly longitudinal or transverse. In future we shall drop the suffix j.

Writing $$C = \frac{h^2}{12\pi^2 m}\int |\operatorname{grad} u_{\mathbf{k}}|^2\, d\tau_0 + \int V |u_{\mathbf{k}}|^2\, d\tau_0, \qquad (9\cdot3\cdot8)$$

the matrix elements of ΔV with respect to the electronic wave functions and associated with the normal mode $\mathbf{q}$ are

$$(\mathbf{k} \pm \mathbf{q} \,|\, \Delta V \,|\, \mathbf{k}) = \pm \frac{iC\,|\,\mathbf{q}\,|}{2(2G+1)^{\frac{1}{2}}} \left(\xi_{\mathbf{q}} \pm \frac{ip_{\mathbf{q}}}{2\pi M \nu_{\mathbf{q}}} \right). \tag{9·3·9}$$

The quantity C is of the same order as the energy of the state $\mathbf{k}$, since the first term is two-thirds of the kinetic energy associated with $u_{\mathbf{k}}$, and the second term is the mean potential energy. C may therefore be expected to lie between 1 and 10 eV. Further, it cannot vary very much with $\mathbf{k}$, and we shall take it to be a constant.

9·31. The lattice wave functions. In § 9·2 we showed that it was possible to reduce the classical Hamiltonian for the lattice vibrations to a sum of squares. If, therefore, we set up the corresponding Schrödinger equation, it will be the equation for a number of independent simple harmonic oscillators, which are distinguished by $\mathbf{q}$ and j. The wave function breaks up into a product

$$\prod_{\mathbf{q},j} \phi(\xi_{\mathbf{q}j}),$$

where
$$\frac{d^2\phi}{d\xi^2} + \frac{8\pi^2 M}{h^2} (E - 2\pi^2 M \nu_{\mathbf{q}j}^2 \xi^2)\,\phi = 0.$$

The solution of this is the well-known one

$$\phi_N = e^{-\frac{1}{2}\lambda^2 \xi^2} H_N(\lambda \xi), \tag{9·31·1}$$

where H_N is the Nth Hermite polynomial ($N = 0, 1, 2, \ldots$), and

$$\lambda^2 = 4\pi^2 M \nu_{\mathbf{q}j}/h,$$

the energy levels being
$$E = (N + \tfrac{1}{2}) h \nu_{\mathbf{q}j}. \tag{9·31·2}$$

Since the perturbing energy ΔV contains the lattice coordinates linearly, only one oscillator at a time can take part in a transition, and, if an electron in the state $\mathbf{k}$ jumps to the state $\mathbf{k} + \mathbf{q}$ (or $\mathbf{k} - \mathbf{q}$), it is the oscillator associated with the longitudinal vibration $\mathbf{q}$ which undergoes a transition.

The recurrence relations for the Hermite polynomials are

$$dH_N(x)/dx = 2N H_{N-1}(x), \quad H_{N+1}(x) - 2x H_N(x) + 2N H_{N-1}(x) = 0, \tag{9·31·3}$$

so that

$$\frac{1}{2}\left(\xi + \frac{1}{\lambda^2} \frac{\partial}{\partial \xi} \right) \phi_N(\xi) = \frac{N}{\lambda} \phi_{N-1}(\xi), \quad \frac{1}{2}\left(\xi - \frac{1}{\lambda^2} \frac{\partial}{\partial \xi} \right) \phi_N(\xi) = \frac{1}{2\lambda} \phi_{N+1}(\xi). \tag{9·31·4}$$

Also
$$\int_{-\infty}^{\infty} \{H_N(x)\}^2\, e^{-x^2}\, dx = \pi^{\frac{1}{2}} 2^N N!, \tag{9·31·5}$$

and the only non-zero matrix elements associated with the factor in (9·3·9) referring to the lattice vibrations are

$$\frac{1}{2}\int\phi_{N-1}\Big(\xi+\frac{1}{\lambda^2}\frac{\partial}{\partial\xi}\Big)\phi_N\,d\xi=\Big(\frac{hN}{8\pi^2Mv}\Big)^{\frac{1}{2}},\quad \frac{1}{2}\int\phi_{N+1}\Big(\xi-\frac{1}{\lambda^2}\frac{\partial}{\partial\xi}\Big)\phi_N\,d\xi=\Big\{\frac{h(N+1)}{8\pi^2Mv}\Big\}^{\frac{1}{2}}.$$

(9·31·6)

9·32. The transition probabilities. If we denote the electronic wave functions generically by $\psi_k(\mathbf{r})$ and the lattice wave functions by $\phi_N(\mathbf{R})$, the general solution of the time-dependent Schrödinger equation is

$$\Psi=\sum_{k,N}c_{kN}(t)\,\psi_k(\mathbf{r})\,\phi_N(\mathbf{R})\,e^{-iE_{kN}t/\hbar},\qquad(9\cdot32\cdot1)$$

and according to the usual perturbation theory the equations determining the coefficients are

$$-i\hbar\dot{c}_{kN}(t)=\sum_{k',N'}c_{k'N'}(t)\,(kN\,|\,\Delta V\,|\,k'N')\,e^{i(E_{kN}-E_{k'N'})t/\hbar}.\qquad(9\cdot32\cdot2)$$

If at $t=0$ only the state kN is excited, the solution of (9·32·2) is

$$-i\hbar c_{k'N'}(t)=(k'N'\,|\,\Delta V\,|\,kN)\frac{e^{i(E_{k'N'}-E_{kN})t/\hbar}-1}{i(E_{k'N'}-E_{kN})/\hbar},$$

and the probability per unit time of a transition $k\to k'$, $N\to N'$ is given by

$$\frac{\partial}{\partial t}|\,c_{k'N'}(t)\,|^2=\frac{2}{\hbar^2}|\,(k'N'\,|\,\Delta V\,|\,kN)\,|^2\,\Omega(E_{k'N'}-E_{kN}),\qquad(9\cdot32\cdot3)$$

where

$$\Omega(x)=\frac{\partial}{\partial t}\frac{1-\cos xt/\hbar}{x^2/\hbar^2}=\frac{\sin xt/\hbar}{x/\hbar}.\qquad(9\cdot32\cdot4)$$

9·33. Collecting together the results of the preceding sections, the transition $\mathbf{k}\to\mathbf{k}'$ of an electron and the simultaneous transition $N_q\to N_q\mp1$ of the qth lattice mode can only occur if $\mathbf{k}-\mathbf{k}'\pm\mathbf{q}=0$, and the corresponding transition probabilities are

$$\frac{\partial}{\partial t}|\,c(\mathbf{k}+\mathbf{q},N_q-1)\,|^2=\frac{C^2\,|\,\mathbf{q}\,|^2\,N_q}{(2G+1)^3\,Mh\nu_q}\,\Omega(E_{\mathbf{k+q}}-E_{\mathbf{k}}-h\nu_q),\qquad(9\cdot33\cdot1)$$

$$\frac{\partial}{\partial t}|\,c(\mathbf{k}-\mathbf{q},N_q+1)\,|^2=\frac{C^2\,|\,\mathbf{q}\,|^2(N_q+1)}{(2G+1)^3\,Mh\nu_q}\,\Omega(E_{\mathbf{k-q}}-E_{\mathbf{k}}+h\nu_q).\qquad(9\cdot33\cdot2)$$

We have also to take into account the thermal distribution of the oscillators, and replace N_q by its average value. In the equilibrium state

$$\text{Av}\,(N_q)=\frac{\Sigma N_q\exp(-hN_q\nu_q/kT)}{\Sigma\exp(-hN_q\nu_q/kT)}=\frac{1}{e^{h\nu_q/kT}-1}.\qquad(9\cdot33\cdot3)$$

We shall write N_q for the average value, since in future only this average occurs.

The number of electrons forced into the state **k** by collisions, per unit time, is

$$\frac{C^2}{(2G+1)^3 Mh} \sum_{\mathbf{q}} \frac{|\mathbf{q}|^2}{\nu_{\mathbf{q}}} [f(\mathbf{k}+\mathbf{q})\{1-f(\mathbf{k})\}(N_{\mathbf{q}}+1)\,\Omega(E_{\mathbf{k}}-E_{\mathbf{k}+\mathbf{q}}+h\nu_{\mathbf{q}})$$
$$+f(\mathbf{k}-\mathbf{q})\{1-f(\mathbf{k})\}N_{\mathbf{q}}\,\Omega(E_{\mathbf{k}}-E_{\mathbf{k}-\mathbf{q}}-h\nu_{\mathbf{q}})], \quad (9\cdot33\cdot4)$$

while the number ejected from **k** per unit time is

$$\frac{C^2}{(2G+1)^3 Mh} \sum_{\mathbf{q}} \frac{|\mathbf{q}|^2}{\nu_{\mathbf{q}}} [f(\mathbf{k})\{1-f(\mathbf{k}+\mathbf{q})\}N_{\mathbf{q}}\,\Omega(E_{\mathbf{k}}-E_{\mathbf{k}+\mathbf{q}}+h\nu_{\mathbf{q}})$$
$$+f(\mathbf{k})\{1-f(\mathbf{k}-\mathbf{q})\}(N_{\mathbf{q}}+1)\,\Omega(E_{\mathbf{k}}-E_{\mathbf{k}-\mathbf{q}}-h\nu_{\mathbf{q}})]. \quad (9\cdot33\cdot5)$$

The change in the distribution function due to collisions is therefore given by

$$\left[\frac{\partial f}{\partial t}\right]_{\text{coll.}} = \frac{C^2}{(2G+1)^3 Mh} \sum_{\mathbf{q}} \frac{|\mathbf{q}|^2}{\nu_{\mathbf{q}}} [\{f(\mathbf{k}+\mathbf{q})(1-f(\mathbf{k}))(N_{\mathbf{q}}+1)$$
$$-f(\mathbf{k})(1-f(\mathbf{k}+\mathbf{q}))N_{\mathbf{q}}\}\,\Omega(E_{\mathbf{k}}-E_{\mathbf{k}+\mathbf{q}}+h\nu_{\mathbf{q}})$$
$$+\{f(\mathbf{k}+\mathbf{q})(1-f(\mathbf{k}))N_{-\mathbf{q}}$$
$$-f(\mathbf{k})(1-f(\mathbf{k}+\mathbf{q}))(N_{-\mathbf{q}}+1)\}\,\Omega(E_{\mathbf{k}}-E_{\mathbf{k}+\mathbf{q}}-h\nu_{\mathbf{q}})]. \quad (9\cdot33\cdot6)$$

In the last set of terms we have written $-\mathbf{q}$ for $\mathbf{q}$ to make the expressions more symmetrical. The frequency $\nu_{\mathbf{q}}$ is an even function of $\mathbf{q}$, but we have not assumed that $N_{\mathbf{q}}=N_{-\mathbf{q}}$, that is, we have not assumed that $N_{\mathbf{q}}$ necessarily has its equilibrium value $(9\cdot33\cdot3)$. In § 8·1 we denoted the transition probabilities by $\mathscr{W}(\mathbf{k},\mathbf{k}')$. From $(8\cdot1\cdot4)$, $(9\cdot33\cdot4)$ and $(9\cdot33\cdot5)$ we see that in the equilibrium state

$$\frac{\mathscr{W}(\mathbf{k}+\mathbf{q},\mathbf{k})}{\mathscr{W}(\mathbf{k},\mathbf{k}+\mathbf{q})} = \frac{N_{\mathbf{q}}+1}{N_{\mathbf{q}}} = e^{h\nu_{\mathbf{q}}/kT}.$$

The factor $\Omega(E_{\mathbf{k}}-E_{\mathbf{k}+\mathbf{q}}\pm h\nu_{\mathbf{q}})$, as is well known, acts as a kind of δ function, which is different from zero only when $E_{\mathbf{k}}-E_{\mathbf{k}+\mathbf{q}}\pm h\nu_{\mathbf{q}}$ is zero, so that

$$\frac{\mathscr{W}(\mathbf{k}+\mathbf{q},\mathbf{k})}{\mathscr{W}(\mathbf{k},\mathbf{k}+\mathbf{q})} = \frac{e^{E_{\mathbf{k}+\mathbf{q}}/kT}}{e^{E_{\mathbf{k}}/kT}},$$

which is the relation already derived in § 8·1. There is an exactly similar relation for $\mathscr{W}(\mathbf{k}-\mathbf{q},\mathbf{k})$, and we see that these symmetry relations are equivalent to the assumption that $N_{\mathbf{q}}$ and $N_{-\mathbf{q}}$ have their equilibrium values $(9\cdot33\cdot3)$. We shall make this assumption until § 9·9, where we shall examine it more carefully.

If we put $f=f_0-\Phi\,\partial f_0/\partial E$ and replace the summation with respect to **q** by the integration

$$\left(\frac{2G+1}{2\pi}\right)^3 \Delta \int d\mathbf{q},$$

where Δ is the volume of the unit cell, we obtain

$$\left[\frac{\partial f}{\partial t}\right]_{\text{coll.}} = \frac{C^2\Delta}{8\pi^3 MhkT} \int \frac{|\mathbf{q}|^2}{\nu_q} N_q [f_0(\mathbf{k})\{1 - f_0(\mathbf{k+q})\}\,\Omega(E_\mathbf{k} - E_\mathbf{k+q} + h\nu_q)$$

$$+ f_0(\mathbf{k+q})\{1 - f_0(\mathbf{k})\}\,\Omega(E_\mathbf{k} - E_\mathbf{k+q} - h\nu_q)]\{\Phi(\mathbf{k+q}) - \Phi(\mathbf{k})\}\,d\mathbf{q}.$$

$$(9\cdot33\cdot7)$$

The most symmetrical form for $[\partial f/\partial t]_{\text{coll.}}$ is that given in equation (8·1·8), namely,

$$\left[\frac{\partial f}{\partial t}\right]_{\text{coll.}} = -\frac{1}{kT} \int V(\mathbf{k}, \mathbf{k}')\{\Phi(\mathbf{k}) - \Phi(\mathbf{k}')\}\,d\mathbf{k}'. \qquad (9\cdot33\cdot8)$$

To transform (9·33·7) into an expression of this type, change the variables of integration from $\mathbf{q}$ to $\mathbf{k}' = \mathbf{k} + \mathbf{q}$. We then obtain

$$V(\mathbf{k}, \mathbf{k}') = \frac{C^2\Delta}{8\pi^3 Mh} \frac{|\mathbf{k} - \mathbf{k}'|^2}{\nu_{|\mathbf{k}-\mathbf{k}'|}} \frac{f_0(\mathbf{k})\,f_0(\mathbf{k}')}{|e^{-(E-\zeta)/kT} - e^{-(E'-\zeta)/kT}|}$$

$$\times \{\Omega(E - E' + h\nu_{|\mathbf{k}-\mathbf{k}'|}) + \Omega(E - E' - h\nu_{|\mathbf{k}-\mathbf{k}'|})\}, \quad (9\cdot33\cdot9)$$

from which the symmetry properties of $V(\mathbf{k}, \mathbf{k}')$ are obvious. In many ways this is the preferable form of the collision operator, but if detailed calculations have to be carried out it is usually necessary at some stage or other to revert to $\mathbf{q}$ as a variable of integration, and it is then simpler to carry through all the calculations starting from (9·33·7).

9.34. The triple integrals occurring in (9·33·7) can always be reduced to double integrals by taking $E_\mathbf{k} - E_\mathbf{k+q} \pm h\nu_q$ as one of the variables of integration, but the resulting expressions are very complicated and cannot in general be simplified further. So far it has only proved possible to carry out the calculations completely when $E_\mathbf{k}$ is a quadratic function of $|\mathbf{k}|$, and therefore has spherical symmetry. It is also necessary to make the further assumptions, which have already been mentioned, that the frequencies of the lattice vibrations are the same as those of an elastic continuum (ν_q a function of $|\mathbf{q}|$ only), and that the values of $|\mathbf{q}|$ lie inside a sphere of radius q_0.

Writing $K = |\mathbf{k}|$ and $q = |\mathbf{q}|$, we have

$$E_\mathbf{k+q} - E_\mathbf{k} \pm h\nu_q = \mathbf{q}\,.\,\text{grad}_\mathbf{k}\,E + \tfrac{1}{2}q^2 d^2 E/dK^2 \pm h\nu_q$$

$$= q\frac{dE}{dK}\cos\vartheta + \tfrac{1}{2}q^2\frac{d^2 E}{dK^2} \pm h\nu_q,$$

where ϑ is the angle between the vectors $\mathbf{k}$ and $\mathbf{q}$. Therefore, introducing polar coordinates q, ϑ, ϖ in the $\mathbf{q}$ space, the polar axis being in the direction of $\mathbf{k}$, the part of the integrals depending on ϑ contains, apart from slowly varying quantities, the expression

$$\int_0^\pi \Omega\left(q\frac{dE}{dK}\cos\vartheta + E_q \pm h\nu_q\right)\sin\vartheta\,d\vartheta.$$

Putting

$$y = \frac{2\pi t}{h}\left(q\frac{dE}{dK}\cos\vartheta + E_q \pm h\nu_q\right),$$

this becomes

$$\frac{h}{2\pi q(dE/dK)}\int_{\vartheta=\pi}^{\vartheta=0}\frac{\sin y}{y}\,dy, \qquad (9\cdot34\cdot1)$$

the upper limit of integration being

$$y = \frac{2\pi t}{h}\left(E_q \pm h\nu_q + q\frac{dE}{dK}\right),$$

and the lower being

$$y = \frac{2\pi t}{h}\left(E_q \pm h\nu_q - q\frac{dE}{dK}\right).$$

If these limits are of opposite sign, then for large values of t the integral in $(9\cdot34\cdot1)$ is effectively

$$\int_{-\infty}^{\infty}\frac{\sin y}{y}\,dy = \pi,$$

while if the limits have the same sign the integral is zero. We must therefore examine these limits carefully. If $E = \lambda K^2$ and $\nu_q = u_0 q/2\pi$, then

$$E_q \pm q\frac{dE}{dK} \pm h\nu_q = \nu_q\left(\frac{4\pi^2\lambda\nu_q}{u_0^2} \pm \frac{4\pi\lambda K}{u_0} \pm h\right).$$

Taking the upper sign for $h\nu_q$, we see that the upper limit is always positive, but that the lower limit becomes positive if

$$\nu_q > \frac{u_0 K}{\pi} - \frac{h u_0^2}{4\pi^2\lambda}. \qquad (9\cdot34\cdot2a)$$

Considering next the lower sign for $h\nu_q$, the upper limit becomes negative if

$$\nu_q < \frac{h u_0^2}{4\pi^2\lambda} - \frac{u_0 K}{\pi}, \qquad (9\cdot34\cdot2b)$$

and the lower limit becomes positive if

$$\nu_q > \frac{h u_0^2}{4\pi^2\lambda} + \frac{u_0 K}{\pi}. \qquad (9\cdot34\cdot2c)$$

The permissible ranges of ν_q and K, which are given by reversing the inequalities $(9\cdot34\cdot2)$, can be put into a more convenient form by considering ν_q to take on negative as well as positive values. We can then write $E_{k+q} = E_k + h\nu_q$ with ν_q positive when an electron absorbs a sound quantum and ν_q negative when a sound quantum is emitted. With this convention the inequalities representing the permitted ranges of ν_q and K can all be written as

$$\frac{u_0 K}{\pi} > \left| \nu_q - \frac{h u_0^2}{4\pi^2\lambda} \right|. \tag{9·34·3}$$

The second term on the right can usually be neglected. For λ is of the order of 10^{-28} so that $h u_0^2/(4\pi^2\lambda)$ is of the order of 10^{10}, while ν_0, the maximum lattice frequency, is of the order of 10^{12}, so that $h u_0^2/(4\pi^2\lambda\nu_q)$ can be neglected except for very long lattice waves.

In addition to the inequality (9·34·3) we must have

$$h\nu_q \leqslant k\Theta = (3/4\pi)^{\frac{1}{3}} h u_0/a. \tag{9·34·4}$$

Now when $E = \zeta$ we have $K = \pi(3n/\pi)^{\frac{1}{3}}$, and hence (9·34·4) is a more restrictive condition on ν_q than (9·34·3) unless $n < \frac{1}{4}n_a$, where $n_a = 1/a^3$ is the number of atoms per unit volume. The restrictions on ν are therefore

$$\left. \begin{array}{ll} h\nu < k\Theta & (n > \frac{1}{4}n_a), \\ h\nu < h u_0 K/\pi & (n < \frac{1}{4}n). \end{array} \right\} \tag{9·34·5}$$

In general, the first inequality is the one to be used, but for semi-metals, and particularly for semi-conductors, the second inequality is the more restrictive.

In order to carry out a second integration it is necessary to put

$$\Phi(\mathbf{k}) = k_1 c(E), \tag{9·34·6}$$

which is the appropriate form when there is an electric field and a temperature gradient along the x-axis. In general, Φ consists of three terms of the above type, but when required they can be obtained at once by replacing k_1 by k_2 or k_3.

We shall now restrict the remaining calculations of this section to the case of free electrons (with an effective mass m^*). If the conduction electrons occupy two or more bands in each of which $E_k = \text{constant} + \lambda K^2$, and if the inter-band transitions are zero, the electrons in each band can be treated as independent of the electrons in the other bands. This case can therefore be considered to be a trivial generalization of the free-electron case, there being a separate distribution function for each band. This applies, for example, to the two-band model discussed extensively in Chapter VIII. (Note, however, that λ can have either sign, and that when $\lambda < 0$ the conduction is most simply thought of as being due to 'positive holes'.) The requirement that there should be no inter-band transitions is unnecessarily restrictive but will be adhered to for the moment. It is considered further in § 9·51.

When we substitute (9·34·6) into (9·33·7), the only quantity depending upon the azimuth ϖ in the q space is q_1, and if θ and θ_1 are the angles between the x-axis and $\mathbf{k}$ and $\mathbf{q}$ respectively, then

$$q_1 = q \cos \theta_1 = q(\cos \theta \cos \vartheta + \sin \theta \sin \vartheta \cos \varpi),$$

so that
$$\int_0^{2\pi} q_1 \, d\varpi = 2\pi q \cos \vartheta \, k_1/K.$$

Now since almost the whole of the integrals involving the Ω's comes from the neighbourhood of $y = 0$ we can replace any slowly varying functions of ϑ by their values when $y = 0$, i.e. we can put

$$\cos \vartheta = - \frac{\tfrac{1}{2}q^2(d^2 E/dK^2) \pm h\nu_q}{q \, dE/dK}, \tag{9·34·7}$$

and in particular

$$\int_0^{2\pi} q_1 \, d\varpi = -\pi k_1 \left(\frac{q^2}{K} \frac{d^2 E/dK^2}{dE/dK} \pm \frac{2h\nu_q}{K \, dE/dK} \right) = -\pi k_1 \left(\frac{q^2}{K^2} \pm \frac{h\nu_q}{E} \right), \tag{9·34·8}$$

since $E = \lambda K^2$. If we use the relations $E_{\mathbf{k}+\mathbf{q}} - E_\mathbf{k} \pm h\nu_q = 0$ to eliminate $E_{\mathbf{k}+\mathbf{q}}$, and assume that N_q has its equilibrium value (9·33·3), (9·33·7) becomes

$$\left[\frac{\partial f}{\partial t} \right]_{\text{coll.}} = - \frac{C^2 \Delta}{8\pi^2 M \, dE/dK} k_1 \frac{\partial f_0}{\partial E}$$

$$\times \int_0^{q_0} \frac{q}{\nu_q} N_q \left[\left\{ \left(1 + \frac{h\nu_q}{2E} - \frac{q^2}{2K^2} \right) c(E + h\nu_q) - c(E) \right\} \frac{f_0(E + h\nu_q)}{f_0(E)} e^{h\nu_q/kT} \right.$$

$$+ \left. \left\{ \left(1 - \frac{h\nu_q}{2E} - \frac{q^2}{2K^2} \right) c(E - h\nu_q) - c(E) \right\} \frac{f_0(E - h\nu_q)}{f_0(E)} \right] q^2 dq. \tag{9·34·9}$$

If we write $\eta = (E - \zeta)/kT$, $z = h\nu_q/kT$ and allow z to take negative as well as positive values, we obtain

$$\left[\frac{\partial f}{\partial t} \right]_{\text{coll.}} \equiv -k_1 \mathscr{L}(c)$$

$$= k_1 \frac{\partial f_0}{\partial E} \frac{(\tfrac{1}{2}m^*)^{\frac{3}{2}}}{\hbar^2 \Lambda E^{\frac{3}{2}}} \left(\frac{T}{\Theta} \right)^3 \int_{-\Theta/T}^{\Theta/T} \left[E c(\eta) \right.$$

$$\left. - c(\eta + z) \left\{ E + \tfrac{1}{2}kTz - D\left(\frac{T}{\Theta} \right)^2 z^2 \right\} \right] \frac{e^\eta + 1}{e^{\eta+z} + 1} \frac{z^2 \, dz}{|1 - e^{-z}|}, \tag{9·34·10}$$

where
$$\Lambda = \left(\frac{4\pi}{3} \right)^{\frac{1}{3}} \frac{4Mak\Theta}{3\hbar^2 C^2}, \qquad D = \frac{(6\pi^2)^{\frac{2}{3}} \hbar^2}{4m^* a^2}, \tag{9·34·11}$$

and where m^* is the effective mass. With this notation, the Boltzmann equation takes the form

$$\mathscr{L}(c) = \frac{\hbar}{m^*} \frac{\partial f_0}{\partial E} \left(\epsilon \mathscr{E} + T \frac{\partial}{\partial x} \frac{\zeta}{T} + \frac{E}{T} \frac{\partial T}{\partial x} \right). \tag{9·34·12}$$

9·35. *The time of relaxation.* We now show that when $(\Theta/T)^2$ is small compared with unity, a time of relaxation can be unambiguously defined for the scattering of the conduction electrons by the lattice vibrations. Since Θ/T is small we can expand the integrand of (9·34·10) in powers of z, obtaining

$$\left[\frac{\partial f}{\partial t}\right]_{\text{coll.}} = k_1 \frac{\partial f_0}{\partial E} \frac{(\tfrac{1}{2}m^*)^{\frac{1}{2}}}{\hbar^2 \Lambda E^{\frac{1}{2}}} \left(\frac{T}{\Theta}\right)^3 \int_0^{\Theta/T} \left[2D\left(\frac{T}{\Theta}\right)^2 c(\eta)\right.$$

$$\left. - kTc'(\eta) - Ec''(\eta) - \frac{e^\eta - 1}{e^\eta + 1}\{\tfrac{1}{2}kTc(\eta) + Ec'(\eta)\} + O(z^2)\right] z^3\,dz. \quad (9·35·1)$$

For sufficiently large values of T/Θ, only the first term in the square brackets need be retained and we have

$$\left[\frac{\partial f}{\partial t}\right]_{\text{coll.}} = k_1 \frac{\partial f_0}{\partial E} \frac{(\tfrac{1}{2}m^*)^{\frac{1}{2}} D T}{2\hbar^2 \Lambda E^{\frac{1}{2}} \Theta} c(\eta) = -\frac{f - f_0}{\tau}, \quad (9·35·2)$$

where

$$\tau = \frac{2\hbar^2 \Lambda}{(\tfrac{1}{2}m^*)^{\frac{1}{2}} D} \frac{\Theta}{T} E^{\frac{1}{2}} \quad (T \gg \Theta). \quad (9·35·3)$$

The time of relaxation can therefore be uniquely defined; it is proportional to $E^{\frac{1}{2}}$ and to T^{-1}, and all the results of Chapter VIII can be applied at once.

9·351. For semi-metals in which the number of conduction electrons is small the limits of integration in (9·34·10) and (9·35·1) must be changed to conform to the second inequality in (9·34·5). The correct limits of integration are obtained by replacing Θ/T by $(2E/D)^{\frac{1}{2}} \Theta/T$, and the time of relaxation is given by

$$\tau = \frac{\hbar^2 D \Lambda}{2(\tfrac{1}{2}m^*)^{\frac{1}{2}}} \frac{\Theta}{T} \frac{1}{E^{\frac{1}{2}}} \quad (T \gg (2E/D)^{\frac{1}{2}} \Theta). \quad (9·351·1)$$

Since only the longest lattice vibrations are capable of scattering the electrons in semi-metals, the time of relaxation in semi-metals must be considerably larger than that in metals (Sondheimer, 1952). This is in agreement with the evidence deduced from the behaviour of thin films and wires of bismuth and from the anomalous skin effect (§ 8·7), which shows that, although bismuth has an abnormally small number of free electrons and a low electrical conductivity, the free path is abnormally large.

9·36. *The conductivity of semi-conductors.* The collision operator for semi-conductors is the same as for semi-metals, but whereas for semi-metals the integral operator only takes a simple form for high temperatures, we can define a time of relaxation for semi-conductors for all temperatures. We put $f_0(E) = e^{-(E-\zeta)/kT}$ in (9·34·9), and restrict $h\nu$ to the range $\pm h u_0 K/\pi$.

We then have

$$\left[\frac{\partial f}{\partial t}\right]_{\text{coll.}} = k_1 \frac{\partial f_0}{\partial E} \frac{(\tfrac{1}{2}m^*)^{\frac{1}{2}}}{\hbar^2 \Lambda E^{\frac{1}{2}}} \left(\frac{T}{\Theta}\right)^3$$

$$\times \int_{-(2E/D)^{\frac{1}{2}}\Theta/T}^{(2E/D)^{\frac{1}{2}}\Theta/T} \left[Ec(\eta) - c(\eta+z)\left\{E + \tfrac{1}{2}kTz - D\left(\frac{T}{\Theta}\right)^2 z^2\right\}\right] \frac{e^{-z}z^2 dz}{|1-e^{-z}|}.$$

$$(9\cdot36\cdot1)$$

Now the maximum value of E that we need consider is of the order of kT and $(2kT/D)^{\frac{1}{2}}\Theta/T = (2k\Theta/D)^{\frac{1}{2}}(\Theta/T)^{\frac{1}{2}}$ is of the order of $\frac{1}{10}(\Theta/T)^{\frac{1}{2}}$. Therefore, except for extremely low temperatures which are of no interest since the conductivity is negligible there, we can expand the integrand of (9·36·1) in powers of z obtaining

$$\left[\frac{\partial f}{\partial t}\right]_{\text{coll.}} = -\frac{f-f_0}{\tau} = -(f-f_0)\frac{2(\tfrac{1}{2}m^*)^{\frac{1}{2}}}{\hbar^2 D\Lambda}\frac{T}{\Theta}E^{\frac{1}{2}}. \qquad (9\cdot36\cdot2)$$

The free path is $$l = \frac{2\pi}{h}\frac{dE}{dK}\tau = \frac{Mh^2k\Theta}{4\pi(6\pi^2)^{\frac{1}{3}}m^{*2}aC^2}\frac{\Theta}{T}, \qquad (9\cdot36\cdot3)$$

if the electrons have effective mass m^*. The conductivity is given by

$$\sigma = \frac{4}{3}\frac{n\epsilon^2 l}{(2\pi m^* kT)^{\frac{1}{2}}} \qquad (9\cdot36\cdot4)$$

and is proportional to $nm^{*-\frac{5}{2}}T^{-\frac{3}{2}}$. The free path is of the same order as that for metals but is independent of the energy. The temperature dependence of the conductivity is, in general, almost entirely determined by the variation in the number of free electrons.

9·361. In polar semi-conductors, that is, in all except the elemental semi-conductors silicon, germanium, grey tin, selenium and tellurium, the optical lattice vibrations are, in certain circumstances, more effective than the acoustical vibrations in scattering the electrons. In a polar solid the optical vibrations produce an electric polarization, and the vibrations can, to a first approximation, be treated as having the same frequency ν_l. The scattering of the electrons can be treated exactly as in §§ 9·3 to 9·34, the main difference being that, since ν_l is a constant instead of being proportional to the wave number q, the distribution function is determined by a difference equation instead of by an integral equation. By a different type of argument Fröhlich and Mott (1939) obtained the order of magnitude of the time of relaxation determining the electrical conductivity and showed that

$$\frac{1}{\tau_p} \propto \frac{m^{*\frac{1}{2}}\epsilon^4}{Ma^3}\frac{1}{(h\nu_l)^{\frac{1}{2}}}\frac{1}{e^{h\nu_l/kT}-1} \qquad (9\cdot361\cdot1)$$

for sufficiently slow electrons, where $2/M$ is the sum of the reciprocals of the masses of the ions. A more exact theory has been given by Howarth and Sondheimer (1953).

In semi-conductors the ionized atoms act as scattering centres having a Coulomb field. The time of relaxation in this case (see § 1·5 and Conwell and Weisskopf, 1950) is given by

$$\frac{1}{\tau_e} = \frac{\pi N \epsilon^4}{2\sqrt{2}\,\kappa^2 m^{*\frac{1}{2}} E^{\frac{3}{2}}} \log(1 + 36\kappa^2 d^2 k^2 T^2 \epsilon^{-4}), \qquad (9\cdot361\cdot2)$$

where N is the number of ionized scattering centres per unit volume, κ is the dielectric constant and $2d$ is the average distance apart of the ions.

When all the above mechanisms of scattering are present in a semi-conductor, the reciprocals of the times of relaxation are additive. Hence if we denote by τ_a the expression for τ associated with the acoustical lattice vibrations, given by (9·36·2), we have

$$\frac{1}{\tau} = \frac{1}{\tau_a} + \frac{1}{\tau_p} + \frac{1}{\tau_e}. \qquad (9\cdot361\cdot3)$$

The energy variation of τ does not therefore follow any simple law unless one of the individual τ's is much smaller than the other two. In addition the partial free path l_i, which is the average of $v\tau_i$ weighted by the distribution function of the electrons, depends upon the energy variation of τ_i and, unlike the case of metals where only one value of the velocity is involved, the reciprocal of the total free path is not the sum of the reciprocals of the separate averages of the $v\tau_i$'s.

At any particular temperature τ can be considered to be approximately of the form μv^p, with μ and p being functions of the temperature. The conductivity is then given by (1·41·4), so that

$$\sigma \propto n\mu(T)\,T^{\frac{1}{2}p(T)}.$$

In general, the time of relaxation will be mainly determined at each temperature by the smallest τ_i, and there will be large temperature ranges where the temperature variation of σ will be comparatively simple. The application of the above formulae has already been discussed in § 5·51.

THE RESIDUAL RESISTANCE

9·4. In general the part of the resistance which is due to impurities is very small, but for alloys the number of 'foreign atoms' is of the same order as the total number of atoms present, and unless a superlattice is formed the resistance due to the irregular distribution of the atoms over the lattice sites is large. We now calculate the scattering of the conduction electrons in an alloy; the scattering by impurities in a normal metallic specimen can be considered to be a special case in which the concentration of one of the constituents is very small. The present discussion is limited to a calculation of the effect of the irregular distribution of the atoms upon the resistance (Nordheim, 1931). Two other effects, the change in the number of free electrons and the change in the interaction constant C with the composition, are neglected.

The potential energy of an electron in the crystal can no longer be treated as periodic, but must be taken to be

$$V = \sum_{\mathbf{g}} V_{\mathbf{g}}(\mathbf{r} - \mathbf{g}_a),$$

where $V_{\mathbf{g}}(\mathbf{r} - \mathbf{g}_a)$ depends on the particular lattice point. The potential energy can, however, be considered to consist of a periodic part and an irregularly fluctuating part. The periodic part is of no interest to us, since it only affects the magnitude of C, and we therefore eliminate it as far as possible. To do this, we subtract from V the average potential energy, which is

$$\sum_{\mathbf{g}} \sum_{s} p_s V_s(\mathbf{r} - \mathbf{g}_a),$$

where p_s is the concentration of the atoms of the type s, and $V_s(\mathbf{r})$ is the potential energy due to one atom s. The average of the perturbing potential energy over the unperturbed states is then zero, and we have now to calculate the transition probability for an electron, the lattice remaining undisturbed. The matrix element which determines the transition is

$$(\mathbf{k'} \,|\, \Delta V \,|\, \mathbf{k}) = \int \psi_{\mathbf{k'}}^* \sum_{\mathbf{g}} [V_{\mathbf{g}}(\mathbf{r} - \mathbf{g}_a) - \sum_{s} p_s V_s(\mathbf{r} - \mathbf{g}_a)] \psi_{\mathbf{k}} d\tau,$$

and

$$|\,(\mathbf{k'} \,|\, \Delta V \,|\, \mathbf{k})\,|^2 = \sum_{\mathbf{g}} \sum_{\mathbf{h}} \int \psi_{\mathbf{k'}} \{V_{\mathbf{g}}(\mathbf{r} - \mathbf{g}_a) - \sum_{s} p_s V_s(\mathbf{r} - \mathbf{g}_a)\} \psi_{\mathbf{k}}^* d\tau$$

$$\times \int \psi_{\mathbf{k'}}^* \{V_{\mathbf{h}}(\mathbf{r} - \mathbf{h}_a) - \sum_{s} p_s V_s(\mathbf{r} - \mathbf{h}_a)\} \psi_{\mathbf{k}} d\tau.$$

Now the distribution of atoms is supposed to be random, so that only the terms with $\mathbf{g} = \mathbf{h}$ contribute anything to the summation, and the summation with respect to $\mathbf{g}$ can be replaced by a summation with respect to the different atoms in the proper proportions and a multiplication by $(2G+1)^3$, the number of atoms in the crystal, so that

$$|\,(\mathbf{k'} \,|\, \Delta V \,|\, \mathbf{k})\,|^2 = (2G+1)^3 \sum_{s} p_s \left| \int \psi_{\mathbf{k'}}^*(V_s - \sum_{t} p_t V_t) \psi_{\mathbf{k}} d\tau \right|^2$$

$$= (2G+1)^{-3} \sum_{s} p_s \left| \int u_{\mathbf{k'}}^*(V_s - \sum_{t} p_t V_t) u_{\mathbf{k}} d\tau_0 \right|^2,$$

u_k being, as usual, normalized in the unit cell, whereas ψ_k is normalized in the whole crystal. Since p_s is a concentration, we have

$$\sum_s p_s = 1,$$

and the matrix element can easily be transformed into

$$|(\mathbf{k'}|\Delta V|\mathbf{k})|^2 = \tfrac{1}{2}(2G+1)^{-3}\sum_{s,t} p_s p_t |V_s - V_t|^2_{\mathbf{kk'}}, \qquad (9\cdot4\cdot1)$$

where

$$|V_s - V_t|^2_{\mathbf{kk'}} = \left|\int (V_s - V_t) u_k^* u_k d\tau_0\right|^2 \qquad (9\cdot4\cdot2)$$

The collision operator is

$$\left[\frac{\partial f}{\partial t}\right]_{\text{coll.}} = (2G+1)^3\frac{\Delta}{\pi h^2}\int [f(\mathbf{k'})\{1-f(\mathbf{k})\}\,\Omega(E_{\mathbf{k}}-E_{\mathbf{k'}})$$
$$-f(\mathbf{k})\{1-f(\mathbf{k'})\}\,\Omega(E_{\mathbf{k'}}-E_{\mathbf{k}})]\,|(\mathbf{k'}|\Delta V|\mathbf{k})|^2 d\mathbf{k'},$$

and can be reduced to a more manageable form by introducing polar coordinates K', ϑ, ϖ, the polar axis being along the direction of $\mathbf{k}$. Instead of proceeding exactly as in §9·34 it is convenient to change the variable from K' to $E_{\mathbf{k'}}$. The collisions are, of course, elastic, since $\Omega(E_{\mathbf{k}}-E_{\mathbf{k'}})$ is effectively different from zero only if $E_{\mathbf{k}}=E_{\mathbf{k'}}$, and if we assume that $(\mathbf{k'}|\Delta V|\mathbf{k})$ is a function of the angle ϑ between $\mathbf{k}$ and $\mathbf{k'}$ we can carry out all the integrations but one for an isotropic body. Putting

$$f = f_0 - k_1 c(E)\,\partial f_0/\partial E$$

and performing the integration with respect to $E_{\mathbf{k'}}$, we have

$$\left[\frac{\partial f}{\partial t}\right]_{\text{coll.}} = -c(E)\frac{\partial f_0}{\partial E}K^2\frac{dK}{dE}\frac{\Delta}{4\pi h}\sum_{s,t} p_s p_t \iint |(V_s - V_t)|^2_{\vartheta}(k_1' - k_1)\sin\vartheta\,d\vartheta\,d\varpi.$$

Also since the collisions are elastic, we have $|\mathbf{k}|=|\mathbf{k'}|=K$, and if θ and α are the angles between the x axis and $\mathbf{k}$ and $\mathbf{k'}$ respectively, we have

$$k_1 = |\mathbf{k}|\cos\theta, \quad k_1' = |\mathbf{k'}|\cos\alpha = |\mathbf{k}|(\cos\theta\cos\vartheta + \sin\theta\sin\vartheta\cos\varpi),$$

so that

$$\int k_1' d\varpi = 2\pi|\mathbf{k}|\cos\theta\cos\vartheta = 2\pi k_1\cos\vartheta.$$

Hence

$$\left[\frac{\partial f}{\partial t}\right]_{\text{coll.}} = -(f-f_0)K^2\frac{dK}{dE}\frac{\Delta}{2h}\sum_{s,t} p_s p_t U_{st}, \qquad (9\cdot4\cdot3)$$

where

$$U_{st} = \int |V_s - V_t|^2_{\mathbf{kk'}}\sin\vartheta(1-\cos\vartheta)\,d\vartheta. \qquad (9\cdot4\cdot4)$$

A time of relaxation therefore always exists and is given by

$$\frac{1}{\tau} = \frac{\Delta}{2h}K^2\frac{dK}{dE}\sum_{s,t} p_s p_t\, U_{st}. \qquad (9\cdot4\cdot5)$$

The free path l is given by

$$\frac{1}{l} = \frac{\Delta}{4\pi} \left(K \frac{dK}{dE} \right)^2 \sum_{s,t} p_s p_t U_{st} = \frac{4\pi^3 \Delta m^{*2}}{h^4} \sum_{s,t} p_s p_t U_{st}, \qquad (9\cdot4\cdot6)$$

the last formula applying only to electrons with an effective mass m^* in a band of standard form.

A similar calculation applies to scattering by impurities, and we shall assume that the time of relaxation is given by

$$\frac{1}{\tau} = \sqrt{\frac{2}{m^*} \frac{E^{\frac{1}{2}}}{l_r}}, \qquad (9\cdot4\cdot7)$$

where the free path l_r is independent of E and of T. The collision operator $\mathscr{L}(c)$ is then given by

$$\mathscr{L}(c) = - \sqrt{\frac{2}{m^*} \frac{E^{\frac{1}{2}}}{l_r} \frac{\partial f_0}{\partial E}} c(E). \qquad (9\cdot4\cdot8)$$

THE ELECTRICAL CONDUCTIVITY AT NORMAL TEMPERATURES

9·5. *Numerical results.* For a completely degenerate electron gas the electrical conductivity is given by

$$\sigma = \frac{ne^2\tau(\zeta)}{m^*} = \frac{2^{\frac{1}{2}}ne^2\hbar^2\Lambda}{m^{*\frac{1}{2}}D} \zeta^{\frac{1}{2}} \frac{\Theta}{T} = \frac{32\sqrt{2}\,e^2\Delta Mk\Theta}{9\pi h^2} \frac{n\zeta^{\frac{1}{2}}}{m^{*\frac{1}{2}}C^2} \frac{\Theta}{T}. \qquad (9\cdot5\cdot1)$$

The resistance is therefore proportional to T, a result which is in good agreement with experiment, since even down to temperatures of the order of $0\cdot2\Theta$ the resistance is a linear function of T. All the constants in the expression $(9\cdot5\cdot1)$ for σ are known with the exception of n, ζ, m^* and C, while for the monovalent elements we can also consider n, ζ and m^* to be known. In principle, we can calculate C by means of the wave functions obtained by the methods of § 3·32, but a simpler procedure is to use the measured conductivities to find C.

For monovalent metals and perfectly free electrons $(9\cdot5\cdot1)$ becomes

$$\sigma = 2\cdot83 \times 10^{-32} \frac{nM_A}{C^2} \frac{\Theta^2}{T} \text{ e.s.u.,} \qquad (9\cdot5\cdot2)$$

where M_A is the atomic weight. In Table IX 1 the calculated values of ζ and C are given, the latter being obtained from the measured values of σ at $0°$ C. It will be seen that, as indicated at the end of § 9·3, C and ζ are of the same order of magnitude. The table also gives the values of $\sigma/(M_A\Theta^2)$.

For multivalent metals it is necessary to consider the electrons to occupy more than one band. We may then use the formula $(8\cdot21\cdot5)$ for σ, namely,

$$\sigma = \left(\frac{n_1\tau_1}{m_1} + \frac{n_2\tau_2}{m_2} \right) e^2,$$

and as a rough approximation we can assume that τ for each band is given by an expression such as (9·35·3). In this case the measured values of $\sigma/(M_A \Theta^2)$ will enable us to determine the order of magnitude of the quantity

$$\frac{1}{n_a} \left(\frac{n_1 \zeta_1^{\frac{4}{3}}}{m_1^{\frac{3}{2}} C_1^2} + \frac{n_2 \zeta_2^{\frac{4}{3}}}{m_2^{\frac{3}{2}} C_2^2} \right).$$

In Tables IX 1 and IX 2 the values of $\sigma/(M_A \Theta^2)$ are shown for a selection of metals including both good and bad conductors, and it will be seen that this quantity does not vary as much as might be expected. This is undoubtedly due to the partial compensation of the factors which affect the conductivity. We should, for example, expect C to be larger for divalent than for monovalent elements, since the electrons are more tightly bound, but ζ is also larger. It is therefore difficult to draw any quantitative conclusions from small variations in the conductivity from element to element, and only the gross differences are of any significance in the present state of the theory.

Table IX 1. *Calculated values of C and ζ in electron volts for monovalent metals*

	M_A	$n \times 10^{-21}$	Θ	σ in e.s.u. $\times 10^{-16}$	C in eV.	ζ in eV.	$\sigma/(M_A \Theta^2)$ in e.s.u. $\times 10^{-10}$
Li	7	48	430	10·5	4·9	4·8	8
Na	23	26	160	20·8	1·7	3·2	36
K	39	14	99	13·3	1·3	2·1	35
Cu	64	85	310	57	6·1	7·0	9
Rb	85	11	59	7·7	1·3	1·8	26
Ag	108	59	220	60	4·6	5·5	12
Cs	133	9	43	5·0	1·3	1·57	20
Au	197	59	185	40	6·35	5·5	6

The difference between the conductivity of the divalent and the monovalent metals can be ascribed almost wholly to the fact that the effective number of conduction electrons in divalent metals is less than in monovalent metals. The same reason accounts for the poor conductivity of antimony and bismuth, except that for these semi-metals the effective number of conduction electrons is of the order of 10^{-3} per atom or less, whereas it is of the order of one-half for the divalent metals.

The low conductivity of the transition metals, on the other hand, cannot be ascribed to there being very few conduction electrons, since all the other evidence points to the number of conduction electrons being of the order of one-half per atom. It is therefore probable that the low conductivity is due to an abnormally small time of relaxation, i.e. to an abnormally large scattering of the electrons by the lattice vibrations. Mott (1935, 1936) has suggested that the major part of the scattering is due to inter-band transitions, and the investigation of these requires refinements in the theory which we now proceed to investigate.

Table IX 2. *Values of $\sigma/(M_A \Theta^2)$ for multivalent metals*

	M_A	Θ	σ in e.s.u. $\times 10^{-16}$	$\sigma/(M_A \Theta^2)$ in e.s.u. $\times 10^{-10}$
Be	9	900	16·2	2·2
Mg	24	330	22·5	8·6
Al	27	410	36	8·0
Ca	40	220	22·5	11·5
Fe	56	355	10	1·4
Co	59	385	14·5	1·7
Ni	59	320	13·6	2·2
Zn	65	240	16·4	4·4
Sr	88	148	3	1·6
Pd	107	270	8·8	1·1
Cd	112	165	13·5	4·4
Sn	119	160	8·6	2·8
Sb	122	140	2·5	1·0
Ba	137	116	1·6	0·85
Ta	181	245	6·5	0·6
W	184	315	18·5	1·0
Ir	193	285	19	1·2
Pt	195	225	9·2	0·95
Pb	207	88	4·6	2·8
Bi	209	80	0·83	0·6

9·51. *The conductivity of the transition metals.* We have already seen in §3·612 that the transition elements are peculiar in that, in the solid state, the valency electrons are divided between an *s*- and a *d*-band, the *s*-band being more or less normal but the *d*-band being very narrow with a corresponding large density of states. Owing to the large effective mass of the *d*-electrons their contribution to the conductivity will be small and can be neglected compared with that due to the *s*-electrons. The vacant levels in the *d*-band have, however, a considerable effect upon the conductivity of the electrons, since the *s*-electrons can be scattered not only into energy levels in the *s*-band (*s-s* transitions) but also into energy levels in the *d*-band (*s-d* transitions). Now the transition probability is proportional to the density of energy levels in the final state, and, since the density is large for the *d*-levels, the *s-d* transitions will give rise to a scattering probability that is much greater than normal.

While the above explanation is undoubtedly qualitatively correct, its quantitative aspects leave much to be desired. Since the exact theory (Wilson, 1938) is somewhat complicated, only a simplified version will be given here, dealing with the main points.

We assume that the energy bands are as shown in fig. II 14 (i) p. 45, i.e. that

$$E_s(\mathbf{k}) = \tfrac{1}{2}\hbar^2 \,|\,\mathbf{k}\,|^2/m_s, \quad E_d(\mathbf{k}) = A - \tfrac{1}{2}\hbar^2 \,|\,\mathbf{k}\,|^2/m_d \quad (A > 0).$$

The calculation of the s-d transitions then proceeds exactly as in § 9·33, the only change being that each initial state $\mathbf{k}$ refers to the s-band while each final state $\mathbf{k}+\mathbf{q}$ refers to the d-band. We therefore have to replace equation (9·33·7) by the equation

$$\left[\frac{\partial f_s}{\partial t}\right]_{sd} = \frac{\varpi_d C_{sd}^2 \Delta}{8\pi^3 MhkT} \int \frac{|\,\mathbf{q}\,|^2}{\nu_{\mathbf{q}}} N_{\mathbf{q}} [f_{0s}(\mathbf{k})\{1 - f_{0d}(\mathbf{k}+\mathbf{q})\}\,\Omega(E_{\mathbf{k},s} - E_{\mathbf{k}+\mathbf{q},d} + h\nu_{\mathbf{q}})$$

$$+ f_{0d}(\mathbf{k}+\mathbf{q})\{1 - f_{0s}(\mathbf{k})\}\,\Omega(E_{\mathbf{k},s} - E_{\mathbf{k}+\mathbf{q},d} - h\nu_{\mathbf{q}})] \{\Phi_d(\mathbf{k}+\mathbf{q}) - \Phi_s(\mathbf{k})\}\,d\mathbf{q},$$

$$(9 \cdot 51 \cdot 1)$$

where $f = f_0 - \Phi\,\partial f_0/\partial E$ and ϖ_d is the weight factor of the d-bands. As before we can carry out one integration by taking polar coordinates q, ϑ, ϖ in the $\mathbf{q}$-space. Then

$$E_{\mathbf{k},s} - E_{\mathbf{k}+\mathbf{q},d} \pm h\nu_{\mathbf{q}} = \frac{\hbar^2}{2}\left(\frac{1}{m_s} + \frac{1}{m_d}\right) k^2 + \frac{\hbar^2}{2m_d}(2kq\cos\vartheta + q^2) - A \pm h\nu_q,$$

and, putting

$$y = 2\pi t(E_{\mathbf{k},s} - E_{\mathbf{k}+\mathbf{q},d} \pm h\nu_{\mathbf{q}})/h,$$

we have

$$\int_0^\pi \Omega(E_{\mathbf{k},s} - E_{\mathbf{k}+\mathbf{q},d} \pm h\nu_{\mathbf{q}})\sin\vartheta\,d\vartheta = \frac{2\pi m_d}{hkq}\int_{\vartheta=\pi}^{\vartheta=0}\frac{\sin y}{y}\,dy, \quad (9 \cdot 51 \cdot 2)$$

the integral being π if the limits of integration are of opposite signs and being zero otherwise. Now since in any transition we must have

$$\mathbf{k} - \mathbf{k}' \pm \mathbf{q} = 0, \quad (9 \cdot 51 \cdot 3)$$

the arrangement of the energy bands introduces severe restrictions on the possible transitions. If k_s and k_d are the wave numbers of the highest occupied levels in the s- and the d-bands (see fig. II 14), it is clear that q must be greater than $|\,k_d - k_s\,|$ and that, if this condition is satisfied, there must be a value of $\cos\vartheta$ for which $E_{\mathbf{k},s} = E_{\mathbf{k}+\mathbf{q},d}$. Therefore if we neglect the small term $\pm h\nu_{\mathbf{q}}$, the integral in (9·51·2) has the value π if $q \geqslant |\,k_d - k_s\,|$. Also $h\nu_q < k\Theta$, and so, for an s-d transition to be possible we must have $k\Theta_E \leqslant h\nu_{\mathbf{q}} \leqslant k\Theta$, where

$$k\Theta_E = \hbar u_0\,|\,k_d - k_s\,|.$$

To proceed further we can put as usual $\Phi = k_1 c(E)$, but to simplify the resulting formulae we can neglect c_d compared with c_s. In this case, since $\Phi_s(\mathbf{k}+\mathbf{q})$ does not occur in (9·51·1) no assumption need be made about the form of Φ_s, and the expression for $[\partial f_s/\partial t]_{sd}$ reduces at once to

$$\left[\frac{\partial f_s}{\partial t}\right]_{sd} = -(f_s - f_{s0})\frac{2^{\frac{1}{2}}\varpi_d m_d}{m_s^{\frac{1}{2}}\hbar^2 \Lambda_{sd} E^{\frac{1}{2}}}\left(\frac{T}{\Theta}\right)^3 \left[\int_{-\Theta/T}^{-\Theta_E/T} + \int_{\Theta_E/T}^{\Theta/T}\frac{e^\eta + 1}{e^{\eta+z} + 1}\frac{z^2\,dz}{|\,1 - e^{-z}\,|}\right],$$

$$(9 \cdot 51 \cdot 4)$$

where

$$\Lambda_{sd} = \left(\frac{4\pi}{3}\right)^{\frac{1}{3}}\frac{4Mak\Theta}{3h^2 C_{sd}^2}.$$

9·511. The quantity multiplying $-(f_s - f_{s0})$ is $1/\tau_{sd}$. It will be seen that it is proportional to $\varpi_d m_d$, which, according to the evidence provided by the electronic specific heats, is about $10m$, so that the probability of an s-electron being scattered into the d-band is about 10 times the normal scattering probability provided that Λ_{ss} and Λ_{sd} are of the same order. This is of the right order of magnitude to account for the poor conductivity of the transition elements at room temperatures. At low temperatures, however, the theoretical expression for $1/\tau_{sd}$ approaches zero like $e^{-\Theta_R/T}$, so that the scattering at low temperatures ought to be predominantly of the s-s type and should therefore be of a normal magnitude, the extra resistance due to the s-d transitions dying out as the temperature is lowered. Measurements on platinum at low temperatures have failed to show any such abnormal temperature variation in the resistance, and therefore, although the theory gives the correct order of magnitude of the resistance at room temperature, it seems to give the temperature variation of the resistance incorrectly.

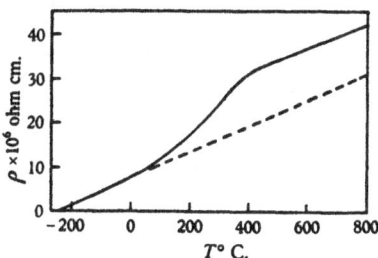

Fig. IX 1. The electrical resistivity of nickel as a function of temperature.

9·512. The above theory should apply, with certain changes, indicated below, to the ferromagnetic metals, which have abnormally high resistances and also abnormal temperature coefficients. The resistance of nickel is shown as a function of the temperature in fig. IX 1, the full curve giving the total resistance and the dotted curve the 'normal resistance', the difference between the two curves representing the 'ferromagnetic part of the resistance'. It will be seen that the ferromagnetic part of the resistance decreases rapidly as the temperature is decreased below the Curie point, and this can be explained qualitatively by making certain assumptions which are plausible but which have not yet so far been incorporated in a quantitative theory.

We assume that, as for the transition elements, the major part of the resistance of a ferromagnetic metal above the Curie point is due to s-d transitions. Below the Curie point, spontaneous magnetization appears and the number of vacant levels in the d-band is a function of the temperature. At the absolute zero all the levels in the d-band with one direction of spin are filled, and the weight factor for the unoccupied levels corresponds to only one direction of spin and is therefore half of what it was above the Curie point. The exact way in which this change in the weight factor with temperature influences the magnitude of the transitions depends upon the effect of temperature on the distribution of the electrons over the levels and upon what happens to the spin of an electron during a transition. If the spin is unchanged during a collision only one-half of the *s*-electrons will undergo *s-d* transitions when all the *d*-levels with one direction of spin are full. But whether this is so or not, it is clear that the relative importance of the ferromagnetic part of the resistance must diminish as the temperature decreases, in accordance with the experimental facts.

9·513. *The conductivity of alloys.* The addition of one metal to another can alter the resistance by (1) changing the number of valency electrons, (2) changing the average field acting on the valency electrons and (3) providing scattering centres when the alloy is in the disordered state. The first two effects can in principle be dealt with by determining the wave functions and energy levels by the methods of Chapters III and IV, but detailed calculations are lacking, and they would not be particularly illuminating. (Some general results can be obtained without calculation. For example, the

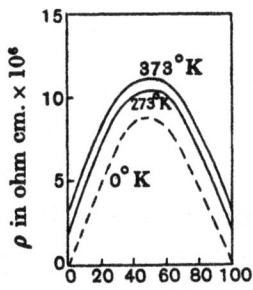

Fig. IX 2. The specific resistance of the silver-gold alloys.

structure of the γ-alloys makes it certain that they must be poor conductors like bismuth.) The third effect is the one discussed in § 9·4 and results in a contribution to the resistance which is independent of the temperature. In general, it can only be disentangled from the other effects when the constituent elements of the alloy have very nearly the same energy levels.

The gold-silver alloys provide the simplest material for a study of the effect of composition upon the conductivity, for gold and silver have the same crystal structure and practically identical atomic volumes, are miscible in all proportions and do not form a superlattice. The resistances of the gold-silver alloys for 373, 273 and 0° K. (extrapolated) are shown in fig. IX 2, and the part of the resistance which is independent of the temperature is proportional to $x(1-x)$, where x is the concentration of one of the components. This is in agreement with equation (9·4·5).

The 50 % alloy has an additional resistance of $9\cdot7 \times 10^{-18}$ e.s.u., which corresponds to a free path of $0\cdot95 \times 10^{-6}$ cm., while the free path in pure silver is $5\cdot2 \times 10^{-6}$ cm. at 0° C., so that irregularities in the lattice are much more efficient scatterers than the lattice vibrations.

9·52. *Higher approximations to the conductivity.* The calculations given in § 9·35 can be extended to include the correction terms of order $(\Theta/T)^2$

and $(kT/\zeta)^2$ and higher (Wilson, 1937). If we write

$$c(E) = \frac{\hbar}{m^*}\left(\epsilon\mathscr{E} + T\frac{\partial}{\partial x}\frac{\zeta}{T}\right)c^{(\frac{1}{2})}(E) + \frac{\hbar}{m^*}\frac{1}{T}\frac{\partial T}{\partial x}c^{(\frac{1}{2})}(E), \qquad (9\cdot52\cdot1)$$

then $c^{(n)}(E)$ satisfies the equation

$$\mathscr{L}(c^{(n)}) = E^{n-\frac{1}{2}}\,\partial f_0/\partial E, \qquad (9\cdot52\cdot2)$$

the operator $\mathscr{L}$ being defined by equation (9·34·10). Also the electric and heat current densities are given by

$$J = -\frac{\epsilon}{4\pi^3}\int v_1 f\,d\mathbf{k} = \mathscr{K}_{\frac{3}{2},\frac{3}{2}}\left(\epsilon^2\mathscr{E} + \epsilon T\frac{\partial}{\partial x}\frac{\zeta}{T}\right) + \mathscr{K}_{\frac{3}{2},\frac{5}{2}}\frac{\epsilon}{T}\frac{\partial T}{\partial x}, \qquad (9\cdot52\cdot3)$$

$$w = \frac{1}{4\pi^3}\int v_1 Ef\,d\mathbf{k} = \mathscr{K}_{\frac{3}{2},\frac{5}{2}}\left(-\epsilon\mathscr{E} - T\frac{\partial}{\partial x}\frac{\zeta}{T}\right) - \mathscr{K}_{\frac{5}{2},\frac{5}{2}}\frac{1}{T}\frac{\partial T}{\partial x}, \qquad (9\cdot52\cdot4)$$

where

$$\mathscr{K}_{m,n} = \frac{16\pi(2m^*)^{\frac{1}{2}}}{3h^3}\int_{-\infty}^{\infty}E^m c^{(n)}(\eta)\frac{\partial f_0}{\partial\eta}\,d\eta. \qquad (9\cdot52\cdot5)$$

If terms of order $(\Theta/T)^2$ compared with the leading term are neglected, the solution of equation (9·52·2) is

$$c_0^{(n)} = -\frac{2\hbar^2\Lambda}{(\frac{1}{2}m^*)^{\frac{1}{2}}D}\frac{\Theta}{T}E^n, \qquad (9\cdot52\cdot6)$$

as found in §9·35. We can therefore solve the equation by successive approximations by putting $c^{(n)} = c_0^{(n)} + c_1^{(n)} + \ldots$, where the ratio of successive terms is of the order of $(\Theta/T)^2$. The correction terms are, however, small and can be obtained more readily by the methods given in the next chapter. We shall therefore only consider the zero-order approximation and use it to find the correction terms of order $(kT/\zeta)^2$. With (9·52·6), $\mathscr{K}_{m,n}$ becomes

$$\mathscr{K}_{m,n} = -\frac{16}{3\pi h}\frac{\Lambda}{D}\frac{\Theta}{T}\int_{-\infty}^{\infty}E^{m+n}\frac{\partial f_0}{\partial\eta}\,d\eta$$

$$= \frac{16}{3\pi h}\frac{\Lambda}{D}\frac{\Theta}{T}\zeta^{m+n}\left\{1 + \frac{(m+n)(m+n-1)}{6}\left(\frac{\pi kT}{\zeta}\right)^2 + \ldots\right\}. \qquad (9\cdot52\cdot7)$$

Hence

$$\sigma = \epsilon^2\mathscr{K}_{\frac{3}{2},\frac{3}{2}} = \frac{16}{3\pi h}\frac{\Lambda}{D}\frac{\Theta}{T}\epsilon^2\zeta^3\left\{1 + \left(\frac{\pi kT}{\zeta}\right)^2 + \ldots\right\}, \qquad (9\cdot52\cdot8)$$

$$\kappa = \frac{\mathscr{K}_{\frac{3}{2},\frac{5}{2}}\mathscr{K}_{\frac{5}{2},\frac{5}{2}} - \mathscr{K}_{\frac{3}{2},\frac{5}{2}}^2}{\mathscr{K}_{\frac{3}{2},\frac{3}{2}}T} = \frac{16\pi}{9h}\frac{\Lambda}{D}\Theta k^2\zeta^3\left\{1 + \frac{6}{5}\left(\frac{\pi kT}{\zeta}\right)^2 + \ldots\right\}, \qquad (9\cdot52\cdot9)$$

$$\mathfrak{S} = \frac{\mathscr{K}_{\frac{3}{2},\frac{5}{2}} - \zeta\mathscr{K}_{\frac{3}{2},\frac{3}{2}}}{\mathscr{K}_{\frac{3}{2},\frac{3}{2}}T} = \frac{\pi^2 k^2 T}{\zeta}\left\{1 - \frac{8}{15}\left(\frac{\pi kT}{\zeta}\right)^2 + \ldots\right\}. \qquad (9\cdot52\cdot10)$$

It will be seen from the above formulae that the Wiedemann-Franz law is no longer exactly obeyed when the correction terms are included. The experimental results at temperatures for which the formulae for κ and $\mathfrak{S}$ are valid are insufficiently accurate for them to be compared with the theoretical predictions. We shall therefore only discuss the behaviour of the electrical conductivity.

The general behaviour of the electrical resistance at high temperatures is shown in fig. IX 3. The resistance normally increases slightly faster than

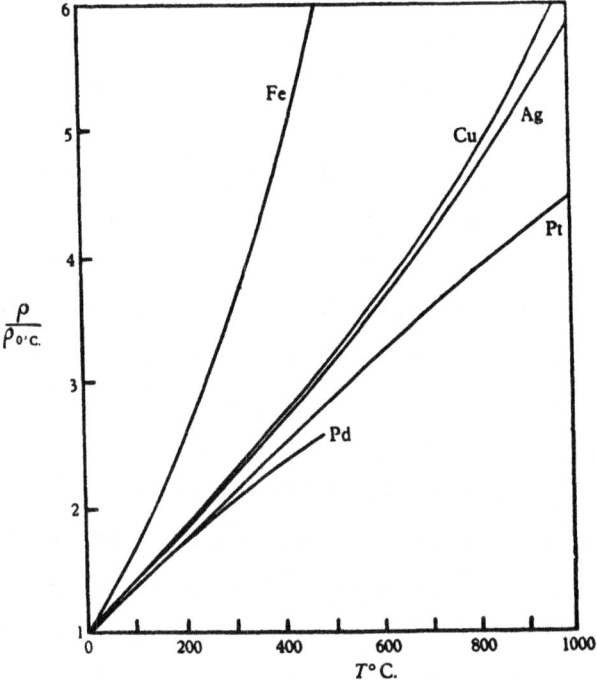

Fig. IX 3. The temperature variation of the relative resistivity of metals at high temperatures.

T, though for palladium and platinum the resistance increases more slowly than T. At the temperatures in question the correction terms in $(\Theta/T)^2$ are negligible, and they only affect the behaviour of σ at temperatures lower than Θ. It is therefore probable that the increase in ρ/T is to be ascribed to the temperature variation of the parameters determining Λ, and in particular to the decrease in Θ due to the thermal expansion of the metal. Numerical calculations to test this hypothesis are lacking, but the variation in Θ required to account for the effect is small and well within the bounds of probability.

The anomalous variation of ρ/T for palladium and platinum has been ascribed (Mott, 1936) to the second-order terms in $(kT/\zeta)^2$ in (9·52·8). If we take into account the temperature variation of ζ $(\zeta = \zeta_0 - \frac{1}{12}\pi^2 k^2 T^2/\zeta_0)$, we obtain

$$\rho = \text{constant } T\{1 - \tfrac{3}{4}(\pi kT/\zeta_0)^2\}. \qquad (9\cdot52\cdot11)$$

The second-order terms therefore always give rise to a decrease in ρ/T with increasing T, but their effect will only be apparent if ζ_0 is sufficiently small. For most metals ζ_0/k is of the order of $50,000°$ K., and the second-order terms, which give a measure of the departure of the electron gas from complete degeneracy, are negligible. The experimental results for palladium and platinum, however, could be made to agree with the theoretical expression (9·52·11) by choosing ζ_0/k for the s-band to be of the order of $5000°$ K. for palladium and $10,000°$ K. for platinum. These values are too small to be reconcilable with the generally accepted structure of the transition elements, according to which there is about $0\cdot6$ electron per atom in the s-band with an effective mass of the order of the normal electron mass. While, therefore, the explanation given above may be the correct one, the quantitative agreement is poor, though the required value of ζ_0/k for the s-band of platinum is just within the bounds of possibility. Rhodes (1950) has suggested that the inclusion of the effect of the excitation of the d-electrons into the s-band might remove the discrepancy.

THE CONDUCTIVITY AT LOW TEMPERATURES

9·6. The general solution of the integral equation for the distribution function for arbitrary temperatures is best carried out by use of a variation principle, and this method is the subject of Chapter X. Solutions for certain special cases can, however, be found by more elementary methods, the most important of which are given below.

9·61. *The electrical conductivity.* At very low temperatures it is possible to find a simple solution of the integral equation which is sufficiently accurate to determine the electrical conductivity, but not the second-order phenomena (Bloch, 1930). If the only external influence is an electric field $\mathscr{E}$ parallel to the x-axis, the integral equation for $c(\eta)$ can be written

$$E\frac{\partial f_0}{\partial E}\int_{-\Theta/T}^{\Theta/T}[c(\eta+z)-c(\eta)]\frac{e^\eta+1}{e^{\eta+z}+1}\frac{z^2 dz}{|1-e^{-z}|} = \frac{2^{\frac{1}{2}}\epsilon\hbar^3}{m^{*\frac{3}{2}}}\mathscr{E}\Lambda\left(\frac{\Theta}{T}\right)^3 E^{\frac{1}{2}}\frac{\partial f_0}{\partial E}$$

$$+\frac{\partial f_0}{\partial E}\int_{-\Theta/T}^{\Theta/T} c(\eta+z)\left\{D\left(\frac{T}{\Theta}\right)^2 z^2 - \tfrac{1}{2}kTz\right\}\frac{e^\eta+1}{e^{\eta+z}+1}\frac{z^2 dz}{|1-e^{-z}|}. \qquad (9\cdot61\cdot1)$$

Now the homogeneous equation obtained by neglecting the right-hand side has the solution $c(\eta) = \text{constant}$, and therefore the non-homogeneous equation can only be solved provided that the right-hand side is orthogonal to the solution of the transposed homogeneous equation, this solution also being $c(\eta) = $ constant. This gives us the condition

$$\frac{2^{\frac{1}{2}}\epsilon\hbar^3}{m^{*\frac{5}{2}}}\mathscr{E}\Lambda\left(\frac{\Theta}{T}\right)^3\int_{-\infty}^{\infty}E^{\frac{5}{2}}\frac{\partial f_0}{\partial\eta}d\eta$$

$$= -\int_{-\infty}^{\infty}\frac{\partial f_0}{\partial\eta}d\eta\int_{-\Theta/T}^{\Theta/T}c(\eta+z)\left\{D\left(\frac{T}{\Theta}\right)^2 z^2 - \tfrac{1}{2}kTz\right\}\frac{e^\eta+1}{e^{\eta+z}+1}\frac{z^2 dz}{|1-e^{-z}|}. \quad (9\cdot61\cdot2)$$

Now if we can treat the operator on the right-hand side of (9·61·1) as small we can solve the equation by successive approximations by putting $c(\eta) = \alpha + \beta(\eta)$, where α is a constant and where $\alpha \gg \beta(\eta)$. If we substitute $c(\eta) = \alpha$ in (9·61·2), we determine α, and $\beta(\eta)$ can then in principle be obtained by solving (9·61·1). When $c(\eta) = \alpha$ the integral with respect to η on the right-hand side of (9·61·2) is an elementary integral. For it is proportional to

$$\int_{-\infty}^{\infty}\frac{d\eta}{(e^{\eta+z}+1)(e^{-\eta}+1)} = \frac{1}{e^z-1}\int_{-\infty}^{\infty}\left(\frac{1}{e^\eta+1} - \frac{1}{e^{\eta+z}+1}\right)d\eta = \frac{z}{e^z-1}. $$

$$(9\cdot61\cdot3)$$

Hence (9·61·2) gives, to the zero order in kT/ζ,

$$\frac{2^{\frac{1}{2}}\epsilon\hbar^3}{m^{*\frac{5}{2}}}\mathscr{E}\Lambda\left(\frac{\Theta}{T}\right)^3\zeta^{\frac{5}{2}} = -2\alpha D\left(\frac{T}{\Theta}\right)^2\int_0^{\Theta/T}\frac{z^5 dz}{(e^z-1)(1-e^{-z})},$$

so that

$$\alpha = -\frac{\epsilon\hbar^3}{2^{\frac{1}{2}}m^{*\frac{5}{2}}}\frac{\mathscr{E}\Lambda\zeta^{\frac{5}{2}}}{D}\left(\frac{\Theta}{T}\right)^5\frac{1}{\mathscr{I}_5(\Theta/T)}, \quad (9\cdot61\cdot4)$$

where

$$\mathscr{I}_n\left(\frac{\Theta}{T}\right) = \int_0^{\Theta/T}\frac{z^n dz}{(e^z-1)(1-e^{-z})}. \quad (9\cdot61\cdot5)$$

To this approximation the distribution function is given by

$$f = f_0 - \alpha k_1 \partial f_0/\partial E,$$

and the conductivity is

$$\sigma = \frac{4\epsilon^2\Lambda\zeta^3}{3\pi\hbar D}\left(\frac{\Theta}{T}\right)^5\frac{1}{\mathscr{I}_5(\Theta/T)} = \frac{8\sqrt{2}\,\epsilon^2\Delta Mk\Theta}{9\pi\hbar^2}\frac{n\zeta^{\frac{5}{2}}}{m^{*\frac{5}{2}}C^2}\left(\frac{\Theta}{T}\right)^5\frac{1}{\mathscr{I}_5(\Theta/T)}. $$

$$(9\cdot61\cdot6)$$

For metals with less than 0·25 electron per atom, the second inequality (9·34·5) holds, and $\mathscr{I}_5(\Theta/T)$ in (9·61·6) must be replaced by $\mathscr{I}_5\{(2\zeta/D)^{\frac{1}{2}}\Theta/T\}$.

9·611. The preceding calculation, though apparently elementary, is not easy to justify directly. In the first place, equation (9·61·1) can be multiplied by any arbitrary function of E and a naïve application of the orthogonality condition would not necessarily lead to equation (9·61·2). To obtain this latter equation we must base the argument upon the properties of integral equations with a symmetrical nucleus, and the calculation then ceases to be an elementary one. In the second place, it is necessary to show that the operators are such that $\alpha \gg \beta(\eta)$, and although it is possible to do this it is simpler to derive the results by other methods, such as that given in § 9·62, or by the use of the variation principle, since they readily lend themselves to a discussion of convergence questions.

9·612. *Comparison with experiment.* The general properties of the integrals $\mathscr{I}_n(x)$ are given in Appendix 5 (iii), but the limiting forms for high and low temperatures can readily be obtained by elementary methods. For small values of x,

$$\mathscr{I}_5(x) = \int_0^x (z^3 - \tfrac{1}{12}z^5 + \ldots)\, dz = \tfrac{1}{4}x^4 - \tfrac{1}{72}x^6 + \ldots,$$

while
$$\mathscr{I}_5(\infty) = 5\int_0^\infty \frac{z^4\, dz}{e^z - 1} = 5!\sum_{s=1}^\infty \frac{1}{s^5} = 124\cdot4.$$

At low temperatures σ is therefore proportional to T^{-5}. It should be noted that, although (9·61·6) has only been derived on the assumption that T/Θ is small, it agrees for large T/Θ with the high-temperature value calculated in § 9·5, and in fact with the second approximation involving terms of the order of $(\Theta/T)^2$ which can be calculated by the method outlined in § 9·52. It therefore has the status of an interpolation formula which is at least approximately valid over the whole temperature range. For the moment we shall assume its correctness, leaving the investigation of its validity until § 10·33.

If T_1 and T_2 are a low and a high temperature respectively, so that $T_2 \gg \Theta \gg T_1$, then

$$\frac{\sigma_2}{\sigma_1} = 497\cdot6\left(\frac{T_1}{\Theta}\right)^4\frac{T_1}{T_2}, \tag{9·612·1}$$

which gives a simple method of determining Θ, and then of calculating the theoretical values of the conductivity of any particular metal. The results for gold (Grüneisen, 1933) are shown in Table IX 3, and the agreement between the calculated and observed values is excellent, so much so that it must be regarded to a certain extent as fortuitous.

There are two main reasons why there should be variations from the theoretical formula (9·61·6). In the first place many simplifications were made in § 9·3, and the scattering of the electrons by the lattice is only approximately given by (9·33·9). In the second place the energy spectrum of the lattice vibrations must differ from that given by the simple Debye

Table IX 3. *The conductivity of gold.* $\Theta = 175^\circ$ K.

T	σ_{273}/σ calculated	σ_{273}/σ observed
273	1	1
87·43	0·2645	0·2551
78·86	0·2276	0·2187
57·8	0·1356	0·1314
20·4	0·00604	0·0058
18·9	0·00346	0·0035
14·3	0·00117	0·00137
12·1	0·00051	0·00048
11·1	0·00033	0·00030
4·2	3×10^{-6}	3×10^{-6}

theory. In particular, the transverse lattice waves must scatter the electrons to some extent, and, while the contribution of the transverse waves to the resistance at very low temperatures is probably small compared with that of the longitudinal waves, at high temperatures the scattering of the electrons by the transverse waves is likely to be by no means negligible. We would therefore expect the ideal resistivity ρ_i to be proportional to T at high temperatures and to T^5 at sufficiently low temperatures, but to differ from $T^5 \mathscr{J}_5(\Theta/T)$ at intermediate temperatures. Further, the values of Θ deduced from the electrical resistance cannot be expected to be the same as those deduced from the specific heat or the elastic constants, since the various phenomena involve different averages of the lattice frequencies. (For gold, Grüneisen chose $\Theta = 175^\circ$ K. to fit the resistance measurements, whereas the Debye Θ is 185° K. and the value deduced from the elastic constants is 158° K. This is an unusually close fit.)

The most detailed measurements at low temperatures are due to MacDonald and Mendelssohn (1950). They found that for sodium there is very close agreement with formula (9·61·6) if Θ is taken as 202° K. (this value of Θ is much higher than the Debye temperature 160° K.). For the other alkalis, however, there are considerable discrepancies. These can be illustrated by analysing the data in a manner similar to that used when discussing the specific heat (§ 6·16). If we fit the results at room temperatures and near a particular temperature T by an expression of the Grüneisen-Bloch type we obtain a value of Θ. If (9·61·6) were exact, Θ would be constant, but, if deviations occur, Θ is a function of T. If we denote this characteristic temperature by Θ_σ and the characteristic temperature which determines the specific heat by Θ_D, the observed variations for lithium are shown in fig. IX 4. The inference is that the Bloch-Grüneisen formula holds approximately, since Θ is a slowly varying function of T, but the total variation in Θ is considerable.

The results for rubidium are distinctly anomalous, as is shown by the apparent values of Θ_σ given in fig. IX 5. When such variations in Θ_σ as these occur it is almost meaningless to calculate Θ_σ from equation (9·612·1) or to apply the Bloch-Grüneisen formula at all. The ideal resistance of rubidium varies as T^2 between 4 and 20° K., but below 2° K. it varies as $T^{4·5}$, so that ultimately it probably obeys the T^5 law, but with a low temperature Θ_σ of the order of 25° K.

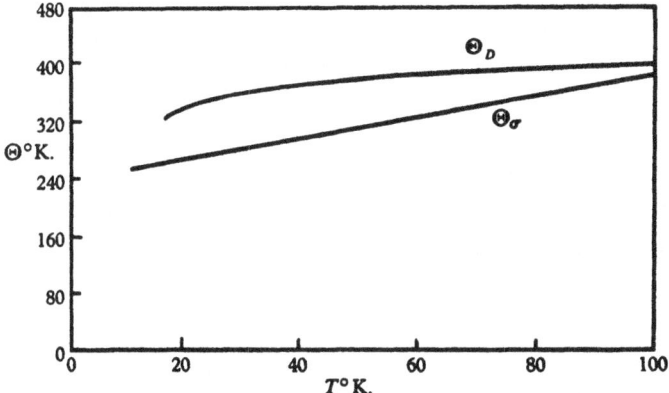

Fig. IX 4. The apparent characteristic temperatures for lithium; Θ_D as deduced from the specific heat, Θ_σ as deduced from the electrical conductivity.

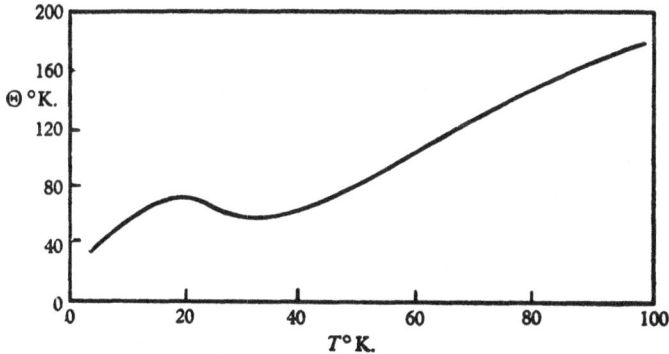

Fig. IX 5. The apparent characteristic temperature Θ_σ for rubidium as deduced from the electrical conductivity.

Similar divergencies from the theoretical behaviour are found for the other alkalis, ρ_i for potassium obeying the T^5 law only below 7° K., while for caesium ρ_i does not vary faster than T^2 down to 4° K.

Various other anomalies may be noted. For the semi-metals such as bismuth the ideal resistance contains the factor $\mathscr{J}_5\{(2\zeta/D)^{\frac{1}{2}}\Theta/T\}$, and according to this the resistance should be a linear function of T down to quite low temperatures, since the effective characteristic temperature is now $(2\zeta/D)^{\frac{1}{2}} \Theta = (4n/n_a)^{\frac{1}{2}} \Theta$, where n_a is the number of atoms per unit volume. Such a behaviour of ρ_i is not observed, which is one more illustration of the difficulties that arise in applying the free electron model to the semi-metals.

At low temperatures the resistance of platinum contains a term in T^2. This has been ascribed by Baber (1937) to the collisions between the s- and d-electrons, but the proposed explanation is not very convincing.

9·613. In view of the approximations made in deriving the conductivity formula (9·61·6), it is not surprising that the agreement between theory and experiment is not entirely satisfactory. It is in fact remarkable that the results for sodium and gold should fit so well. It is, moreover, clear how to obtain theoretical formulae of greater validity, and a method for doing so is given in § 10·6. The difficulties in following out the programme outlined there are formidable, but the difficulties are in the details of the calculations and not in the physical principles involved. There are, however, discrepancies which seem to be of a more fundamental character.

It was found by de Haas, de Boer and van den Berg in 1934 (see van den Berg, 1938) that the resistance of some very pure specimens of gold passed through a minimum as the temperature decreased (see fig. IX 6). The temperature at which the minimum occurs depends upon the impurities present, the minimum shifting to lower temperatures as the metal is purified. The effect, though small, is unmistakable, and has been reported as occurring in other metals also, though, apart from the measurements on gold, the only reliable results seem to be those of MacDonald and Mendelssohn for magnesium (fig. IX 7), and of Mendoza and Thomas (1951) for copper and silver. It is not known whether the resistance curve flattens out again as the absolute zero is reached, but if it does not do so the behaviour of an ideally pure (non-superconducting) metal would be highly anomalous near $T = 0$. The cause of the minimum is so far entirely obscure, and constitutes a most striking departure from Matthiessen's rule, according to which the ideal and residual resistances are additive. The rule is in fact only an approximation, and its validity is discussed in § 10·4. The calculations given there show that deviations from the rule are to be expected when the ideal and residual resistances are comparable, but they give no indication of a minimum in the resistance, so that some new physical principle seems to be involved.

9·62. *The second-order phenomena.* The method given in the preceding section is inapplicable to the second-order phenomena. An elementary method (Wilson, 1937) of solving the integral equation at low temperatures consists in considering simultaneously the scattering due to both the

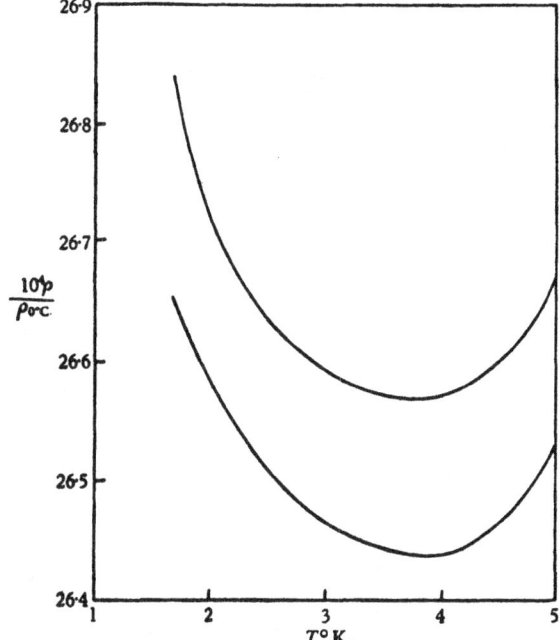

Fig. IX 6. The minimum in the electrical resistance of gold.

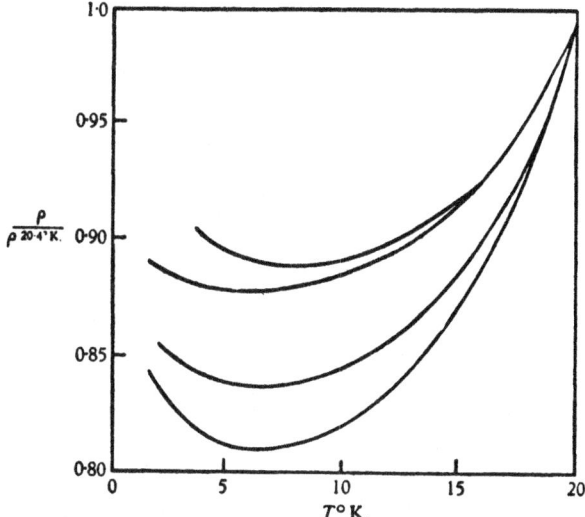

Fig. IX 7. The minimum in the electrical resistance of
magnesium for four specimens.

impurities and to the lattice vibrations. In this case the integral equation
(9·52·2) for $c^{(n)}$ takes the form

$$\sqrt{\frac{2}{m^*}} \frac{E^2}{l_r} c(\eta) + \frac{(\tfrac{1}{2}m^*)^{\tfrac{1}{2}}}{\hbar^2 \Lambda} \left(\frac{T}{\Theta}\right)^3$$

$$\times \int_{-\Theta/T}^{\Theta/T} \left[Ec(\eta) - c(\eta+z) \left\{ E + \tfrac{1}{2}kTz - D\left(\frac{T}{\Theta}\right)^2 z^2 \right\} \right] \frac{e^\eta + 1}{e^{\eta+z} + 1} \frac{z^2 dz}{|1 - e^{-z}|} = -E^n.$$

$$(9·62·1)$$

Now for small values of T/Θ, the first term on the left is much larger than
the second, and we can obtain a solution $c_0^{(n)} + c_1^{(n)} + \dots$ by treating the
integral operator as small. The first two approximations are

$$c_0^{(n)} = -(\tfrac{1}{2}m^*)^{\tfrac{1}{2}} l_r E^{n-2}, \qquad (9·62·2)$$

$$c_1^{(n)} = \frac{m^{*\tfrac{3}{2}} l_r^2}{2^{\tfrac{3}{2}} \hbar^2 \Lambda} \left(\frac{T}{\Theta}\right)^3 \int_{-\Theta/T}^{\Theta/T} \left[E^{n-3} \right.$$

$$\left. - \frac{(E+kTz)^{n-2}}{E^2} \left\{ E + \tfrac{1}{2}kTz - D\left(\frac{T}{\Theta}\right)^2 z^2 \right\} \right] \frac{e^\eta + 1}{e^{\eta+z} + 1} \frac{z^2 dz}{|1 - e^{-z}|}. \quad (9·62·3)$$

Since the maximum value of kTz, namely $k\Theta$, is small compared with ζ for
a degenerate gas, it is sufficient to expand quantities such as $(E + kTz)^n$ in
powers of kTz/E. If we then substitute for $c^{(n)}(\eta)$ in the expression (9·52·5)
for $\mathscr{K}_{m,n}$, the integrals with respect to η can all be reduced to special cases
of the integral

$$\int_{-\infty}^{\infty} \frac{F(\eta)}{(e^{\eta+z} + 1)(e^{-\eta} + 1)} d\eta,$$

which is evaluated in Appendix 5 (ii). After a straightforward calculation
we find

$$\mathscr{K}_{\frac{1}{2},\frac{1}{2}} = -\frac{16\pi m^* l_r}{3h^3} \int_{-\infty}^{\infty} \left[E - \frac{m^* l_r}{2\hbar^2 \Lambda} \left(\frac{T}{\Theta}\right)^3 \left\{ \frac{D}{E} \left(\frac{T}{\Theta}\right)^2 \left(2\mathscr{J}_5 + \tfrac{5}{12}\mathscr{J}_7 \frac{k^2 T^2}{E^2}\right) \right. \right.$$

$$\left. \left. - \tfrac{1}{4}\mathscr{J}_5 \frac{k^2 T^2}{E^2} - \tfrac{7}{64}\mathscr{J}_7 \frac{k^4 T^4}{E^4} \right\} \right] \frac{\partial f_0}{\partial \eta} d\eta, \quad (9·62·4)$$

$$\mathscr{K}_{\frac{3}{2},\frac{1}{2}} = -\frac{16\pi m^* l_r}{3h^3} \int_{-\infty}^{\infty} \left[E^2 - \frac{m^* l_r}{2\hbar^2 \Lambda} \left(\frac{T}{\Theta}\right)^3 \left\{ D\left(\frac{T}{\Theta}\right)^2 \left(2\mathscr{J}_5 + \tfrac{1}{4}\mathscr{J}_7 \frac{k^2 T^2}{E^2}\right) \right. \right.$$

$$\left. \left. - E\left(\tfrac{1}{4}\mathscr{J}_5 \frac{k^2 T^2}{E^2} + \tfrac{13}{192}\mathscr{J}_7 \frac{k^4 T^4}{E^4}\right) \right\} \right] \frac{\partial f_0}{\partial \eta} d\eta, \quad (9·62·5)$$

$$\mathscr{K}_{\frac{5}{2},\frac{1}{2}} = -\frac{16\pi m^* l_r}{3h^3} \int_{-\infty}^{\infty} \left[E^3 - \frac{m^* l_r}{2\hbar^2 \Lambda} \left(\frac{T}{\Theta}\right)^3 \left\{ ED\left(\frac{T}{\Theta}\right)^2 \left(2\mathscr{J}_5 - \tfrac{1}{4}\mathscr{J}_7 \frac{k^2 T^2}{E^2}\right) \right. \right.$$

$$\left. \left. + E^2\left(\tfrac{3}{4}\mathscr{J}_5 \frac{k^2 T^2}{E^2} + \tfrac{1}{64}\mathscr{J}_7 \frac{k^4 T^4}{E^4}\right) \right\} \right] \frac{\partial f_0}{\partial \eta} d\eta. \quad (9·62·6)$$

On evaluating the integrals in powers of kT/ζ and retaining only the lowest powers required to give a non-zero result, we find

$$\sigma = \sigma_r\left[1 - \sigma_r\frac{3\pi hD}{4\epsilon^2\Lambda\zeta^3}\left(\frac{T}{\Theta}\right)^5\mathscr{J}_5\right], \tag{9·62·7}$$

$$\kappa = \kappa_r\left[1 - \kappa_r\frac{27h}{8\pi^3k^2\Lambda\zeta^2\Theta}\left(\frac{T}{\Theta}\right)^2\left\{\mathscr{J}_5 + \frac{D}{\zeta}\left(\frac{T}{\Theta}\right)^2(\tfrac{2}{3}\pi^2\mathscr{J}_5 - \tfrac{1}{3}\mathscr{J}_7)\right\}\right], \tag{9·62·8}$$

and
$$\mathfrak{S} = \frac{\pi^2k^2T}{3\zeta}\left[1 + \frac{2\pi^2m^*l_rD}{h^2\Lambda\zeta^2}\left(\frac{T}{\Theta}\right)^5\left(4\mathscr{J}_5 + \frac{1}{2\pi^2}\mathscr{J}_7\right)\right], \tag{9·62·9}$$

where $1/\sigma_r = 3h^3/(16\pi m^*\epsilon^2 l_r\zeta)$ is the residual electrical resistivity and $\kappa_r = \tfrac{1}{3}(\pi k/\epsilon)^2\sigma_r T$ is the corresponding residual thermal conductivity.

9·621. We can now obtain the ideal conductivities if we assume that the ideal and residual resistances are additive. If we assume that

$$\frac{1}{\sigma} = \frac{1}{\sigma_r} + \frac{1}{\sigma_i} \tag{9·621·1}$$

and compare this with (9·62·7), which only holds for small values of σ_r/σ_i, we see that

$$\frac{1}{\sigma_i} = \frac{3\pi hD}{4\epsilon^2\Lambda\zeta^3}\left(\frac{T}{\Theta}\right)^5\mathscr{J}_5\left(\frac{\Theta}{T}\right). \tag{9·621·2}$$

This is the same expression for σ_i as given by (9·61·6), and the present method gives a justification of the interpolation formula for σ_i provided that the ideal and residual resistances are additive. This assumption is, however, only approximately true, and the rigorous treatment of the problem given in Chapter X shows that correction terms must be added to (9·621·2), but that these correction terms are small.

To determine the ideal thermal conductivity κ_i we assume that

$$\frac{1}{\kappa} = \frac{1}{\kappa_r} + \frac{1}{\kappa_i} \tag{9·621·3}$$

and compare it with (9·62·8). We obtain

$$\frac{1}{\kappa_i} = \frac{27h}{8\pi^3k^2\Lambda\zeta^2\Theta}\left(\frac{T}{\Theta}\right)^2\left\{\mathscr{J}_5 + \frac{D}{\zeta}\left(\frac{T}{\Theta}\right)^2(\tfrac{2}{3}\pi^2\mathscr{J}_5 - \tfrac{1}{3}\mathscr{J}_7)\right\}, \tag{9·621·4}$$

$$= \frac{1}{\kappa_\infty}\frac{6\zeta}{\pi^2D}\left(\frac{T}{\Theta}\right)^2\left\{\mathscr{J}_5 + \frac{D}{\zeta}\left(\frac{T}{\Theta}\right)^2(\tfrac{2}{3}\pi^2\mathscr{J}_5 - \tfrac{1}{3}\mathscr{J}_7)\right\}, \tag{9·621·5}$$

where κ_∞ is the value of κ_i for a fully degenerate gas for $T \gg \Theta$ (see equation (9·52·9)). This formula for κ_i, though derived for low temperatures only, passes over into the correct formula for high temperatures if we make T/Θ

sufficiently large, and we shall therefore consider it to be an interpolation formula valid at least approximately over the whole temperature range. We shall examine its validity, and the corrections to be made to it, in Chapter X.

If the above formulae are applied to semi-metals in which the number of electrons per atom is less than 0·25, the permissible lattice vibrations are determined by the second of the inequalities (9·34·5), and $\mathscr{J}_n(\Theta/T)$ must be replaced by

$$\mathscr{J}_n\left\{\left(\frac{2\zeta}{D}\right)^{\frac{1}{3}}\frac{\Theta}{T}\right\} = \mathscr{J}_n\left\{\left(\frac{4n}{n_a}\right)^{\frac{1}{3}}\frac{\Theta}{T}\right\}, \tag{9·621·6}$$

where n_a is the number of atoms per unit volume.

9·63. *The thermoelectric power.* The formula (9·62·9) for $\mathfrak{S}$ can be written as

$$\mathfrak{S} = \frac{\pi^2 k^2 T}{3\zeta}\left[1 + \frac{2\sigma_r}{\sigma_i}\left(1 + \frac{1}{8\pi^2}\frac{\mathscr{J}_7}{\mathscr{J}_5}\right)\right]. \tag{9·63·1}$$

An interpolation formula, approximately valid over all temperature ranges and all ratios of σ_r to σ_i, has been obtained by Sondheimer (1947) by a consideration of various limiting cases, including the second approximation to $\mathfrak{S}$ at high temperatures. The formula is

$$\mathfrak{S} = \frac{\pi^2 k^2 T}{3\zeta}\frac{\dfrac{1}{\sigma_r} + \dfrac{3}{\sigma_i}\left\{1 + \dfrac{\mathscr{J}_7}{4\pi^2\mathscr{J}_5} + \dfrac{\zeta}{2\pi^2 D}\left(\dfrac{\Theta}{T}\right)^2\right\}}{\dfrac{1}{\sigma_r} + \dfrac{1}{\sigma_i}\left\{1 + \dfrac{\mathscr{J}_7}{2\pi^2\mathscr{J}_5} + \dfrac{3\zeta}{2\pi^2 D}\left(\dfrac{\Theta}{T}\right)^2\right\}}. \tag{9·63·2}$$

If the effect of the thermal vibrations is negligible compared with the residual resistance, $\mathfrak{S}$ is always $\frac{1}{3}\pi^2 k^2 T/\zeta$, while, if $\rho_r = 0$, $\mathfrak{S}$ decreases from $\pi^2 k^2 T/\zeta$ to $\frac{1}{3}\pi^2 k^2 T/\zeta$ as T decreases from a high value to near the absolute zero. Some calculated and observed values of $-\mathfrak{S}/\epsilon$ for sodium of two different purities are shown in fig. IX 8, both above and below the melting-point (100° C.), and it will be seen that the agreement is reasonable but by no means perfect. For most metals, however, the thermoelectric phenomena at low temperatures are very complicated, and sign reversals occur which cannot be explained by a model based upon free or nearly free electrons.

THE THERMAL CONDUCTIVITY

9·7. The thermal resistance, like the electric resistance, consists of a part due to impurities and an ideal part characteristic of the pure metal, and to a first approximation these are independent of one another. For collisions of the electrons with impurities, a universal time of relaxation always exists, and so the Wiedemann-Franz law holds. The 'residual thermal resistance' $1/\kappa_r$ can therefore be deduced from the residual electric resistance $1/\sigma_r$ and is $1/(L_n \sigma_r T)$, where L_n is the normal value $\frac{1}{3}(\pi k/\epsilon)^2$ of the Lorenz number. If the residual and ideal resistivities are additive, the total thermal resistivity is given by

$$\frac{1}{\kappa} = \frac{1}{\kappa_r} + \frac{1}{\kappa_i} = \frac{1}{L_n \sigma_r T} + \frac{1}{\kappa_i}. \qquad (9\cdot7\cdot1)$$

The simplest way of determining κ_i experimentally, therefore, is to plot T/κ against T and thus to determine the limiting values of T/κ as T tends to zero. This gives the residual thermal resistance and, by subtraction, the ideal resistance. A more indirect method, but one which does not require

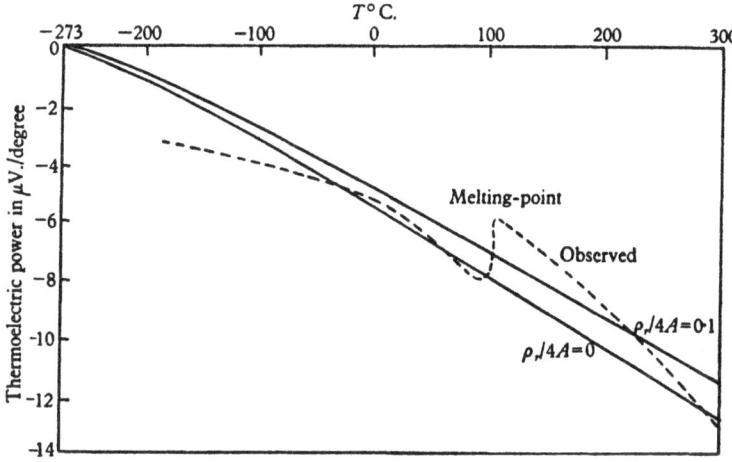

Fig. IX 8. The absolute thermoelectric power of sodium.

measurements of κ at very low temperatures, was adopted by Grüneisen and Goens (1927), based on the 'law of isothermal lines'. Grüneisen and Goens found that for different specimens of a metal at the same temperature the thermal resistance is a linear function of the electrical resistance, i.e.

$$\kappa^{-1} = a + b\sigma^{-1}, \qquad (9\cdot7\cdot2)$$

where a and b are functions of the temperature. By determining a and b experimentally, κ_i can be found by substituting the known values of σ_i in (9·7·2). By comparing (9·7·2) with (9·7·1) it is seen that

$$b = 1/(L_n T), \quad a + 1/(L_n \sigma_i T) = 1/\kappa_i. \qquad (9\cdot7\cdot3)$$

The theoretical thermal conductivity is given by (9·7·1), with κ_i given by (9·621·4). If we write

$$\rho_r = \frac{1}{\sigma_r}, \quad \rho_i = \frac{1}{\sigma_i} = 4A\left(\frac{T}{\Theta}\right)^5 \mathscr{J}_5\!\left(\frac{\Theta}{T}\right), \qquad (9\cdot7\cdot4)$$

then $\dfrac{1}{\kappa} = \dfrac{\rho_r}{L_n T} + \dfrac{4A}{L_n T}\left(\dfrac{T}{\Theta}\right)^5 \left[\left\{1 + \dfrac{3}{2\pi^2}\dfrac{\zeta}{D}\left(\dfrac{\Theta}{T}\right)^2\right\}\mathscr{I}_5\left(\dfrac{\Theta}{T}\right) - \dfrac{1}{2\pi^2}\mathscr{I}_7\left(\dfrac{\Theta}{T}\right)\right]$

$$= \dfrac{\rho_r}{L_n T} + \dfrac{4A}{L_n T}\chi\left(\dfrac{T}{\Theta}, \dfrac{D}{\zeta}\right); \qquad (9.7.5)$$

i.e. $$\kappa = \dfrac{\Theta}{4A}\dfrac{L_n T/\Theta}{\tfrac{1}{4}\rho_r/A + \chi(T/\Theta, D/\zeta)}. \qquad (9.7.6)$$

The value of the parameter D/ζ is given by

$$D/\zeta = 2^{-\frac{1}{3}}(n_a/n)^{\frac{2}{3}}, \qquad (9.7.7)$$

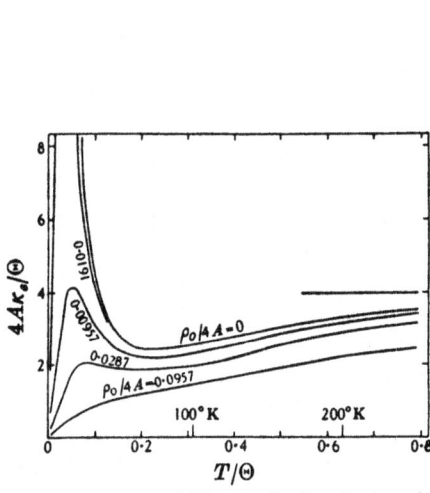

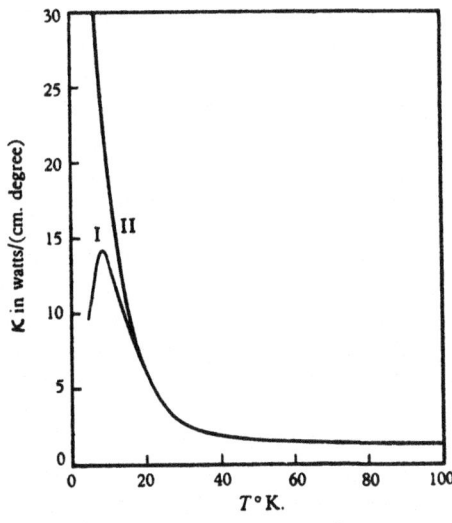

Fig. IX 9. Theoretical electronic thermal conductivity for a monovalent metal, showing the effect of impurity. The temperatures marked correspond to copper ($\Theta = 315°$ K.).

Fig. IX 10. The thermal conductivity of two specimens of sodium. Specimen I has a residual electrical resistivity $\rho_r = 9.5 \times 10^{-9}$ ohm cm., while specimen II has $\rho_r = 1.3 \times 10^{-9}$ ohm cm.

where n_a is the number of atoms per unit volume. Hence D/ζ is $1/\sqrt[3]{2}$ for monovalent metals. Curves showing the behaviour of κ as a function of T/Θ have been computed by Makinson (1938) for different values of the parameter $\rho_r/(4A)$. His results for monovalent metals are given in fig. IX 9.

For semi-metals, $\mathscr{I}_n(\Theta/T)$ must be replaced by (9.621·6), and κ is then constant down to quite low temperatures. It must, however, be borne in mind that, since the theoretical formula for σ is at variance with the facts for semi-metals, there is no reason for the formula for κ to have any greater validity.

9.71. *Comparison with experiment.* Some experimental results for sodium (Berman and MacDonald, 1951) are shown in fig. IX 10, and they are qualitatively in agreement with the theoretical formula for κ. For small values of T, the observed thermal conductivity can be expressed as

$$T/\kappa = \alpha T^3 + \beta, \qquad (9 \cdot 71 \cdot 1)$$

in agreement with (9·7·5), and the predicted maximum in κ is observed for the less pure specimen. There are, however, a number of discrepancies between theory and experiment. In the first place, as shown in fig. IX 9 and more particularly in fig. X 1, p. 309, the calculated conductivity curve has a fairly pronounced minimum near $T = 0 \cdot 25\Theta$, provided that the amount of impurity present is not too great, and κ always decreases initially as T is decreased below Θ. This minimum is not observed experimentally. A more serious discrepancy is in the value of α. According to the theoretical formulae we have

$$\alpha = 95 \cdot 3(n/n_a)^{\frac{2}{3}}/(\kappa_\infty \Theta^2), \quad \beta = \rho_r/L_n, \qquad (9 \cdot 71 \cdot 2)$$

and it is found that, whereas the calculated and observed values of β are in excellent agreement, the theoretical values of α are very much too large. The same discrepancy has been observed by Berman and MacDonald (1952) for copper and by Hulm (1950) for a number of other metals. It is of course always possible to obtain the correct value of α by adjusting n and Θ, as shown in Table IX 4, where we give the Debye Θ's and the values of $\Theta_\kappa(n_a/n)^{\frac{1}{3}}$ required to fit (9·71·1) and (9·71·2) to the experimental results.

Table IX 4. *Thermal conductivity coefficients at low temperatures*

	$\alpha \times 10^4$ observed	Θ_D in °K.	Apparent $\Theta_\kappa(n_a/n)^{\frac{1}{3}}$
Na	3·8	160	440
Cu	0·23	310	1000
Hg	200	96	200
In	19	100	315
Sn	3·9	160	770
Pb	22	88	370
W	0·88	315	950

It will be seen that either Θ_κ is much larger than Θ_D, or that n is of the order of one thirtieth of n_a or less; but it must be borne in mind that when $n/n_a < \frac{1}{4}$ equation (9·7·5) for κ is no longer valid. Since the latter hypothesis must be ruled out, we must conclude that the theory given above considerably over-emphasizes the interaction between the conduction electrons and the lattice vibrations at low temperatures. On the other hand, the theory of the residual resistivities seems entirely satisfactory. There are a number of obvious ways in which the theory could be modified in order to produce a larger value of κ_i at low temperatures, but most of them would also result in a larger value of σ_i as well, which would disturb the excellent agreement between theory and experiment found for the electrical conductivity of sodium by MacDonald and Mendelssohn. The discrepancy will therefore not be easy to resolve. It is, however, worth noting that the formulae derived in § 10·63 for σ and κ for a more general model than the free-electron model contain one multiplying constant in the formula for σ and two constants in that for κ. A suitable choice for these constants could produce the right kind of relations between the high and low temperature conductivities, but in the absence of numerical estimates based upon a particular model this suggestion is a very tentative one.

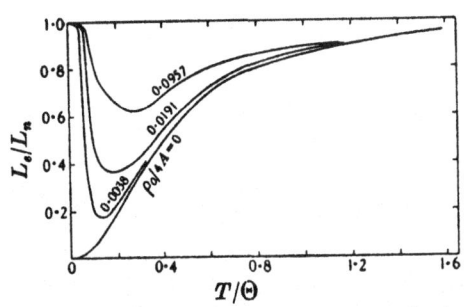

Fig. IX 11. The ratio L_e/L_n for monovalent metals.

9.72. *The Lorenz number.* By combining equations (9·7·4) and (9·7·5) we see that the Lorenz number is given by

$$\frac{L}{L_n} = \frac{\tfrac{1}{4}\rho_r/A + (T/\Theta)^5 \, \mathscr{J}_5(\Theta/T)}{\tfrac{1}{4}\rho_r/A + \chi(T/\Theta, D/\zeta)}. \tag{9·72·1}$$

This quantity is shown in fig. IX 11 for various values of $\tfrac{1}{4}\rho_r/A$ and for $D/\zeta = 1/\sqrt[3]{2}$. It will be seen that $L = L_n$ for very high and for very low temperatures except for ideally pure metals, but that L differs very much from L_n at intermediate temperatures in a manner which depends strongly upon the amount of impurity present.

The experimental curves for sodium are shown in fig. IX 12 together with two theoretical curves. There is general qualitative agreement, and, as predicted by the theory, L first diminishes as T decreases from room temperature and finally increases to the normal value L_n. The experimental curves are, however, steeper than the calculated ones round about 20° K., and the observed minima are higher than the theoretical. Some further experimental curves for L/L_n are shown in fig. IX 13, and it will be seen that the results for copper, platinum and lead are in qualitative agreement with the theoretical predictions.

A strict interpretation of the theoretical formulae results in a constant value of L/L_n for semi-metals such as polycrystalline bismuth (treated as

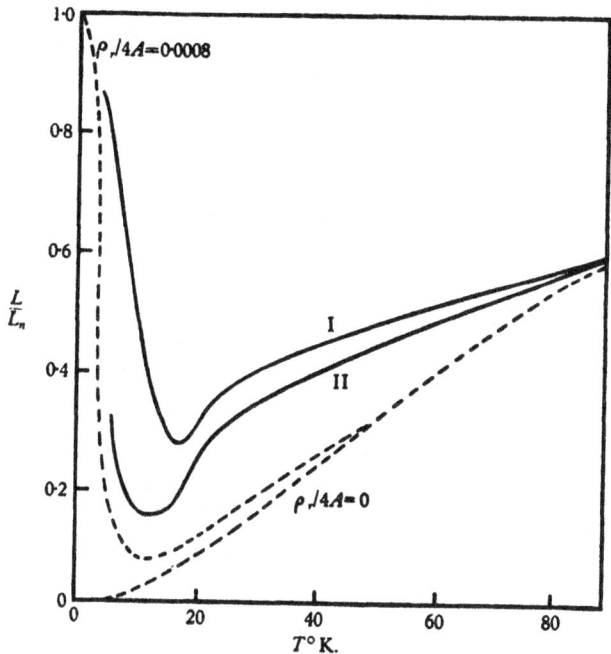

Fig. IX 12. Experimental and theoretical curves for the Lorenz numbers of two specimens of sodium. Sample I has $\rho_r/4A = 0.0008$, sample II has $\rho_r/4A = 0.0001$. Experimental curves ———, theoretical curves · · · ·.

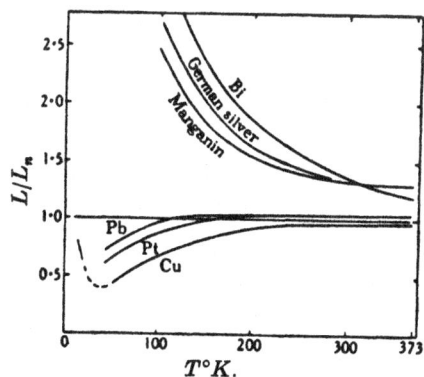

Fig. IX 13. Experimental curves for the ratio L/L_n.

isotropic) down to very low temperatures. Whether one believes this or assumes that L/L_n behaves more or less as for monovalent metals, it is clear that the experimental curve for bismuth shown in fig. IX 13 is highly anomalous. The discrepancy is due to the neglect of the thermal conductivity of the lattice.

THE LATTICE THERMAL CONDUCTIVITY

9·8. Since both insulators and metals conduct heat, the lattice vibrations in a metal, as well as the free electrons, must be responsible for some heat transfer. In good metals the free electrons are such good carriers of heat that they swamp the effect of the lattice vibrations, but in semi-metals, or in metals containing large amounts of impurities, the electronic thermal conductivity can be so low that the lattice conductivity must be taken into account.

The free electrons and the lattice vibrations provide two different mechanisms of heat transfer, and we can therefore write

$$\kappa = \kappa_e + \kappa_g, \tag{9·8·1}$$

where κ_e is the electronic and κ_g the lattice conductivity, the two heat currents being additive but not independent since in a metal the free electrons provide a mechanism for scattering the lattice vibrations which is absent in an insulator.

9.81. To determine the thermal conductivity of the lattice we have to set up the Boltzmann equation for the distribution function $N(\mathbf{q})$ of the lattice vibrations. If the temperature gradient is in the x-direction, the velocity component of the lattice wave in this direction is $u_0 q_1/q$, and hence the Boltzmann equation can be written as

$$\left[\frac{\partial N}{\partial t}\right]_{\text{coll.}} = u_0 \frac{q_1}{q} \frac{\partial N}{\partial x} = u_0 \frac{q_1}{q} \frac{\partial N}{\partial T} \frac{\partial T}{\partial x}. \tag{9·81·1}$$

The change in N due to collisions (in the wide sense) is due to four principal causes:

(i) anharmonic terms in the equations of motion of the lattice vibrations, which cause scattering of the waves by one another,

(ii) irregularities of atomic dimensions, due to impurities and lattice defects,

(iii) grain boundaries and the boundaries of single crystal specimens,

(iv) interaction with the conduction electrons.

The first three causes determine the thermal conductivity of an insulator, and it is sufficient to assume that their effects can be represented by appropriate free paths. We therefore write for the anharmonic scattering

$$\left[\frac{\partial N}{\partial t}\right]_a = -\frac{N - N_0}{L_a(q, T)} u_0. \tag{9·81·2}$$

The function $L_a(q, T)$ has been calculated by Peierls (1929) for certain limiting cases, and he finds

$$L_a(q, T) = \text{constant } T^{-1} \quad (T \gg \Theta)$$

and
$$L_a(q, T) = \text{constant } e^{\frac{1}{2}\Theta/T} \quad (\Theta \gg T). \tag{9·81·3}$$

The scattering by irregularities of atomic dimensions follows the Rayleigh λ^{-4} law, so that

$$\left[\frac{\partial N}{\partial t}\right]_i = -\frac{N - N_0}{L_i(q)} u_0, \quad L_i(q) = \frac{1}{\mathscr{I}}\left(\frac{q_0}{q}\right)^4, \tag{9·81·4}$$

where $\mathscr{I}$ is a constant, while, for the boundary scattering,

$$\left[\frac{\partial N}{\partial t}\right]_b = -\frac{N - N_0}{L_b} u_0, \tag{9·81·5}$$

where L_b is independent of q and T and is determined by the configuration of the boundaries.

The scattering of the lattice waves by the free electrons is determined by the transition probabilities already calculated in § 9·33, but they now have to be summed over $\mathbf{k}$ instead of over $\mathbf{q}$, so that

$$\left[\frac{\partial N_j}{\partial t}\right]_{\text{electrons}} = \frac{C_j^2}{(2G+1)^3 Mh\nu_q} \frac{q^2}{}$$
$$\times \sum_{\mathbf{k}} \Omega(E_{\mathbf{k}} - E_{\mathbf{k+q}} + h\nu_q)\left[f(\mathbf{k}+\mathbf{q})\{1 - f(\mathbf{k})\}(N_q + 1) - f(\mathbf{k})\{1 - f(\mathbf{k}+\mathbf{q})\}N_q\right], \tag{9·81·6}$$

where j denotes the direction of polarization of the lattice waves ($j = 1$ for longitudinal and $j = 2, 3$ for transverse waves).

If we now put

$$f = f_0, \quad N = N_0 - q_1 \frac{b(z)}{kT} \frac{\partial N_0}{\partial z}, \tag{9·81·7}$$

where $z = h\nu/kT$, the summation over $\mathbf{k}$, when replaced by an integration, can be carried out exactly as in § 9·34 by taking the direction of $\mathbf{q}$ as the polar axis. The result is

$$\left[\frac{\partial N_j}{\partial t}\right]_{\text{electrons}} = -\frac{q_1 b_j(z)}{e^z - 1} \frac{C_j^2 \Delta}{2\pi M u_0 kT} \int_0^\infty \frac{dK}{dE} \frac{K^2 dK}{(e^\eta + 1)(e^{-\eta - z} + 1)} \tag{9·81·8}$$

$$= -\frac{q_1 z}{(e^z - 1)(1 - e^{-z})} b_j(z) \frac{C_j^2 m^{*2} a^3}{2\pi\hbar^4 M u_0}. \tag{9·81·9}$$

If we substitute the above expressions into (9·81·1) we find

$$b_j(z) = -\frac{1}{T}\frac{\partial T}{\partial x}\frac{k\Theta}{q_0}\bigg/\left[\frac{1}{L_a(q, T)} + \frac{1}{L_b} + \mathscr{D}_j\frac{q}{q_0} + \mathscr{I}\left(\frac{q}{q_0}\right)^4\right], \tag{9·81·10}$$

where

$$\mathscr{D}_j = 6\pi^2\left(\frac{4\pi}{3}\right)^{\frac{1}{3}}\frac{m^{*2}aC_j^2}{h^2 k\Theta M}. \tag{9·81·11}$$

The thermal current density is

$$w_g = \sum_{q,j} \hbar q_1 u_0^2 N_j(q), \tag{9·81·12}$$

since the velocity of the lattice waves in the x direction is $q_1 u_0/q$ and the energy is $h\nu_q = \hbar q u_0$. Hence, substituting from (9·81·7) and (9·81·10), integrating over the angles in q space and changing the variable of integration from q to z, we obtain

$$\kappa_g = \frac{w_g}{-\partial T/\partial x} = \sum_j \left(\frac{4\pi}{3}\right)^{\frac{1}{3}} \frac{k^2 \Theta}{ha^2} \left(\frac{T}{\Theta}\right)^3 \int_0^{\Theta/T} \frac{L_j(z,T) z^4 dz}{(e^z - 1)(1 - e^{-z})}, \tag{9·81·13}$$

where

$$\frac{1}{L_j(z,T)} = \frac{1}{L_a(z,T)} + \frac{1}{L_b} + \mathcal{D}_j \frac{zT}{\Theta} + \mathcal{I} z^4 \frac{T^4}{\Theta^4}. \tag{9·81·14}$$

9.82. In the calculations made in the preceding section it is assumed that the lattice conductivity could be calculated assuming that the free electrons have their equilibrium distribution. In principle this is not correct, and the electronic and lattice conductivities should be calculated by taking into account simultaneously the departure of both $f(\mathbf{k})$ and $N(\mathbf{q})$ from their equilibrium values. This means that $[\partial f/\partial t]_{\text{coll.}}$ and $[\partial N/\partial t]_{\text{coll.}}$ are linear functions of $c(\eta)$ and $b(z)$, and the two Boltzmann equations for f and N lead to two simultaneous linear integral equations for $c(\eta)$ and $b(z)$. Makinson (1938) has, however, shown that the correction terms introduced by this more exact calculation are in general small, and we shall neglect them here. The reader is referred to Makinson's paper for details of the complicated calculations.

9.83. It will be seen by comparing (6·11·3) and (9·81·13) that we can write

$$\kappa_g = \tfrac{1}{3} C_v u_0 \bar{L}, \tag{9·83·1}$$

but unless the free path L is constant this equation is effectively only the definition of the mean free path $\bar{L}$, and in general it is necessary to evaluate (9·81·13) explicitly. At sufficiently low temperatures, the effect of the grain boundaries is predominant and we then have $\kappa_g = \tfrac{1}{3} C_v u_0 L_b$, so that κ_g tends to zero like T^3. At higher temperatures, but still much less than Θ, in an insulator the impurity scattering becomes important, the conductivity reaches a maximum and then decreases roughly as $1/T$ (exactly as $1/T$ if no other scattering mechanisms are important, as is seen by substituting (9·81·14) into (9·81·13)). At still higher temperatures the anharmonic scattering predominates, and κ_g behaves like $1/T$.

In a metal the scattering of the lattice waves by the free electrons produces a diminution in κ_g which is significant in the region in which κ_g

would otherwise be large. In this region we can obtain an upper limit for κ_g by neglecting all the terms in (9·81·14) except that involving $\mathscr{D}_j$. This gives

$$\kappa_g = \sum_j \frac{hk^3\Theta^2 M}{6\pi^2 m^{*2}a^3 C_j^2} \left(\frac{T}{\Theta}\right)^2 \mathscr{J}_3\left(\frac{\Theta}{T}\right). \qquad (9\cdot83\cdot2)$$

The general form of κ_g is shown in fig. IX 14.

9·84. In the poorly conducting metals for which κ_g is comparable with κ_e in some temperature range, the Lorenz numbers will be abnormally high, and we can conclude from the curves of fig. IX 13 that the lattice conductivity in bismuth, German silver and manganin is large. When we try to compare theory and experiment quantitatively, however, we immediately come up against the difficulty that the observations give the

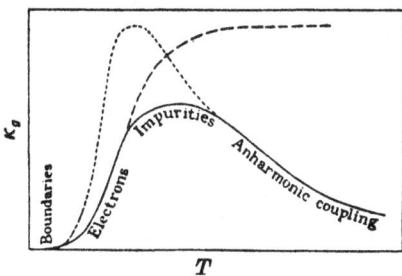

Fig. IX 14. The theoretical general form of κ_g. The dotted line shows the form for an insulator and the dashed line the form for a metal if only electrons scattered the lattice waves.

total κ, and we can only deduce κ_g if we can calculate κ_e. But the discussion of § 9·71 shows that the theoretical expression for κ_e has only a qualitative validity, and reliable values for κ_g are therefore impossible to obtain. An alternative method of deducing κ_g has been used by Reddemann (1934) for metals whose electrical conductivity is greatly decreased by a magnetic field. If the thermal and electrical conductivities are measured as functions of the magnetic field and if it is assumed that κ_e and σ are reduced in the same ratio, and that κ_g is unaffected, then κ_g is easily calculated. The former assumption is, however, not in general true (see § 10·514), and the method is subject to uncertainties of the same order as occur in the more direct method. We cannot therefore expect to obtain results of a high order of accuracy except at high and very low temperatures where the Lorenz number is accurately known.

The lattice conductivity can only be important in relatively poor conductors, where, either because the number of free electrons is small or because the impurities are numerous, the electronic thermal conductivity is abnormally small. Bismuth falls into the first category, and it will be seen by comparing figs. IX 14 and IX 15 that the observed trend of κ_g is in agreement with the theoretical predictions. Measurements by Hulm (1950) on a tin specimen containing 0·33 % of impurities show, however, that, at least at low temperatures, the quantitative agreement is poor. His results are shown in fig. IX 16. Since the values of T/κ lie well below the value ρ_r/L_n, there must be a considerable lattice conductivity, which can be estimated by putting $\kappa = \kappa_g + L_n T/\rho_r$. It is found that the lattice conductivity is about ten times larger than the theoretical upper limit for κ_g given by (9·83·2), which is obtained by assuming that the only scattering mechanism operative is that due to the free electrons. This discrepancy is of the same order as that occurring in the theory of the electronic conductivity, and this strengthens the conclusion that at low temperatures the theoretical expressions derived hitherto over-estimate the effect on the thermal conductivity of the interaction between the free electrons and the lattice vibrations.

The Steady State of the Lattice

9.9. It was pointed out by Peierls (1930) that, as a consequence of the interference conditions $\mathbf{k}' = \mathbf{k} \pm \mathbf{q}$, the quantity

$$\mathbf{S} = \sum_{\mathbf{k}} f(\mathbf{k})\,\mathbf{k} + \sum_{\mathbf{q}} N(\mathbf{q})\,\mathbf{q} \qquad (9\cdot9\cdot1)$$

is conserved in the absence of external influences. (It will be seen from (9·33·1) and (9·33·2) that, when an electron jumps from $\mathbf{k}$ to $\mathbf{k}+\mathbf{q}$, $N(\mathbf{q})$ diminishes by 1, and that $N(\mathbf{q})$ increases by 1 when an electron jumps from $\mathbf{k}$ to $\mathbf{k}-\mathbf{q}$.) The existence of the invariant $\mathbf{S}$ is due to the fact that the states in an infinite lattice are degenerate with respect to a reversal in the direction of motion, and it is for this reason that we chose to describe the lattice vibrations by progressive waves. Now, if we have an electric field $\mathscr{E}_x$ acting on the metal, the rate of change of S_x per unit volume is given by

$$\frac{\partial S_x}{\partial t} = \sum_{\mathbf{k}} k_1 \frac{\partial f}{\partial t} = \frac{2\pi e \mathscr{E}_x}{h} \sum_{\mathbf{k}} k_1 \frac{\partial f}{\partial k_1} = \frac{e\mathscr{E}_x}{2\pi^2 h} \int k_1 \frac{\partial f}{\partial k_1}\, d\mathbf{k} = -\frac{2\pi}{h} n e \mathscr{E}_x, \quad (9\cdot9\cdot2)$$

by equation (2·83·3) giving the rate of change of f due to an electric field. But S_x cannot be affected by collisions of the electrons with the lattice, and unless there are other processes tending to reduce S_x it is impossible for a stationary state to exist. We have so far implicitly assumed that these processes are the interactions of the lattice vibrations with one another (due to the anharmonic forces) or with the crystal imperfections. If, however, we put $f = f_0 + f_1$, $N = N_0 + N_1$ and set up the equations for a steady state of both the

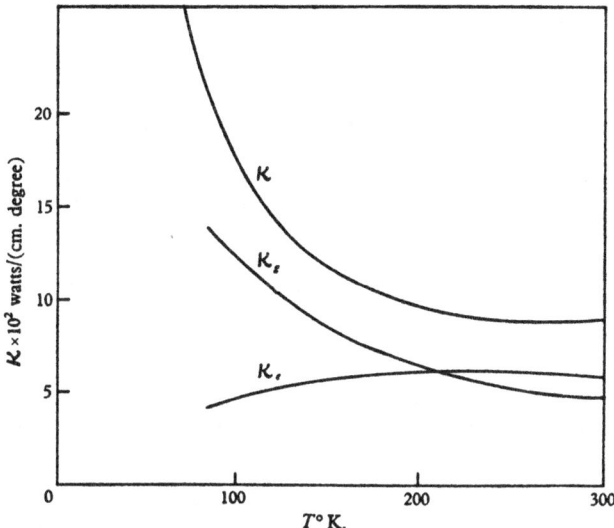

Fig. IX 15. The electronic and lattice conductivities of bismuth.

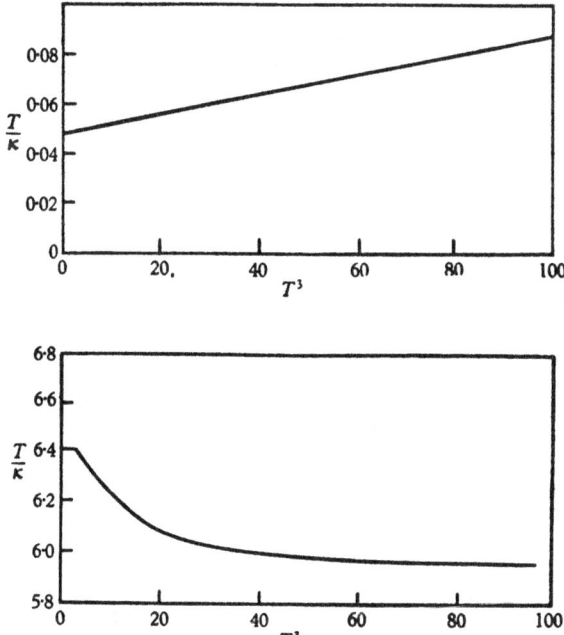

Fig. IX 16. T/κ as a function of T^3 for two specimens of tin. The purer specimen has $\rho_r/L_n = 0.045$ and shows very little lattice conductivity. The less pure specimen has $\rho_r/L_n = 6.72$, and has a large lattice conductivity. κ is measured in watts/(cm. degree).

electrons and the lattice, we run into difficulties which have been discussed in detail by Peierls. Peierls concludes that the interaction between the lattice vibrations is too small to maintain equilibrium at low temperatures, and that unless modifications are made to the theory the electrical conductivity will go to infinity much more rapidly than T^{-5} as $T \to 0$.

Peierls's suggestion (1932) to clear up the difficulty is to include the collisions between an electron and the lattice in which

$$\mathbf{k} - \mathbf{k'} \pm \mathbf{q} = 2\pi \mathbf{g}_b \quad (\mathbf{g}_b \neq 0) \qquad (9\cdot9\cdot3)$$

(see equation ($9\cdot3\cdot3$)), which we have neglected so far, and he shows that, if these more general processes are included, it is possible to set up a steady state and that many of the preceding results can be re-established. If, however, the collisions satisfying ($9\cdot9\cdot3$) are to be the dominant ones at low temperatures, they can only arise from electrons which are near a Brillouin zone boundary, and Peierls has shown that the surface of the Fermi distribution must be far from spherical and must be multiply-connected.

The assumptions made by Peierls have been criticized by Klemens (1951), who maintains that the calculations show that the anharmonic lattice forces are sufficiently strong to maintain a steady state if a more elaborate theory of their interaction is used than that considered by Peierls. This would avoid the difficulty of having to postulate a multiply-connected Fermi surface for the monovalent metals, which is distinctly unpalatable, but the subject is still very obscure.

REFERENCES

Baber, W. G. (1937). The contribution to the electrical resistance of metals from collisions between electrons. *Proc. Roy. Soc.* A, **158**, 383.

Bardeen, J. (1937). The conductivity of monovalent metals. *Phys. Rev.* **52**, 688.

van den Berg, G. J. (1938). Thesis, Leiden.

Berman, R. and MacDonald, D. K. C. (1951). The thermal and electrical conductivity of sodium at low temperatures. *Proc. Roy. Soc.* A, **209**, 368.

Berman, R. and MacDonald, D. K. C. (1952). The thermal and electrical conductivity of copper at low temperatures. *Proc. Roy. Soc.* A, **211**, 122.

Bloch, F. (1928). The quantum mechanics of electrons in crystal lattices. *Z. Phys.* **52**, 555.

Bloch, F. (1930). The electrical resistivity law at low temperatures. *Z. Phys.* **59**, 208.

Conwell, E. and Weisskopf, V. (1950). The theory of impurity scattering in semi-conductors. *Phys. Rev.* **77**, 388.

Fröhlich, H. and Mott, N. F. (1939). The mean free path of electrons in polar crystals. *Proc. Roy. Soc.* A, **171**, 496.

Grüneisen, E. (1933). The temperature dependence of the electrical resistance of pure metals. *Ann. Phys., Lpz.* (5), **16**, 530.

Grüneisen, E. and Goens, E. (1927). The electrical and thermal conductivity of metals of the regular system. *Z. Phys.* **44**, 615.

de Haas, W. J., de Boer, J. H. and van den Berg, G. J. (1934). The electrical resistance of gold, copper and lead at low temperatures. *Physica*, **1**, 1115.

Howarth, D. J. and Sondheimer, E. H. (1953). The theory of electronic conduction in polar semi-conductors. *Proc. Roy. Soc.* A, **219**, 53.

Hulm, J. K. (1950). The thermal conductivity of tin, mercury, indium and tantalum at liquid helium temperatures. *Proc. Roy. Soc.* A, **204**, 98.

Klemens, P. G. (1951). Electrical conductivity of metals at low temperatures. *Proc. Phys. Soc.* A, **64**, 1030.

MacDonald, D. K. C. and Mendelssohn, K. (1950). Resistivity of pure metals at low temperatures. *Proc. Roy. Soc.* A, **202**, 103 and 523.

Makinson, R. E. B. (1938). The thermal conductivity of metals. *Proc. Camb. Phil. Soc.* 34, 474.

Mendoza, E. and Thomas, J. G. (1951). The electrical resistance of copper, silver and gold at low temperatures. *Phil. Mag.* **42**, 291.

Mott, N. F. (1935). A discussion of the transition metals. *Proc. Phys. Soc.* 47, 571.

Mott, N. F. (1936). The electrical conductivity of transition metals. *Proc. Roy. Soc.* A, **153**, 699.

Nordheim, L. (1931). The electron theory of metals. *Ann. Phys., Lpz.* (5), **9**, 607.

Peierls, R. (1929). The kinetic theory of the thermal conductivity of crystals. *Ann. Phys., Lpz.* (5), **3**, 1055.

Peierls, R. (1930). The theory of the electrical and thermal conductivity of metals. *Ann. Phys., Lpz.* (5), **4**, 121.

Peierls, R. (1932). The electrical resistivity law at low temperatures. *Ann. Phys., Lpz.* (5), **12**, 154.

Reddemann, E. (1934). The change of the thermal and electrical resistance of a single crystal of bismuth in a magnetic field. *Ann. Phys., Lpz.* (5), **20**, 441.

Rhodes, P. (1950). The Bloch integral equation and electrical conductivity. *Proc. Roy. Soc.* A, **160**, 207.

Sondheimer, E. H. (1947). The theory of the thermoelectric power of metals. *Proc. Camb. Phil. Soc.* **43**, 571.

Sondheimer, E. H. (1952). A note on the theory of conduction in metals. *Proc. Phys. Soc.* A, **65**, 561.

Wilson, A. H. (1937). The second order electrical effects in metals. *Proc. Camb. Phil. Soc.* 33, 371.

Wilson, A. H. (1938). The electrical conductivity of the transition metals. *Proc. Roy. Soc.* A, **167**, 580.

Chapter X

APPLICATION OF THE VARIATION PRINCIPLE
TO CONDUCTION PHENOMENA

INTRODUCTION

10·1. In Chapter IX the fundamental integral equation for the distribution function was obtained and solved in certain special cases, but all the solutions obtained have limited ranges of validity. We now investigate a more powerful but more complicated method of solving the integral equation, which depends upon the solution of a variational problem (Kohler, 1948, 1949a; Sondheimer, 1950).

The most general form of the Boltzmann equation is given in § 8·1, and, if terms involving squares and products of the electric fields and temperature gradients are neglected, it is a linear integral equation. Hence, if the distribution function is written as

$$f = f_0 - \Phi \, \partial f_0/\partial E,$$

where

$$\Phi = \sum_{i=1}^{3} \left[\left(e\mathscr{E}_i + T \frac{\partial}{\partial x_i} \frac{\zeta}{T} \right) \phi_i + \frac{1}{T} \frac{\partial T}{\partial x_i} \psi_i \right], \tag{10·1·1}$$

then equations (8·1·2) (with $H = 0$) and (8·1·8) can be replaced by

$$L(\phi_i) = v_i \, \partial f_0/\partial E \tag{10·1·2}$$

and

$$L(\psi_i) = v_i E \, \partial f_0/\partial E, \tag{10·1·3}$$

where L is the integral operator defined by

$$L(\Phi) = \frac{1}{kT} \int V(\mathbf{k}, \mathbf{k}') \{ \Phi(\mathbf{k}) - \Phi(\mathbf{k}') \} \, d\mathbf{k}'. \tag{10·1·4}$$

The solution of the equations (10·1·2) and (10·1·3) can be obtained by means of the following principle.

Let F be the solution of the equation

$$L(F) = v_i E^n \, \partial f_0/\partial E, \tag{10·1·5}$$

and let G be any function satisfying the relation

$$(G, G) \equiv \frac{1}{4\pi^3} \int G L(G) \, d\mathbf{k} = \frac{1}{4\pi^3} \int G v_i E^n \frac{\partial f_0}{\partial E} \, d\mathbf{k}. \tag{10·1·6}$$

Then of all the functions G satisfying the above relation, F is the one which makes (G, G) a maximum.

The proof of this statement depends upon the relations $(F, G) = (G, F)$ and $(G, G) \geqslant 0$, which are consequences of the symmetry property

$$V(\mathbf{k}, \mathbf{k}') = V(\mathbf{k}', \mathbf{k})$$

and of the fact that $V(\mathbf{k}, \mathbf{k}')$, being a transition probability, is essentially positive. To establish these relations we write

$$(F, G) = \frac{1}{4\pi^3 kT} \int\!\!\int V(\mathbf{k}, \mathbf{k}')\, F(\mathbf{k})\, \{G(\mathbf{k}) - G(\mathbf{k}')\}\, d\mathbf{k}\, d\mathbf{k}'$$

$$= \frac{1}{4\pi^3 kT} \int\!\!\int V(\mathbf{k}, \mathbf{k}')\, F(\mathbf{k}')\, \{G(\mathbf{k}') - G(\mathbf{k})\}\, d\mathbf{k}\, d\mathbf{k}',$$

since $V(\mathbf{k}, \mathbf{k}') = V(\mathbf{k}', \mathbf{k})$. Hence, by addition,

$$(F, G) = \frac{1}{8\pi^3 kT} \int\!\!\int V(\mathbf{k}, \mathbf{k}')\, \{F(\mathbf{k}) - F(\mathbf{k}')\}\, \{G(\mathbf{k}) - G(\mathbf{k}')\}\, d\mathbf{k}\, d\mathbf{k}'.$$

$$(10\cdot1\cdot7)$$

Since this is symmetrical in F and G, it establishes the relation

$$(F, G) = (G, F). \tag{10·1·8}$$

The inequality
$$(G, G) \geqslant 0 \tag{10·1·9}$$

follows at once from $(10\cdot1\cdot7)$ by putting $F = G$, since the integrand is then essentially positive.

To prove the variation principle we proceed as follows. If we multiply the equation $(10\cdot1\cdot5)$ by G and integrate with respect to $\mathbf{k}$ we find that

$$(G, F) = \frac{1}{4\pi^3} \int G v_i E^n \frac{\partial f_0}{\partial E}\, d\mathbf{k} = (G, G) \tag{10·1·10}$$

by $(10\cdot1\cdot6)$. Now $(F - G, F - G) \geqslant 0$, and so $(F, F) - 2(F, G) + (G, G) \geqslant 0$. Hence, on using $(10\cdot1\cdot10)$, we have

$$(G, G) \leqslant (F, F), \tag{10·1·11}$$

which proves the maximum principle.

An alternative statement is the following:

Let G be any function for which (G, G) exists. Then of all such functions, the solution F of equation $(10\cdot1\cdot5)$ is the one which makes

$$(G, G) \Big/ \left[\frac{1}{4\pi^3} \int G v_i E^n \frac{\partial f_0}{\partial E}\, d\mathbf{k} \right]^2 \tag{10·1·12}$$

a minimum.

To prove this we note that

$$(\lambda F + \mu G, \lambda F + \mu G) = \lambda^2 (F, F) + 2\lambda\mu (F, G) + \mu^2 (G, G) \geqslant 0$$

for all real λ and μ. The discriminant is therefore positive, i.e.

$$(F, F)\, (G, G) \geqslant (F, G)^2. \tag{10·1·13}$$

Now if F satisfies equation (10·1·5), we have

$$(F, F) = \frac{1}{4\pi^3} \int F v_i E^n \frac{\partial f_0}{\partial E} d\mathbf{k}, \quad (G, F) = \frac{1}{4\pi^3} \int G v_i E^n \frac{\partial f_0}{\partial E} d\mathbf{k}. \quad (10\cdot1\cdot14)$$

Hence dividing the inequality (10·1·13) by $(F, F)(F, G)^2$, which is positive, we have

$$\frac{(G, G)}{(F, G)^2} \geqslant \frac{1}{(F, F)},$$

which, by (10·1·14), can be written as

$$(G, G) \Big/ \left[\frac{1}{4\pi^3} \int G v_i E^n \frac{\partial f_0}{\partial E} d\mathbf{k} \right]^2 \geqslant (F, F) \Big/ \left[\frac{1}{4\pi^3} \int F v_i E^n \frac{\partial f_0}{\partial E} d\mathbf{k} \right]^2,$$
$$(10\cdot1\cdot15)$$

which proves the minimum principle.

THE VARIATION PRINCIPLE FOR ISOTROPIC METALS

10·2. We may consider the variation principle for the particular transport equations discussed in Chapter IX to be a special case of the general variation principle, or we can derive it independently. The direct proof is as follows.

If we consider scattering of the electrons both by impurities and by the lattice vibrations, we have, from (9·34·10), (9·34·12) and (9·4·8),

$$f = f_0 - k_1 c(E) \partial f_0 / \partial E \quad (10\cdot2\cdot1)$$

and

$$\left[\frac{\partial f}{\partial t} \right]_{\text{coll.}} = -k_1 \mathscr{L}(c) = -k_1 \frac{\hbar}{m^*} \frac{\partial f_0}{\partial E} \left(\epsilon \mathscr{E} + T \frac{\partial}{\partial x} \frac{\zeta}{T} + \frac{E}{T} \frac{\partial T}{\partial x} \right), \quad (10\cdot2\cdot2)$$

where

$$\mathscr{L}(c) = -\sqrt{\frac{2}{m^*} \frac{E^{\frac{1}{2}}}{l_r} \frac{\partial f_0}{\partial E} c(E)} - \frac{\partial f_0}{\partial E} \frac{(\frac{1}{2} m^*)^{\frac{1}{2}}}{\hbar^2 \Lambda E^{\frac{1}{2}}} \left(\frac{T}{\Theta} \right)^3$$

$$\times \int_{-\Theta/T}^{\Theta/T} \left[E c(\eta) - c(\eta + z) \left\{ E + \frac{1}{2} k T z - D \left(\frac{T}{\Theta} \right)^2 z^2 \right\} \right] \frac{e^\eta + 1}{e^{\eta+z} + 1} \frac{z^2 dz}{|1 - e^{-z}|}.$$
$$(10\cdot2\cdot3)$$

Then, if $c^{(n)}(E)$ is the solution of

$$\mathscr{L}(c^{(n)}) = E^{n-\frac{3}{2}} \partial f_0 / \partial E, \quad (10\cdot2\cdot4)$$

the solution of (10·2·2) is given by

$$c(E) = \frac{\hbar}{m^*} \left(\epsilon \mathscr{E} + T \frac{\partial}{\partial x} \frac{\zeta}{T} \right) c^{(\frac{3}{2})}(E) + \frac{\hbar}{m^*} \frac{1}{T} \frac{\partial T}{\partial x} c^{(\frac{5}{2})}(E). \quad (10\cdot2\cdot5)$$

The variation principle in the form in which we shall use it can now be stated as follows:

The solution of equation (10·2·4) *is such as to make the integral*

$$(c^{(n)}, c^{(n)}) = kT \int_{-\infty}^{\infty} E^{\frac{1}{2}} c^{(n)} \mathscr{L}(c^{(n)}) \, d\eta, \tag{10·2·6}$$

where $\eta = (E - \zeta)/kT$, *a maximum, subject to the subsidiary condition*

$$kT \int_{-\infty}^{\infty} E^{\frac{1}{2}} c^{(n)} \mathscr{L}(c^{(n)}) \, d\eta = \int_{-\infty}^{\infty} E^n c^{(n)} \frac{\partial f_0}{\partial \eta} \, d\eta. \tag{10·2·7}$$

To prove the variation principle we have merely to establish the relations $(c, d) = (d, c)$ and $(c, c) \geqslant 0$, and the proof is then exactly the same as for the general case. The first relation follows at once by a change of variable. For the only term in (c, d) which is not obviously symmetrical is that involving

$$\int_{-\infty}^{\infty} \int_{-\Theta/T}^{\Theta/T} c(\eta) \, d(\eta + z) \left\{ E + \tfrac{1}{2} kTz - D \left(\frac{T}{\Theta} \right)^2 z^2 \right\} \frac{z^2 \, dz}{|1 - e^{-z}|} \frac{d\eta}{(e^{\eta + z} + 1)(e^{-\eta} + 1)}.$$

By using the identity

$$\int_{-\infty}^{\infty} \left[\frac{F(\eta + z)}{e^{\eta + z} + 1} - \frac{F(\eta)}{e^{\eta} + e^z} \right] \frac{d\eta}{e^{-\eta} + 1} = 0, \tag{10·2·8}$$

this can be transformed into

$$\int_{-\infty}^{\infty} \int_{-\Theta/T}^{\Theta/T} c(\eta - z) \, d(\eta) \left\{ E - \tfrac{1}{2} kTz - D \left(\frac{T}{\Theta} \right)^2 z^2 \right\} \frac{z^2 \, dz}{|1 - e^{-z}|} \frac{d\eta}{(e^{\eta} + e^z)(e^{-\eta} + 1)}.$$

If we now put $-z$ for z, we obtain the preceding expression with $c(\eta) \, d(\eta + z)$ replaced by $c(\eta + z) \, d(\eta)$.

The inequality $(c, c) \geqslant 0$ is more difficult to establish. We have

$$(c, c) = - \sqrt{\frac{2}{m^*}} \int_{-\infty}^{\infty} \frac{E^2}{l_r} \frac{\partial f_0}{\partial \eta} \{c(E)\}^2 \, d\eta$$

$$+ \frac{(\tfrac{1}{2} m^*)^{\frac{1}{2}}}{\hbar^2 \Lambda} \left(\frac{T}{\Theta} \right)^3 \int_{-\infty}^{\infty} \int_{-\Theta/T}^{\Theta/T} \left[E c(\eta) \, c(\eta) - c(\eta) \, c(\eta + z) \left\{ E + \tfrac{1}{2} kTz - D \left(\frac{T}{\Theta} \right)^2 z^2 \right\} \right]$$

$$\times \frac{z^2 \, dz}{|1 - e^{-z}|} \frac{d\eta}{(e^{\eta + z} + 1)(e^{-\eta} + 1)}. \tag{10·2·9}$$

Now, by using the identity (10·2·8) and replacing z by $-z$, twice the quantity in square brackets can be replaced by

$$E c(\eta) \, c(\eta) - 2 \{ E + \tfrac{1}{2} kTz - D(T/\Theta)^2 z^2 \} c(\eta) \, c(\eta + z) + (E + kTz) \, c(\eta + z) \, c(\eta + z),$$

and if this is a positive definite form every term of (10·2·9) is positive, so that we must have $(c, c) \geqslant 0$. We have therefore to prove that

$$\{ E + \tfrac{1}{2} kTz - D(T/\Theta)^2 z^2 \}^2 - E(E + kTz) \leqslant 0.$$

On substituting the values (9·34·4) and (9·34·11) for Θ and D, and putting $E = h^2 K^2/(8\pi^2 m^*)$, $z = h\nu/kT$, the inequality becomes, after some simplification,

$$\frac{hK}{4\pi m^* u_0} \geqslant \left| \frac{h\nu}{4m^* u_0^2} - \frac{1}{2} \right|,$$

and this is always satisfied, since it is the condition (9·34·3) for the transition probabilities to have non-zero values. We therefore see that every term in (10·2·9) is essentially positive and therefore that $(c, c) \geqslant 0$.

10·21. The variation principle has been stated in the form most suited to the case in which the integrals with respect to η can be taken between the limits $\pm \infty$ with negligible error. That is, we have assumed that the asymptotic formula for evaluating integrals involving the Fermi function is applicable. If this is not true, the lower limits must everywhere be $-\zeta/kT$, and it is then simpler to use E directly rather than η. In this case, the electrons cannot be treated as forming a degenerate gas and the limits occurring in the integral equation need amendment. The variation principle and its proof must be modified to take these more complicated limits into account, and it is somewhat simpler to start from the general variation principle and evaluate the integrals directly rather than to evaluate the collision operator first and then employ the variation principle. Such calculations have not so far been carried out, but there is no difficulty in doing so if they should be required.

SOLUTION OF THE VARIATIONAL PROBLEM

10·3. We can obtain a convenient solution of equation (10·2·4) as a power series in E or in $\eta = (E - \zeta)/kT$, and the latter is slightly more convenient. We therefore put

$$c^{(n)}(\eta) = \sum_{r=0}^{\infty} c_r^{(n)} \eta^r \qquad (10·3·1)$$

and

$$\alpha_r^{(n)} = \int_{-\infty}^{\infty} E^n \eta^r \frac{\partial f_0}{\partial \eta} d\eta, \quad d_{rs} = (\eta^r, \eta^s) = kT \int_{-\infty}^{\infty} E^{\frac{3}{2}} \eta^r \mathscr{L}(\eta^s) \, d\eta. \quad (10·3·2)$$

Then

$$(c^{(n)}, c^{(n)}) = \sum_{r=0}^{\infty} \sum_{s=0}^{\infty} d_{rs} c_r^{(n)} c_s^{(n)} \qquad (10·3·3)$$

must be a maximum, subject to the condition

$$C \equiv \sum_{r=0}^{\infty} \sum_{s=0}^{\infty} d_{rs} c_r^{(n)} c_s^{(n)} - \sum_{r=0}^{\infty} \alpha_r^{(n)} c_r^{(n)} = 0. \qquad (10·3·4)$$

If we maximize $(\lambda - 1)(c^{(n)}, c^{(n)}) + C$, where λ is a Lagrange undetermined multiplier, the equations satisfied by the c's are $2\lambda \Sigma d_{rs} c_r^{(n)} - \alpha_s^{(n)} = 0$, and, if we multiply the rth equation by $c_r^{(n)}$, sum with respect to r and use (10·3·4), we find $2\lambda = 1$. Hence the coefficients $c_r^{(n)}$ satisfy the infinite set of linear equations

$$\sum_{s=0}^{\infty} d_{rs} c_s^{(n)} - \alpha_r^{(n)} = 0 \quad (r = 0, 1, \ldots). \tag{10·3·5}$$

The current densities can now be found in the usual way. They are

$$J = -\frac{\epsilon}{4\pi^3} \int v_1 f d\mathbf{k} = \frac{(2m^*)^{\frac{3}{2}} \epsilon}{3\pi^2 \hbar^4} \int_{-\infty}^{\infty} E^{\frac{3}{2}} c(\eta) \frac{\partial f_0}{\partial \eta} d\eta$$

$$= \mathcal{K}_{\frac{3}{2},\frac{3}{2}} \left(\epsilon^2 \mathscr{E} + \epsilon T \frac{\partial}{\partial x} \frac{\zeta}{T} \right) + \mathcal{K}_{\frac{3}{2},\frac{5}{2}} \frac{\epsilon}{T} \frac{\partial T}{\partial x} \tag{10·3·6}$$

and

$$w = \frac{1}{4\pi^3} \int v_1 E f d\mathbf{k} = \mathcal{K}_{\frac{3}{2},\frac{5}{2}} \left(-\epsilon \mathscr{E} - T \frac{\partial}{\partial x} \frac{\zeta}{T} \right) - \mathcal{K}_{\frac{5}{2},\frac{5}{2}} \frac{1}{T} \frac{\partial T}{\partial x}, \tag{10·3·7}$$

where

$$\mathcal{K}_{m,n} = \frac{16\pi (2m^*)^{\frac{1}{2}}}{3h^3} \int_{-\infty}^{\infty} E^m c^{(n)}(\eta) \frac{\partial f_0}{\partial \eta} d\eta = \frac{16\pi (2m^*)^{\frac{1}{2}}}{3h^3} \sum_{r=0}^{\infty} \alpha_r^{(m)} c_r^{(n)} = \mathcal{K}_{n,m}. \tag{10·3·8}$$

10·31. It will be noticed that the set of equations (10·3·5) could have been obtained by multiplying the equation (10·2·4) by $E^{\frac{1}{2}} \eta^r$ ($r = 0, 1, 2, \ldots$) and integrating with respect to η. This was the method adopted by Kroll (1933), who first derived the equations, but he gave no justification for his procedure.

The main theoretical difficulties that arise in applying the above equations are in ensuring that the set of functions used is complete and that the formal series obtained converges to the solution of the transport equations. The equations are, however, so complex that we are limited to finding an approximate solution whose form is suggested by physical considerations.

In all the methods of solution discussed in Chapter IX the validity of the approximate solutions obtained must be taken for granted, and they do not lend themselves to a discussion of the errors involved in stopping the approximation at any particular point. The variation principle, on the other hand, gives rise to a monotonic sequence of approximations to the conductivities, and the effect of the higher approximations can easily be calculated, although it turns out that the first approximation is usually the dominant one.

A direct numerical solution of the integral equation has been obtained by Rhodes (1950), his method being primarily an iterative one. This method has the advantage of giving not only the conductivity but also the distribution function to a reasonable degree of accuracy. The calculations are, however, laborious and have only been applied to the electrical conductivity.

10·32. The infinite set of equations (10·3·5) can easily be solved formally, and on substituting the result into (10·3·8) we find

$$\mathscr{K}_{m,n} = -\frac{16\pi(2m^*)^{\frac{1}{2}}}{3h^3}\frac{\mathscr{D}_{m,n}}{\mathscr{D}}, \qquad (10\cdot32\cdot1)$$

where $\mathscr{D} = |d_{rs}|$ is the determinant of the d_{rs}'s, and $\mathscr{D}_{m,n}$ is the determinant formed by bordering $\mathscr{D}$ with $0, \alpha_0^{(m)}, \alpha_1^{(m)}, \ldots$ and $0, \alpha_0^{(n)}, \alpha_1^{(n)}, \ldots$, i.e.

$$\mathscr{D}_{m,n} = \begin{vmatrix} 0 & \alpha_0^{(m)} & \alpha_1^{(m)} & \cdots \\ \alpha_0^{(n)} & d_{00} & d_{01} & \cdots \\ \alpha_1^{(n)} & d_{10} & d_{11} & \cdots \\ \vdots & \vdots & \vdots & \end{vmatrix}, \qquad (10\cdot32\cdot2)$$

Neither $\mathscr{D}_{m,n}$ nor $\mathscr{D}$ is in general convergent, and we therefore have to deal directly with their ratio. We proceed to evaluate σ, κ and $\mathfrak{S}$, treating each separately.

10·33. *The electrical conductivity.* The electrical conductivity is given by

$$\sigma = \epsilon^2 \mathscr{K}_{\frac{1}{2},\frac{1}{2}} = -\frac{16\pi(2m^*)^{\frac{1}{2}}\epsilon^2 \mathscr{D}_{\frac{1}{2},\frac{1}{2}}}{3h^3}\frac{\mathscr{D}_{\frac{1}{2},\frac{1}{2}}}{\mathscr{D}}. \qquad (10\cdot33\cdot1)$$

Let $\mathscr{D}^{(n)}$ denote the determinant obtained by taking only the first n rows and columns of $\mathscr{D}$, let $\mathscr{D}_{\alpha\alpha}^{(n)}$ be the determinant of $n+1$ rows and columns formed by bordering $\mathscr{D}^{(n)}$ in the $(n+1)$th row and column with $\alpha_0^{(\frac{3}{2})}, \alpha_1^{(\frac{3}{2})}, \ldots,$ $\alpha_{n-1}^{(\frac{3}{2})}, 0$, and let $\mathscr{D}_\alpha^{(n-1)}$ be the determinant of n rows and columns formed by replacing the last row (or column) of $\mathscr{D}^{(n)}$ by $\alpha_0^{(\frac{3}{2})}, \alpha_1^{(\frac{3}{2})}, \ldots, \alpha_{n-1}^{(\frac{3}{2})}$. Consider the adjoint $\mathscr{D}_{\alpha\alpha}^{(n)\prime}$ (the determinant formed by the minors of $\mathscr{D}_{\alpha\alpha}^{(n)}$). Then, if M and M' are corresponding m-rowed minors of any determinant D and its adjoint D' respectively, M' is equal to D^{m-1} times the algebraic complement of M. We apply this well-known theorem to the two-rowed minor formed by the four elements at the bottom right-hand corner of $\mathscr{D}_{\alpha\alpha}^{(n)\prime}$. It gives

$$\begin{vmatrix} \mathscr{D}_{\alpha\alpha}^{(n-1)} & \mathscr{D}_\alpha^{(n-1)} \\ \mathscr{D}_\alpha^{(n-1)} & \mathscr{D}^{(n)} \end{vmatrix} = \mathscr{D}_{\alpha\alpha}^{(n)}\mathscr{D}^{(n-1)},$$

which can be written

$$-\frac{\mathscr{D}_{\alpha\alpha}^{(n)}}{\mathscr{D}^{(n)}} + \frac{\mathscr{D}_{\alpha\alpha}^{(n-1)}}{\mathscr{D}^{(n-1)}} = \frac{(\mathscr{D}_\alpha^{(n-1)})^2}{\mathscr{D}^{(n-1)}\mathscr{D}^{(n)}} \quad (n \geqslant 2).$$

Also

$$-\frac{\mathscr{D}_{\alpha\alpha}^{(1)}}{\mathscr{D}^{(1)}} = \frac{(\alpha_0^{(\frac{3}{2})})^2}{d_{00}}$$

Hence, by addition, we have

$$\sigma = -\frac{16\pi(2m^*)^{\frac{1}{2}}}{3h^3}\epsilon^2 \lim_{n\to\infty}\frac{\mathscr{D}_{\alpha\alpha}^{(n)}}{\mathscr{D}^{(n)}} = \frac{16\pi(2m^*)^{\frac{1}{2}}}{3h^3}\epsilon^2\left[\frac{(\alpha_0^{(\frac{3}{2})})^2}{d_{00}} + \sum_{n=2}^{\infty}\frac{(\mathscr{D}_\alpha^{(n-1)})^2}{\mathscr{D}^{(n-1)}\mathscr{D}^{(n)}}\right],$$

$$(10\cdot33\cdot2)$$

which is the most convenient form for σ. Since the d_{rs}'s are the coefficients of a positive definite quadratic form, all the $\mathscr{D}^{(n)}$'s are positive, and hence the successive approximations to σ form a monotonically increasing sequence.

To find the first approximation to σ we merely need $\alpha_0^{(\frac{3}{2})}$ and d_{00}, and these only to the zero order in kT/ζ. Now $\alpha_0^{(\frac{3}{2})} = -\zeta^{\frac{3}{2}}$, and

$$d_{00} = -\sqrt{\frac{2}{m^*}}\frac{1}{l_r}\int_{-\infty}^{\infty} E^2 \frac{\partial f_0}{\partial \eta}\, d\eta$$

$$+ \frac{(\frac{1}{2}m^*)^{\frac{1}{2}}}{\hbar^2\Lambda}\left(\frac{T}{\Theta}\right)^3 \int_{-\infty}^{\infty}\int_{-\Theta/T}^{\Theta/T}\left[D\left(\frac{T}{\Theta}\right)^2 z^2 - \tfrac{1}{2}kTz\right]\frac{z^2\,dz}{|1-e^{-z}|}\frac{d\eta}{(e^{\eta+z}+1)(e^{-\eta}+1)}.$$

$$(10\cdot33\cdot3)$$

The integral with respect to η in the second term is a particular case of Appendix (A 5·11) (it is, in fact, an elementary integral), and on carrying out the integrations we find

$$d_{00} = \sqrt{\frac{2}{m^*}}\frac{\zeta^2}{l_r} + \frac{(2m^*)^{\frac{1}{2}}D}{\hbar^2\Lambda}\left(\frac{T}{\Theta}\right)^5 \mathscr{J}_5\left(\frac{\Theta}{T}\right) = \frac{16\pi(2m^*)^{\frac{1}{2}}\epsilon^2\zeta^3}{3h^3}\left(\frac{1}{\sigma_r}+\frac{1}{\sigma_i}\right),$$

$$(10\cdot33\cdot4)$$

where

$$\frac{1}{\sigma_r} = \rho_r = \frac{3h^3}{16\pi m^*\epsilon^2 l_r\zeta} \qquad (10\cdot33\cdot5)$$

and

$$\frac{1}{\sigma_i} = \rho_i = \frac{3\pi hD}{4\epsilon^2\Lambda\zeta^3}\left(\frac{T}{\Theta}\right)^5 \mathscr{J}_5\left(\frac{\Theta}{T}\right) \qquad (10\cdot33\cdot6)$$

are the expressions found in §§ 9·62 and 9·621 for the residual and ideal resistances. Hence

$$\sigma = \sigma^{(0)} + \frac{16\pi(2m^*)^{\frac{1}{2}}\epsilon^2}{3h^3}\sum_{n=2}^{\infty}\frac{(\mathscr{D}_\alpha^{(n-1)})^2}{\mathscr{D}^{(n-1)}\mathscr{D}^{(n)}}, \qquad \frac{1}{\sigma^{(0)}} = \frac{1}{\sigma_r}+\frac{1}{\sigma_i}, \qquad (10\cdot33\cdot7)$$

where σ_r and σ_i are given by (10·33·5) and (10·33·6).

By using (A 5·11) Sondheimer has obtained explicit expressions for the elements of the determinants $\mathscr{D}_{m,n}$ and $\mathscr{D}$ (or, alternatively, of $\mathscr{D}_\alpha^{(n)}$ and $\mathscr{D}^{(n)}$). The successive approximations to σ can be calculated to any desired accuracy, and the convergence of the expansion (10·33·7) can readily be investigated. The same is true for the expressions (10·34·3) and (10·35·2) given below for κ and $\mathfrak{S}$.

10·34. The thermal conductivity.

The thermal conductivity is given by

$$\kappa = \frac{\mathscr{K}_{\frac{1}{2},\frac{1}{2}}\mathscr{K}_{\frac{3}{2},\frac{3}{2}} - \mathscr{K}_{\frac{1}{2},\frac{3}{2}}^2}{\mathscr{K}_{\frac{1}{2},\frac{1}{2}}T} = \frac{16\pi(2m^*)^{\frac{1}{2}}}{3h^3\Theta}\frac{\Theta}{T}\left(-\frac{\mathscr{D}_{\frac{3}{2},\frac{3}{2}}}{\mathscr{D}} + \frac{\mathscr{D}_{\frac{1}{2},\frac{3}{2}}^2}{\mathscr{D}_{\frac{1}{2},\frac{1}{2}}\mathscr{D}}\right). \qquad (10\cdot34\cdot1)$$

To transform this into a simpler form, consider

$$\mathscr{D}_{\frac{3}{2},\frac{3}{2},\frac{5}{2},\frac{5}{2}} = \begin{vmatrix} 0 & 0 & \alpha_0^{(\frac{3}{2})} & \alpha_1^{(\frac{3}{2})} & \cdots \\ 0 & 0 & \alpha_0^{(\frac{5}{2})} & \alpha_1^{(\frac{5}{2})} & \cdots \\ \alpha_0^{(\frac{3}{2})} & \alpha_0^{(\frac{5}{2})} & d_{00} & d_{01} & \cdots \\ \alpha_1^{(\frac{3}{2})} & \alpha_1^{(\frac{5}{2})} & d_{10} & d_{11} & \cdots \\ \vdots & \vdots & \vdots & \vdots & \end{vmatrix}, \qquad (10\cdot34\cdot2)$$

and consider the two-rowed minor in the left-hand top corner of the adjoint determinant. It is

$$\begin{vmatrix} \mathscr{D}_{\frac{3}{2},\frac{3}{2}} & \mathscr{D}_{\frac{3}{2},\frac{5}{2}} \\ \mathscr{D}_{\frac{5}{2},\frac{3}{2}} & \mathscr{D}_{\frac{5}{2},\frac{5}{2}} \end{vmatrix},$$

and, by the properties of adjoint determinants, it is equal to $\mathscr{D}\mathscr{D}_{\frac{3}{2},\frac{3}{2},\frac{5}{2},\frac{5}{2}}$. Hence

$$\kappa = -\frac{16\pi(2m^*)^{\frac{1}{2}}}{3h^3\Theta}\frac{\Theta}{T}\frac{\mathscr{D}_{\frac{3}{2},\frac{3}{2},\frac{5}{2},\frac{5}{2}}}{\mathscr{D}_{\frac{3}{2},\frac{3}{2}}}. \qquad (10\cdot34\cdot3)$$

Also, since $\mathscr{D}_{\frac{3}{2},\frac{3}{2},\frac{5}{2},\frac{5}{2}}$ is the bordered determinant of $\mathscr{D}_{\frac{3}{2},\frac{3}{2}}$, we can apply the same expansion as was used in § 10·33 to calculate σ; and the zero-order approximation is obtained by taking four rows of $\mathscr{D}_{\frac{3}{2},\frac{3}{2},\frac{5}{2},\frac{5}{2}}$ and three rows of $\mathscr{D}_{\frac{3}{2},\frac{3}{2}}$. This gives

$$\kappa^{(0)} = \frac{16\pi(2m^*)^{\frac{1}{2}}}{3h^3\Theta}\frac{\Theta}{T}\frac{(\alpha_0^{(\frac{3}{2})}\alpha_1^{(\frac{5}{2})} - \alpha_0^{(\frac{5}{2})}\alpha_1^{(\frac{3}{2})})^2}{(\alpha_0^{(\frac{3}{2})})^2 d_{11} - 2\alpha_0^{(\frac{3}{2})}\alpha_1^{(\frac{3}{2})}d_{01} + (\alpha_1^{(\frac{3}{2})})^2 d_{00}}. \qquad (10\cdot34\cdot4)$$

It is sufficient to evaluate this to the smallest non-vanishing power of $(kT/\zeta)^2$. The integrals are of standard type and the d's can be evaluated by the formula given in Appendix (A 5·11). The result is that

$$\frac{1}{\kappa^{(0)}} = \frac{1}{\kappa_r} + \frac{1}{\kappa_i}, \qquad (10\cdot34\cdot5)$$

where $1/\kappa_r = 3(\epsilon/\pi k)^2\rho_r/T$ is the residual thermal resistance and where $1/\kappa_i$ is the expression given by the interpolation formula (9·621·4) for the ideal thermal resistance.

The higher order correction terms have been calculated by Sondheimer and are found to be much larger than for the electrical conductivity. In particular, as $\kappa_r \to 0$, the conductivity does not tend to the formula for κ_i given by (9·621·4), and there are considerable deviations at low temperatures. At sufficiently low temperatures κ_i is proportional to T^{-2}, but the proportionality factor is larger than is given by (9·621·4). (For an ideally pure metal the coefficient of T^{-2} must be increased by a factor 1·33.) Some typical results are shown in fig. X 1. (The parameter D/ζ is equal to $2^{-\frac{1}{3}}$

309

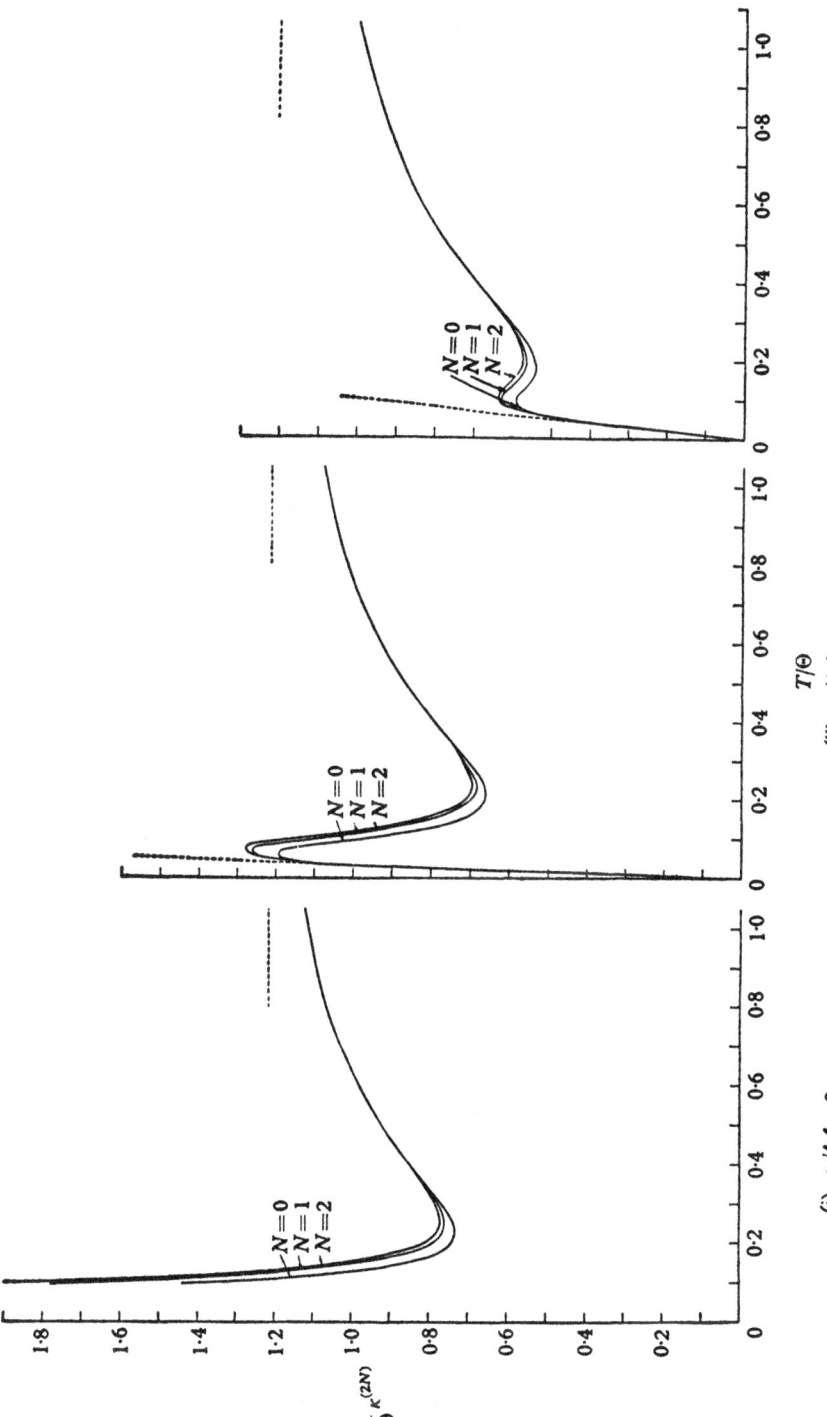

(i) $\rho_r/4A = 0$ (ii) $\rho_r/4A = 0.01$ (iii) $\rho_r/4A = 0.03$

T/Θ

$\dfrac{12A}{\pi^2 L_n \Theta} \kappa^{(2N)}$

Fig. X 1. Successive approximations to the thermal conductivity of monovalent metals.

for a monovalent metal, and $\kappa^{(2N)}$ is the value of κ obtained by retaining N terms in the infinite series for κ similar to (10·33·2). The parameter $\frac{1}{4}\rho_r/A$, defined in (9·7·4), is equal to $\frac{1}{4}T/\Theta$ times ρ_r/ρ_i for $T \gg \Theta$.) Although the deviations from the interpolation formula are considerable, they are insufficient to explain the discrepancies discussed in § 9·71, and the theory requires substantial modification to bring it into line with the experimental data.

10·35. *The thermoelectric power.* The absolute thermoelectric force per degree is given by $d\Theta/dT = -\mathfrak{S}/\epsilon$, where

$$\mathfrak{S} = \frac{\mathcal{X}_{\frac{1}{2},\frac{3}{2}} - \zeta\mathcal{X}_{\frac{1}{2},\frac{1}{2}}}{\mathcal{X}_{\frac{1}{2},\frac{1}{2}}T} = \frac{\mathcal{D}_{\frac{1}{2},\frac{3}{2}} - \zeta\mathcal{D}_{\frac{1}{2},\frac{1}{2}}}{\mathcal{D}_{\frac{1}{2},\frac{1}{2}}T}. \tag{10·35·1}$$

Now $\alpha_r^{(\frac{1}{2})} - \zeta\alpha_r^{(\frac{1}{2})} = kT\alpha_{r+1}^{(\frac{1}{2})}$, and hence

$$\mathfrak{S} = kE_{\frac{1}{2},\frac{1}{2}}/\mathcal{D}_{\frac{1}{2},\frac{1}{2}}, \tag{10·35·2}$$

where

$$E_{m,n} = \begin{vmatrix} 0 & \alpha_0^{(m)} & \alpha_1^{(m)} & \cdots \\ \alpha_1^{(n)} & d_{00} & d_{01} & \cdots \\ \alpha_2^{(n)} & d_{10} & d_{11} & \cdots \\ \vdots & \vdots & \vdots & \end{vmatrix}. \tag{10·35·3}$$

The ratios $E_{\frac{1}{2},\frac{1}{2}}/\mathcal{D}$ and $\mathcal{D}_{\frac{1}{2},\frac{1}{2}}/\mathcal{D}$ can be expanded in series just as in the discussion of the conductivity in § 10·33, but the convergence is not so rapid.

If we restrict the determinants to three rows and three columns we obtain an expression for $\mathfrak{S}$ which resembles (9·63·2) but which differs from it by having $-\mathcal{J}_7$ instead of $\mathcal{J}_7$. This makes practically no difference to the numerical results, but it shows that incorrect results may be obtained by stopping the approximations at too early a stage.

Matthiessen's Rule

10·4. The results of the preceding section show that the interpolation formula for σ derived in § 9·61 by elementary methods gives a lower limit to the exact conductivity. Sondheimer has calculated two further approximations to σ and finds that the corrections to be applied are extremely small. A more interesting question is the validity or otherwise of Matthiessen's rule, according to which the resistance of a metal is separable into an ideal part, characteristic of the pure metal, and a residual part due to impurities, etc. It is obvious that in the first approximation Matthiessen's rule must hold, but the form of (10·33·7) is such that the rule cannot be true when higher approximations are taken into account. The deviations from

Matthiessen's rule can only be significant when the ideal and residual resistances are of the same order of magnitude, and the detailed calculations show that the deviations are positive, in agreement with the observations, but are much too small. Sondheimer's numerical results are given in Table X 1, which demonstrates both the rapid convergence of the series (10·33·7) and the smallness of the calculated deviations from Matthiessen's rule. The value of σ obtained by retaining N terms in the series (10·33·2) is denoted by $\sigma^{(2N)}$, and A is defined by $\rho_i = AT/\Theta$ $(T \gg \Theta)$. The deviation $\Delta^{(2N)}$ from Matthiessen's rule is defined to be $\rho^{(2N)} - (\rho_i^{(2N)} + \rho_r)$.

Table X 1. *The electrical resistance of monovalent metals*

	$100\rho_r/4A = 0$			$100\rho_r/4A = 3$				
T/Θ	$100\,\dfrac{\rho_i^{(0)}}{4A}$	$100\,\dfrac{\rho_i^{(2)}}{4A}$	$100\,\dfrac{\rho_i^{(4)}}{4A}$	$100\,\dfrac{\rho^{(0)}}{4A}$	$100\,\dfrac{\rho^{(2)}}{4A}$	$100\,\dfrac{\rho^{(4)}}{4A}$	$100\,\dfrac{\Delta^{(2)}}{4A}$	$100\,\dfrac{\Delta^{(4)}}{4A}$
0·01	$1{\cdot}2443 \times 10^{-4}$	$1{\cdot}2429 \times 10^{-4}$	$1{\cdot}2429 \times 10^{-4}$	—	—	—	—	—
0·025	$1{\cdot}2152 \times 10^{-4}$	$1{\cdot}2068 \times 10^{-4}$	$1{\cdot}2065 \times 10^{-4}$	—	—	—	—	—
0·05	$3{\cdot}8882 \times 10^{-3}$	$3{\cdot}7821 \times 10^{-3}$	$3{\cdot}7786 \times 10^{-3}$	3·0039	3·0039	—	0·0001	—
0·0769	0·033166	0·031171	0·031069	3·0332	3·0322	3·0322	0·0010	0·0011
0·1	0·11638	0·10674	0·10601	3·1164	3·1102	3·1102	0·0035	0·0042
0·125	0·30968	0·28090	0·27807	3·3097	3·2886	3·2879	0·0077	0·0098
0·167	0·91141	0·83202	0·82217	3·9114	3·8481	3·8423	0·0161	0·0201
0·2	1·6084	1·4881	1·4721	4·6085	4·5098	4·4984	0·0217	0·0263
0·25	2·8797	2·7192	2·6941	5·8796	5·7458	5·7258	0·0266	0·0317
0·333	5·2554	5·0847	5·0536	8·2556	8·1110	8·0846	0·0263	0·0310
0·5	10·0915	9·9746	9·9487	13·0916	12·9907	12·9679	0·0161	0·0192
0·667	14·747	14·677	14·660	17·747	17·686	17·670	0·0091	0·010
0·833	19·253	19·210	19·198	22·253	22·214	—	0·004	—
1	23·662	23·634	23·626	26·662	26·637	—	0·004	—
1·25	30·165	30·150	30·145	33·165	33·151	—	0·001	—

The deviations from Matthiessen's rule will in general be greater than those calculated by Sondheimer for free electrons. This is most easily seen by considering a two-band model, the electrons in each band being supposed to behave independently of those in the other band. If Matthiessen's rule applies separately to each band we have $\rho_1 = \rho_{i1} + \rho_{r1}$, $\rho_2 = \rho_{i2} + \rho_{r2}$, but since $\sigma = \sigma_1 + \sigma_2$ we clearly cannot have $\rho = \rho_i + \rho_r$. To find the deviations from Matthiessen's rule the procedure is as follows. Put

$$\rho_{r1} = a, \quad \rho_{r2} = b, \quad \rho_{i1} = x, \quad \rho_{i2} = y. \tag{10·4·1}$$

Then
$$\rho_1 = a + x, \quad \rho_2 = b + y \tag{10·4·2}$$

and
$$\rho = \frac{(a+x)(b+y)}{a+b+x+y}. \tag{10·4·3}$$

Now define ρ_r as the resistivity when $x = y = 0$, and ρ_i as the resistivity when $a = b = 0$, that is,

$$\rho_r = \frac{ab}{a+b}, \quad \rho_i = \frac{xy}{x+y}. \tag{10·4·4}$$

Then the deviation Δ from Matthiessen's rule is

$$\Delta = \rho - (\rho_r + \rho_i) = \frac{(bx - ay)^2}{(a+b)(x+y)(a+b+x+y)}. \tag{10·4·5}$$

Having found ρ_i by measurements on the ideally pure substance, the apparent residual resistivity of any specimen is $\rho_r' = \rho - \rho_i = \rho_r + \Delta$. Equation (10·4·5) shows that Δ is always positive and is strongly temperature dependent.

To obtain a qualitative picture of the temperature variation of Δ it is sufficient to put $\lambda a = b$ and $\mu x = y$. Then

$$\Delta = \frac{ax(\lambda - \mu)^2}{(1+\lambda)(1+\mu)\{a(1+\lambda) + x(1+\mu)\}}.$$

At low temperatures x is small compared with a, so that Δ is proportional to x, that is, to ρ_i, while at high temperatures, where x is large compared with a, Δ is proportional to a, that is, to ρ_r. It is therefore seen that Δ is zero when $T = 0$; it increases as T increases and finally reaches a constant value proportional to ρ_r. For very pure metals the constancy of Δ holds down to very low temperatures, and the constancy of the apparent residual resistance is no guarantee of the validity of Matthiessen's rule unless the measurements are carried out at temperatures where ρ_i is of the same order of magnitude as ρ_r.

When ρ_i and ρ_r are of the same order of magnitude, then according to equations (10·4·4) and (10·4·5) we have

$$\Delta = \frac{\alpha \rho_i \rho_r}{\beta \rho_i + \gamma \rho_r}, \tag{10·4·6}$$

where α, β and γ are quantities, possibly temperature dependent, of the order of unity. This is in general agreement with the results of Grüneisen (1933) and other workers, but we have no means of calculating α, β and γ exactly.

Kohler (1949b) has given a general proof that the deviations from Matthiessen's rule are always positive. His proof is based upon the minimum principle (10·1·15), namely,

$$(G, G) \Big/ \left[\frac{\epsilon}{4\pi^3} \int G v_i \frac{\partial f_0}{\partial E} d\mathbf{k} \right]^2 \geqslant \rho.$$

Now, if Φ is the correct solution for the distribution function corresponding to the operator L, we have

$$\rho = \frac{1}{4\pi^3} \int \Phi L(\Phi) \, d\mathbf{k} \Big/ \left[\frac{\epsilon}{4\pi^3} \int \Phi v_i \frac{\partial f_0}{\partial E} d\mathbf{k} \right]^2$$

But the collision operator L is linear and is the sum of two operators L_i and L_r referring to the ideal metal and the impurities respectively. Hence if we apply the minimum principle to L_i and L_r separately we have

$$\frac{1}{4\pi^3} \int \Phi L_i(\Phi)\,d\mathbf{k} \Big/ \left[\frac{\epsilon}{4\pi^3} \int \Phi v_i \frac{\partial f_0}{\partial E}\,d\mathbf{k}\right]^2 \geqslant \rho_i,$$

and similarly for L_r, so that $\qquad \rho \geqslant \rho_i + \rho_r.$

The equality can only occur when $\Phi = \alpha \Phi_i = \beta \Phi_r$, where α and β are constants. In particular, Matthiessen's rule is exact if times of relaxation can be unambiguously defined for both the ideal and residual resistances, and if their ratio is independent of $\mathbf{k}$.

The Magneto-resistance Effects

10·5. Expressions were obtained in §§ 8·523 and 8·531 for the electrical and thermal conductivities of a metal containing two conduction bands of standard form in the particular case in which a time of relaxation exists. It can be shown (Sondheimer and Wilson, 1947; Kohler, 1949c)† that the same formulae give the zero-order approximations to σ and κ at temperatures such that a universal time of relaxation cannot be assumed to exist. The proofs are long and complicated and will not be given here.

The formulae for σ and κ can be put into various forms which are equivalent so long as a universal time of relaxation τ exists, but which lead to very different results in the general case. Care must therefore be taken to choose the correct generalizations, which are as follows. The electrical conductivity σ is given in terms of the conductivity σ_0 in zero field by (8·523·4), where σ_0 in the general case is given by (9·621·1) and (9·621·2). Similarly, the thermal conductivity κ is given in terms of κ_0 by (8·531·4), where κ_0 in the general case is given by (9·621·3) and (9·621·4). If, therefore, the formulae of §§ 8·523 and 8·531 are written in such a form that, on eliminating τ, only electrical quantities occur in the formulae for σ and only thermal quantities occur in the formulae for κ, we obtain formulae which hold, at least as interpolation formulae, for all temperatures.

---

† Sondheimer and Wilson obtained exact solutions for a single band of standard form at high and low temperatures and for the limit of infinite H. They then set up interpolation formulae, and showed that σ, R, κ and B_{RL} are practically independent of H. (They also determined the correction terms.) If it is assumed that σ, R, κ and B_{RL} for one band are independent of H, the required formulae for two bands follow at once from (8·525·3), (8·525·4), (8·533·1) and (8·533·2). Kohler transformed the integral equation, without special justification, into an infinite set of linear equations similar to (10·3·5). The solution then follows as in § 10·3.

10·51. *Discussion of the experimental results.* The pioneering experiments of Kapitza (1929), who used magnetic fields up to 300,000 gauss and worked at temperatures down to those of liquid air, demonstrated the most important features of the magneto-resistance effect. The work of Meissner and Scheffers (1929) at very low temperatures and with moderate magnetic fields of the order of 10,000 gauss supplemented the results obtained by Kapitza, and since that time a very large amount of data has been accumulated by many workers. Kapitza's results for cadmium wires are shown in

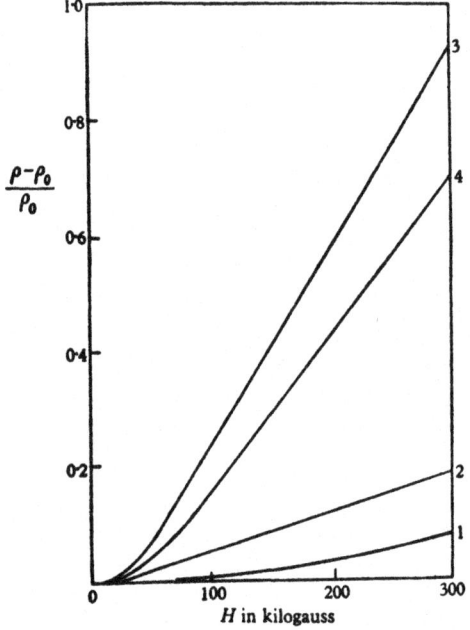

Fig. X 2. Magneto-resistance curves for cadmium wires.

Curves 1, 2, 3; H transverse, $T = 290, 190, 88°$ K.
Curve 4; H longitudinal, $T = 88°$ K.

fig. X 2, and it will be seen that $\Delta\rho/\rho_0$ is proportional to H^2 for small fields and to H^p for large fields where $p < 2$. It will also be seen that $\Delta\rho/\rho_0$ increases considerably for fixed H as the temperature diminishes, and that there is a longitudinal effect as well as a transverse effect, though a transverse field is more effective than a longitudinal field in increasing the resistance.

The change in resistance is considerably reduced by the presence of impurities, especially at low temperatures, but it has been shown by Justi and Scheffers (1938) that if $\Delta\rho/\rho_0$ is plotted against $H/\rho_{red.}$, where $\rho_{red.}$ is the reduced resistivity defined by $\rho_{red.} = \rho_0(T)/\rho_0(\Theta)$, a single curve is obtained for any given metal, whatever the values of H, T and the residual resistivity ρ_r may be. The experimental results for polycrystalline metals are therefore of a comparatively simple nature, but any theory which completely neglects anisotropic effects can only give the order of magnitude of the transverse effect and can give no explanation of the longitudinal effect. The two-band model used in deriving equations (8·523·4) and (8·531·4) is therefore too specialized to cover all aspects of the magneto-resistance effects, but it has not so far been possible to carry out the calculations for any more complicated models. There is a further difficulty in applying the formulae (8·523·4) and (8·531·4), in that, in general, we have no reliable method of deciding what values to give to n_1 and n_2, the number of electrons (or number of vacant levels) in the energy bands. This applies particularly to the monovalent metals where, for most purposes but not for our present discussion, it is sufficient to consider the valency electrons as being free, which is equivalent to putting $n_2 = 0$. For the transition metals, for bismuth and the divalent cubic metals all the evidence goes to show that $n_1 = n_2$, and in such cases we are on much safer ground since the formulae contain fewer undetermined parameters.

A brief description has already been given in § 8·523 of the magneto-resistance effects at high temperatures, and we shall therefore confine the present discussion to the low-temperature region.

10·511. *The electrical conductivity.* If in the general formula

$$\frac{\Delta\rho}{\rho_0} = \frac{\rho - \rho_0}{\rho_0} = \frac{\sigma_0 - \sigma}{\sigma} = \frac{\left(\dfrac{H}{\epsilon c}\right)^2 \left(\dfrac{\sigma_{01}}{n_1} + \dfrac{\sigma_{02}}{n_2}\right)^2 \dfrac{\sigma_{01}\sigma_{02}}{(\sigma_{01} + \sigma_{02})^2}}{1 + \left(\dfrac{H}{\epsilon c}\right)^2 \dfrac{(n_1 - n_2)^2}{n_1^2 n_2^2} \dfrac{\sigma_{01}^2 \sigma_{02}^2}{(\sigma_{01} + \sigma_{02})^2}}, \quad (10\cdot511\cdot1)$$

we put $\rho_1 = \lambda\rho_2$ and assume that λ is constant, we see that $\Delta\rho/\rho_0$ has the functional form

$$\frac{\Delta\rho}{\rho_0} = F\left(\frac{H^2}{\rho_0^2 \epsilon^2 c^2}\right), \quad (10\cdot511\cdot2)$$

so that the magneto-resistance effect is a function of H/ρ_0, as is found in practice. The experimental curves for a number of metals are shown in fig. X 3, which has been called the reduced diagram by Kohler, $\Delta\rho/\rho_0$ being plotted on a double logarithmic scale against $H/\rho_{\text{red.}}$, where $\rho_{\text{red.}}$ is the reduced resistance defined above.

It will be seen that the metals can be roughly divided into two classes, those giving straight lines in the diagram and those giving resistance curves which are convex upwards. Metals in the first group have $\Delta\rho/\rho_0$ roughly proportional to H^2 for all values of H and comprise the bismuth group of metals, all the divalent and quadrivalent metals and the transition metals tungsten and platinum. The second group of metals comprises the monovalent metals, the trivalent metals aluminium and indium, and rhodium. For these metals $\Delta\rho/\rho_0$ increases more slowly than H^2 for moderate values of H and shows signs of saturating for sufficiently large values of H. (The proportionality of $\Delta\rho/\rho_0$ to H^2 is only a fairly crude approximation for metals of the first group. In fact $\Delta\rho/\rho_0 \propto H^p$, where p lies between 1 and 2, the index tending to decrease as H increases. The main criterion is that a metal is placed in group 1 if $\Delta\rho/\rho_0$ increases as H^p with $p \geqslant 1$, and into

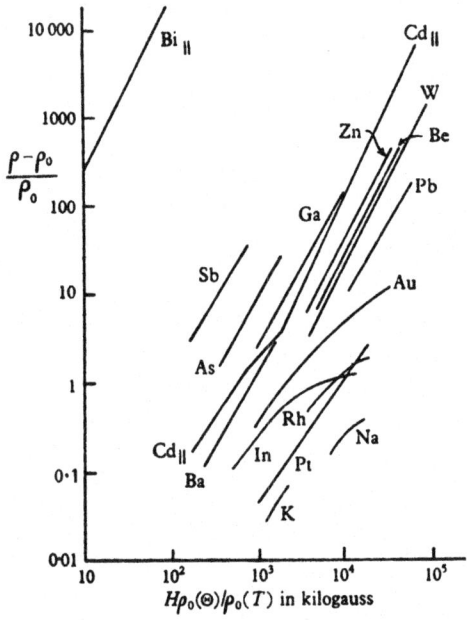

Fig. X 3. The reduced diagram for the relative change in resistance in a transverse magnetic field.

group 2 if $\Delta\rho/\rho_0$ increases as H^p with $p < 1$ for large H, and in particular if it shows signs of tending to a limiting value as H tends to infinity.) This division of metals into two groups is in accordance with the predictions of equation (10·511·1). If $n_1 = n_2$, $\Delta\rho/\rho_0$ is proportional to H^2 for all values of the magnetic field, whereas if $n_1 \neq n_2$ the resistance reaches a saturation value for very large fields. Now although equation (10·511·1) has been derived for a very special model and cannot be expected to reproduce the detailed behaviour of real metals, a more general model must give rise to an equation for σ which is more general than (10·511·1) and which therefore has at least the properties discussed above, though the physical interpretation of the parameters may be different. Surprisingly enough, the division of the

metals into the groups which do or do not show a tendency towards saturation more or less follows the predictions of the two-band model, though this
may be fortuitous and due to the experimental data being inadequate for
a proper test of the theory. But, in all the cases so far examined, the metals
which have $\Delta\rho/\rho_0$ roughly proportional to H^2 for large fields have structures
such that the valency electrons are capable of filling an energy band, so that,
when the electrons are distributed over two bands, the number of electrons
in the upper band is equal to the number of vacant levels in the lower band.

The theory is not only in qualitative, but also in reasonable quantitative
agreement with the experimental data. The absolute magnitude of the
magneto-resistance effect has been discussed in §§ 8·524 and 8·561, where it
was shown that at room temperatures reasonable values of the parameters
could be chosen so as to reproduce the behaviour of divalent metals and
bismuth-type metals for small fields. The reduced resistance diagram then
shows that the behaviour for large values of H and small values of T is in
agreement with the theoretical predictions.

10·512. The behaviour of the resistivity at low temperatures, for constant H, shows some apparent anomalies which can be explained by the
theoretical formulae. In order to obtain a simple qualitative picture it is
sufficient to put $\rho_{01} = \rho_{02} = 2\rho_0$ and to consider the case $n_1 = n_2 = n$. We
then have

$$\rho = \rho_0 + \left(\frac{H}{2nec}\right)^2 \frac{1}{\rho_0}, \qquad (10·512·1)$$

and if ρ is treated as a function of ρ_0 for constant H it has a minimum value
$\rho_{\min.} = H/(nec)$, which occurs at the temperature at which $\rho_0 = H/(2nec)$.
Now since ρ_0 can never be less than the residual resistivity ρ_r, the minimum
can only occur if $H > 2nec\rho_r$. Hence, if $H < 2nec\rho_r$, the resistivity ρ has
the value

$$\rho_r + \left(\frac{H}{2nec}\right)^2 \frac{1}{\rho_r} \qquad (10·512·2)$$

at $T = 0$ and increases steadily with T. At high temperatures ρ differs
negligibly from ρ_0. On the other hand, if $H > 2nec\rho_r$, the resistivity near
$T = 0$ at first decreases from the value (10·512·2) as T increases, reaches the
minimum value $H/(nec)$ and then increases steadily with T. Similar results
are obtained in the general case except that if $(n_1 - n_2)^2$ is sufficiently large
a minimum cannot occur for any values of H and T.

The minimum in the $\rho - T$ curve has been observed by Milner (1937) in
specimens of cadmium. In one specimen ρ_0 was $4·5 \times 10^{-21}$ e.s.u. at
$T = 2·35°$ K., while ρ was $1·4 \times 10^{-18}$ e.s.u. for $H = 2 \times 10^4$ gauss. A minimum value of the resistivity of $2·6 \times 10^{-19}$ e.s.u. was observed in a
field of $2·28 \times 10^4$ gauss at a temperature $T = 15°$ K., corresponding to

$\rho_0 = 6 \times 10^{-20}$ e.s.u. Inserting these values into equation (10·512·1), we find $n = 8\cdot5 \times 10^{21}$ per c.c., which is of the right order of magnitude since the number of atoms is $4\cdot7 \times 10^{22}$ per c.c. Further, the calculated minimum value of ρ is $1\cdot8 \times 10^{-19}$ e.s.u. at the temperature where $\rho_0 = 0\cdot9 \times 10^{-19}$ e.s.u., and these values are in reasonable agreement with the observed values given above.

10·513. The magneto-resistance effect in real metals depends strongly upon the anisotropy of the crystalline field. A formal expression for ρ can be obtained from the formal solution of the Boltzmann equation derived in § 8·55, on the assumption that a time of relaxation exists, and this calculation could be generalized. Unfortunately, it has not proved possible to evaluate numerically the resulting expressions for any model more general than those already discussed, namely, those for which E is a quadratic function of $\mathbf{k}$, or for which the departure from spherical symmetry is small. In the latter case the calculation is restricted to terms of order H^2. The solution is therefore of no assistance except in a very general way in interpreting the experimental data, which are of great complexity. In the first place the longitudinal magneto-resistance is not zero even for cubic metals, in which the resistance in the absence of a magnetic field is isotropic, and secondly, the transverse magneto-resistance effect in single crystals depends in all cases on the orientation of the magnetic field relative to the crystal axes. The latter effect is particularly marked when the relative change in resistance is large, as is shown by the curve in fig. X 4 (Justi and Scheffers, 1936) which gives $\Delta\rho/\rho_0$ for a rod of gold, whose axis of symmetry is a crystal axis, as a function of the angle between the transverse magnetic field and the crystal axes perpendicular to the length of the rod. (The axis of rotation of the magnetic field and the crystal axis were not quite parallel, which accounts for the apparent lack of symmetry between the four quadrants.) It does not seem likely that any model simple enough to be tractable theoretically would give a magneto-resistance curve of the complexity of those actually observed.

10·514. *The thermal conductivity.* The experimental results on the thermal conductivity are usually given in terms of the Lorenz number $L = \kappa/\sigma T$. The theoretical value for L is obtained by combining (8·523·4) and (8·531·4). If for simplicity we put $\sigma_{01} = \sigma_{02} = \frac{1}{2}\sigma_0$, $\kappa_{01} = \kappa_{02} = \frac{1}{2}\kappa_0$, L is given by

$$L = L_0 \frac{1 + \frac{1}{8}\left(\frac{H}{ec}\right)^2\left(\frac{1}{n_1^2} + \frac{1}{n_2^2}\right)\sigma_0^2}{1 + \frac{1}{8}\left(\frac{H}{ec}\right)^2\left(\frac{1}{n_1^2} + \frac{1}{n_2^2}\right)\frac{L_0^2}{L_n^2}\sigma_0^2} \frac{1 + \frac{1}{16}\left(\frac{H}{ec}\right)^2\frac{(n_1-n_2)^2}{n_1^2 n_2^2}\frac{L_0^2}{L_n^2}\sigma_0^2}{1 + \frac{1}{16}\left(\frac{H}{ec}\right)^2\frac{(n_1-n_2)^2}{n_1^2 n_2^2}\sigma_0^2},$$

$$(10\cdot514\cdot1)$$

where $L_0 = \kappa_0/\sigma_0 T$ and where $L_n = \tfrac{1}{3}(\pi k/\epsilon)^2$ is the 'normal' value of the Lorenz number. For small values of H, equation (10·514·1) becomes

$$L = L_0\left\{1 + \frac{1}{16}\left(\frac{H}{\epsilon c}\right)^2 \frac{(n_1 + n_2)^2}{n_1^2 n_2^2}\left(1 - \frac{L_0^2}{L_n^2}\right)\sigma_0^2\right\}, \qquad (10·514·2)$$

while for very large values of H we have $L = L_0$ if $n_1 \neq n_2$ and $L = L_n^2/L_0$ if $n_1 = n_2$. In general we have $L_0 \leqslant L_n$, and hence, according to (10·514·2), L should be increased by a magnetic field. Also if $n_1 \neq n_2$, L passes through a maximum as H is increased and finally tends to L_0 again. If, however,

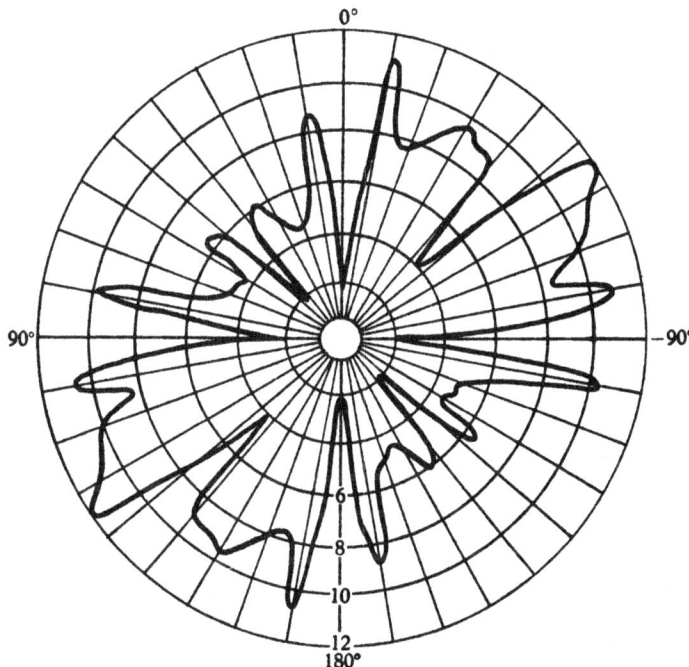

Fig. X 4. $(\sigma_0 - \sigma)/\sigma$ at 4·2° K. for a single crystal rod of gold in a magnetic field of 18600 gauss, which is perpendicular to one of the cubic axes, as a function of the orientation of the field relative to the other two axes.

$n_1 = n_2$, L increases steadily from L_0 to L_n^2/L_0. The experimental data are meagre (Grüneisen and Adenstadt, 1938, copper, silver, platinum, tungsten and beryllium; Grüneisen and Erfling, 1940, beryllium; de Nobel, 1949, tungsten), but they are qualitatively in agreement with the theoretical predictions. Some results for tungsten are shown in fig. X 5, where the total Lorenz number and the electronic Lorenz number are shown, there being an appreciable lattice thermal conductivity. It will be seen that L_e is increased by the magnetic field and that it passes through a maximum as H is increased, in accordance with equation (10·514·1) for $n_1 \neq n_2$. Tungsten, however, is one of the metals whose electrical resistance shows no evidence of saturation in large magnetic fields, so that the hypothesis $n_1 \neq n_2$ is extremely doubtful, and further measurements are required before one can say whether the formula (10·514·1) provides an explanation of de Nobel's results or whether his numerical estimates of the split of κ between κ_e and κ_g are at fault.

If we put $n_1 = n_2 = n$, we can obtain an estimate of n for beryllium by using the results of Grüneisen and Erfling, according to which (in gaussian units) $\rho_0 = 8\cdot4 \times 10^{-21}$ and $L_0 = 1\cdot42 \times 10^{-13}$ at $T = 22\cdot6°\,\mathrm{K.}$, while $\rho = 7\cdot9 \times 10^{-19}$

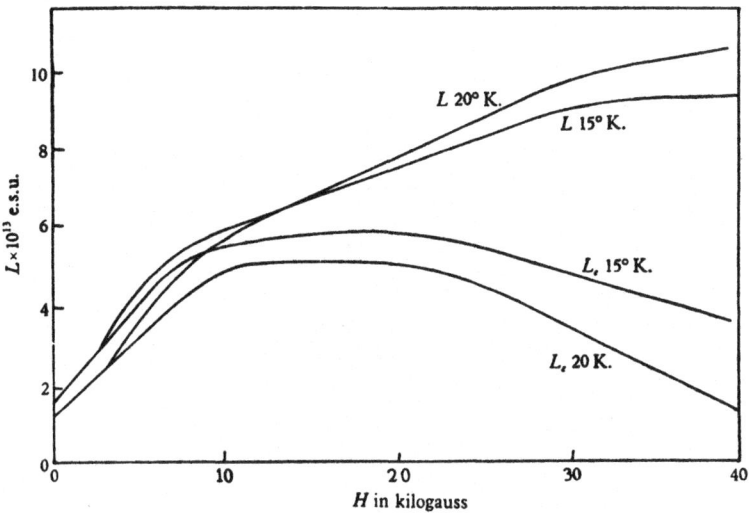

Fig. X 5. The influence of a magnetic field on the Lorenz number of tungsten.

and $L = 2\cdot92 \times 10^{-13}$ for $H = 1\cdot17 \times 10^4$ gauss. Equation (10·514·1) then gives $n = 3\cdot1 \times 10^{22}$ per c.c., while equation (10·512·1), which involves the electrical conductivity alone, gives $n = 5 \times 10^{21}$ per c.c. The agreement is reasonable in view of the crude nature of the model on which the calculations are based and also in view of the inadequacy of the theoretical expressions for the ideal thermal conductivity in the low-temperature region (§ 9·71).

CONDUCTION IN ANISOTROPIC MEDIA

10·6. In the general case of anisotropic media it is impossible to reduce the Boltzmann equation to an integral equation in one variable, and a direct solution of the integral equation seems to be out of the question. However, the fact that the general conduction problem can be reduced to a variation problem means that the physical quantities are insensitive to the exact form of the distribution function, and that reasonable first approximations to σ, κ, etc., can be obtained by assuming simple forms for the functions to be inserted in the variation integrals. We should therefore expect that all the important physical phenomena can be described, as regards order of magnitude, by the formal theory of conduction in which the existence of a time of relaxation is assumed, and that the exact theory will only change the values of the constants without introducing any new physical phenomena.

The calculations of the preceding sections show that, for ideally isotropic metals, we can obtain first approximations to σ and κ by assuming that the distribution function is such that $c^{(n)}(\eta) = c_0^{(n)}$ in calculating σ, and $c^{(n)}(\eta) = c_0^{(n)} + c_1^{(n)}\eta$ in calculating κ, where $c_0^{(n)}$ and $c_1^{(n)}$ are constants. If we assume that correspondingly simple distribution functions are sufficient for general anisotropic media, we can readily calculate σ and κ. For free electrons, the validity of the above assumptions can be checked by calculating the second and higher approximations and investigating their convergence. In the general case of anisotropic media, however, the results are not so firmly based, since no model of an anisotropic metal has yet been proposed for which even the first approximation can be obtained in numerical form; and until a number of approximations can be calculated it is impossible to assess the validity of the results. We shall therefore only give a brief account of the application of the variation principle to conduction in anisotropic media, making the simplest possible assumptions. In practice it will probably be necessary to use more complicated distribution functions, but no general rules can be laid down, and the functions to be used in any particular case will be determined by the complexity of the energy levels and of the transition probabilities of the model being investigated.

Since all metals have at least one axis of symmetry the conductivity tensors reduce to two components if one of the axes, say $O1$, is taken along the axis of the crystal. We then need only consider an electric field and a temperature gradient along the axis $O1$, and the equations determining the distribution function are

$$f = f_0 - \Phi \, \partial f_0/\partial E, \quad \hbar v_1 = \partial E/\partial k_1, \tag{10·6·1}$$

$$L(\Phi) = \left(\epsilon \mathscr{E}_1 + T\frac{\partial}{\partial x_1}\frac{\zeta}{T}\right) v_1 \frac{\partial f_0}{\partial E} + \frac{E}{T}\frac{\partial T}{\partial x_1} v_1 \frac{\partial f_0}{\partial E}. \tag{10·6·2}$$

10·61. To calculate the electrical conductivity we put $\partial \zeta/\partial x_1 = 0$, $\partial T/\partial x_1 = 0$, and then Φ must be such as to make the integral (Φ, Φ) a maximum, where

$$(\Phi, \Phi) = \frac{1}{4\pi^3}\int \Phi L(\Phi)\, dk, \tag{10·61·1}$$

subject to the condition

$$(\Phi, \Phi) = \frac{\epsilon \mathscr{E}_1}{4\pi^3}\int v_1 \Phi \frac{\partial f_0}{\partial E}\, d\mathbf{k}. \tag{10·61·2}$$

Now if we put $\Phi = a_0 \partial E / \partial k_1$, (10·61·3)

the only condition that a_0 has to satisfy is (10·61·2), and hence it is given by

$$a_0 \left(\frac{\partial E}{\partial k_1}, \frac{\partial E}{\partial k_1} \right) = \frac{\epsilon \mathscr{E}_1}{2\pi^2 h} \int \left(\frac{\partial E}{\partial k_1} \right)^2 \frac{\partial f_0}{\partial E} d\mathbf{k}. \qquad (10·61·4)$$

Also

$$\sigma_\parallel = \frac{J_1}{\mathscr{E}_1} = \frac{\epsilon}{4\pi^3 \mathscr{E}_1} \int v_1 \Phi \frac{\partial f_0}{\partial E} d\mathbf{k} = \frac{a_0 \epsilon}{2\pi^2 h \mathscr{E}_1} \int \left(\frac{\partial E}{\partial k_1} \right)^2 \frac{\partial f_0}{\partial E} d\mathbf{k},$$

and so

$$\sigma_\parallel = \frac{\epsilon^2}{4\pi^4 h^2} \left[\int \left(\frac{\partial E}{\partial k_1} \right)^2 \frac{\partial f_0}{\partial E} d\mathbf{k} \right]^2 \Big/ \left(\frac{\partial E}{\partial k_1}, \frac{\partial E}{\partial k_1} \right). \qquad (10·61·5)$$

Similarly, $\sigma_\perp$ is obtained by replacing $\partial E / \partial k_1$ by $\partial E / \partial k_2$.

The numerator can be evaluated as usual by writing

$$d\mathbf{k} = dE \, dS / |\operatorname{grad}_k E|,$$

where dS is the element of area of the surface $E = $ constant. We then find, to the zero order in kT/ζ,

$$\sigma_\parallel = \frac{\epsilon^2}{4\pi^4 h^2} \left[\int_{E=\zeta} \frac{(\partial E / \partial k_1)^2 dS}{|\operatorname{grad}_k E|} \right]^2 \Big/ \left(\frac{\partial E}{\partial k_1}, \frac{\partial E}{\partial k_1} \right). \qquad (10·61·6)$$

10·62. The simplest function that can be used to calculate the thermal conductivity is

$$\Phi = (a_0 + a_1 \eta) \, \partial E / \partial k_1. \qquad (10·62·1)$$

Then instead of calculating the quadratic form (Φ, Φ) and determining its maximum, we may adopt the equivalent procedure of substituting (10·62·1) into (10·6·2), multiplying successively by $\partial E / \partial k_1$ and $\eta \, \partial E / \partial k_1$ and integrating with respect to $\mathbf{k}$. In this way we obtain two simultaneous equations for a_0 and a_1, and, when they are solved, κ can be obtained in a form equivalent to (10·34·3) except that the determinants are now finite ones. The result is that

$$-\hbar^2 T \kappa_\parallel = \frac{\begin{vmatrix} \dfrac{1}{4\pi^3} \int \left(\dfrac{\partial E}{\partial k_1} \right)^2 \dfrac{\partial f_0}{\partial E} d\mathbf{k} & \dfrac{1}{4\pi^3} \int \eta \left(\dfrac{\partial E}{\partial k_1} \right)^2 \dfrac{\partial f_0}{\partial E} d\mathbf{k} \\[4mm] \dfrac{1}{4\pi^3} \int E \left(\dfrac{\partial E}{\partial k_1} \right)^2 \dfrac{\partial f_0}{\partial E} d\mathbf{k} & \dfrac{1}{4\pi^3} \int E \eta \left(\dfrac{\partial E}{\partial k_1} \right)^2 \dfrac{\partial f_0}{\partial E} d\mathbf{k} \end{vmatrix}^2}{\begin{vmatrix} 0 & \dfrac{1}{4\pi^3} \int \left(\dfrac{\partial E}{\partial k_1} \right)^2 \dfrac{\partial f_0}{\partial E} d\mathbf{k} & \dfrac{1}{4\pi^3} \int \eta \left(\dfrac{\partial E}{\partial k_1} \right)^2 \dfrac{\partial f_0}{\partial E} d\mathbf{k} \\[4mm] \dfrac{1}{4\pi^3} \int \left(\dfrac{\partial E}{\partial k_1} \right)^2 \dfrac{\partial f_0}{\partial E} d\mathbf{k} & \left(\dfrac{\partial E}{\partial k_1}, \dfrac{\partial E}{\partial k_1} \right) & \left(\dfrac{\partial E}{\partial k_1}, \eta \dfrac{\partial E}{\partial k_1} \right) \\[4mm] \dfrac{1}{4\pi^3} \int \eta \left(\dfrac{\partial E}{\partial k_1} \right)^2 \dfrac{\partial f_0}{\partial E} d\mathbf{k} & \left(\eta \dfrac{\partial E}{\partial k_1}, \dfrac{\partial E}{\partial k_1} \right) & \left(\eta \dfrac{\partial E}{\partial k_1}, \eta \dfrac{\partial E}{\partial k_1} \right) \end{vmatrix}}.$$

$$(10·62·2)$$

The numerator can readily be calculated since integrals of the form $\int_{-\infty}^{\infty} \phi(E)\, \eta^m \dfrac{\partial f_0}{\partial \eta}\, d\eta$ can be evaluated to the second order in kT/ζ by the method given in the Appendix A 5. The denominator can be simplified if it is assumed that $(\partial E/\partial k_1,\ \partial E/\partial k_1)$ and $(\eta\, \partial E/\partial k_1,\ \eta\, \partial E/\partial k_1)$ are of the same order in kT/ζ. This is true for any obvious interaction functions, but a general proof seems difficult. If we make the above assumption then (10·62·2) can be reduced to

$$\kappa_{\parallel} = \frac{k^2 T}{36 h^2} \left[\int_{E=\zeta} \frac{(\partial E/\partial k_1)^2\, dS}{|\,\mathrm{grad}_k E\,|} \right]^2 \Big/ \left(\eta\, \frac{\partial E}{\partial k_1},\ \eta\, \frac{\partial E}{\partial k_1} \right). \tag{10·62·3}$$

10·63. The expressions for σ and κ have been evaluated approximately by Sondheimer (unpublished) at low temperatures where it is only the differential geometry of the energy surfaces that affects the results. (It is assumed that $V(\mathbf{k}, \mathbf{k}')$ has the form (9·33·9) and that the lattice frequencies are those of a continuum.) He finds the following expressions:

$$\frac{1}{\sigma} = \frac{f_3}{f_1^2} \frac{4\pi^2 h}{e^2 \Lambda} \left(\tfrac{3}{4}\pi^2 n_a\right)^{\frac{4}{3}} \left(\frac{T}{\Theta}\right)^5 \mathscr{I}_5, \tag{10·63·1}$$

$$\frac{1}{\kappa} = \frac{f_2}{f_1^2} \frac{18 h}{\pi^2 k^2 \Lambda \Theta} \left(\frac{T}{\Theta}\right)^2 \left\{ \mathscr{I}_5 + \left(\tfrac{3}{4}\pi^2 n_a\right)^{\frac{2}{3}} \left(\frac{T}{\Theta}\right)^2 \frac{f_3}{f_2} \left(\tfrac{2}{3}\pi^2 \mathscr{I}_5 - \tfrac{1}{3}\mathscr{I}_7\right) \right\}, \tag{10·63·2}$$

where n_a is the number of atoms per unit volume,

$$f_1(\zeta) = \int \frac{(\partial E/\partial k_1)^2\, dS}{|\,\mathrm{grad}_k E\,|}, \quad f_2(\zeta) = \int \frac{(\partial E/\partial k_1)^2\, dS}{|\,\mathrm{grad}_k E\,|^2}, $$

$$f_3(\zeta) = 2 \int \frac{\partial E/\partial k_1}{|\,\mathrm{grad}_k E\,|^3} \sum_i \frac{\partial E}{\partial k_i} \frac{\partial^2 E}{\partial k_1 \partial k_i} P_k\, dS, \tag{10·63·3}$$

and where

$$P_k = \frac{1}{2} \sum_j \frac{\partial}{\partial k_j} \frac{\partial E/\partial k_j}{|\,\mathrm{grad}_k E\,|} \tag{10·63·4}$$

is the mean curvature of the surface $E = \zeta$.

For free electrons we have

$$f_1 = 16\sqrt{2}\,\pi^2 m^{*\frac{3}{2}} \zeta^{\frac{3}{2}}/3h, \quad f_2 = 32\pi^3 m^* \zeta/3h^2, \quad f_3 = \tfrac{8}{3}\pi, \tag{10·63·5}$$

and we regain the interpolation formulae for σ and κ. In the general case the formulae for σ and κ have the same general form as for free electrons but with different constants, provided that the temperature is sufficiently low. At moderate and high temperatures, however, there is no reason to suppose that (10·63·1) and (10·63·2) will continue to hold except for very special models, since the electrons are then deflected through large angles and $V(\mathbf{k}, \mathbf{k}')$ no longer depends only upon the local properties of the energy surfaces. (An elementary example of the divergences from the simple theory is provided by the discussion of the conductivity of the transition metals given in § 9·51, where it is shown that $\rho = AT^5 \mathscr{J}_5 + B e^{-\Theta_s/T}$.)

It is possible to derive an expression for the thermoelectric power similar to (10·63·1) and (10·63·2). It is a complicated function depending upon f_1, f_2, f_3 and their logarithmic derivatives with respect to ζ, but it is interesting in that, by a suitable choice of f_1, f_2, f_3, it is possible to obtain a thermoelectric power which changes its sign at low temperature. Unless, however, the required behaviour of the f's can be derived from a plausible model, this merely remains an interesting suggestion regarding the anomalous variation of the thermoelectric effects at low temperatures.

The method outlined in the preceding paragraphs is, in principle, of sufficient generality to provide the solution of all conduction problems in which magnetic fields are absent. Whether significant results can be obtained by it remains to be seen, and this will depend upon the choice of suitable models and the calculation, probably by numerical methods, of a sufficient number of approximations to ensure that the convergence is adequate. The discussion of the formulae derived for the free electron model has shown that, while this model gives a fair qualitative description of the various phenomena, there are quantitative discrepancies, some small and some serious, even for the monovalent metals. It is therefore important to know whether the divergences between the experimental data and the theoretical formulae are due to the crudeness of the approximations used or whether new physical principles are involved.

REFERENCES

Grüneisen, E. (1933). The temperature dependence of the electrical resistance of pure metals. *Ann. Phys., Lpz.* (5), **16**, 530.

Grüneisen, E. and Adenstedt, H. (1938). The influence of transverse magnetic fields on the electrical and thermal conductivities of metals at low temperatures. *Ann. Phys., Lpz.* (5), **31**, 714.

Grüneisen, E. and Erfling, H. D. (1940). The electrical and thermal resistances of beryllium crystals in a magnetic field. *Ann. Phys., Lpz.* (5), **38**, 399.

Justi, E. and Scheffers, H. (1936). The electrical resistance of gold at low temperatures in a transverse magnetic field. *Phys. Z.* **37**, 383, 475.

Justi, E. and Scheffers, H. (1938). The change in resistance of aluminium single crystals in a strong magnetic field at low temperatures. *Phys. Z.* **39**, 105.

Kapitza, P. (1929). The change in electrical conductivity in strong magnetic fields. *Proc. Roy. Soc.* A, **123**, 292.

Kohler, M. (1948). The treatment of non-equilibrium phenomena by means of a variation principle. *Z. Phys.* **124**, 772.

Kohler, M. (1949a). Transport phenomena in an electron gas. *Z. Phys.* **125**, 679.

Kohler, M. (1949b). The general theory of the deviations from Matthiessen's rule. *Z. Phys.* **126**, 495.

Kohler, M. (1949c). The theory of the magneto-resistance effects in metals. *Ann. Phys., Lpz.* (6), **6**, 18.

Kroll, W. (1933). The theory of the thermoelectric effects. *Z. Phys.* **80**, 50 and **81**, 425.

Meissner, W. and Scheffers, E. (1929). The electrical resistivity of gold in magnetic fields at low temperatures. *Phys. Z.* **30**, 827.

Milner, C. J. (1937). The magneto-resistance effect in cadmium at low temperatures. *Proc. Roy. Soc.* A, **160**, 207.

de Nobel, J. (1949). The thermal and electrical resistance of tungsten at low temperatures and high magnetic fields. *Physica*, **15**, 532.

Rhodes, P. (1950). The Bloch integral equation and electrical conductivity. *Proc. Roy. Soc.* A, **202**, 466.

Sondheimer, E. H. (1950). The theory of the transport phenomena in metals. *Proc. Roy. Soc.* A, **203**, 75.

Sondheimer, E. H. and Wilson, A. H. (1947). The theory of the magneto-resistance effects in metals. *Proc. Roy. Soc.* A, **190**, 435.

Appendix

THE FERMI-DIRAC STATISTICS

THE DISTRIBUTION FUNCTION

A 1. Suppose we have N electrons having the same spin in a volume V. The electrons must move practically independently of each other, but there must be a slight coupling between their motions so as to ensure thermal equilibrium. If the energy levels for any electron are $E_1, E_2, ..., E_s, ...,$ a microscopic state of the assembly is defined by the numbers of electrons $n_1, n_2, ..., n_s, ...$ in the different energy levels. In addition there are the two conditions

$$N = \sum_s n_s, \qquad (\text{A } 1 \cdot 1)$$

$$U = \sum_s n_s E_s, \qquad (\text{A } 1 \cdot 2)$$

expressing the fact that the number of electrons and the total energy of the system are fixed. According to the Pauli principle the numbers n_s can only have the values 0 and 1, and we make the assumption that all such sets of numbers satisfying the conditions (A 1·1) and (A 1·2) have the same *a priori* probability. As a detailed knowledge of the microscopic states is usually out of the question, we have to be content with average values.

Consider the energy levels to be grouped together, all the levels in any one group having approximately the same energy, and let M_s be the number of levels in the group with energy E_s. We shall now suppose that our macroscopic measurements are insufficient to determine the numbers n_s, but will tell us the number of electrons N_s occupying the M_s levels with energy E_s. The number of arrangements of these N_s (indistinguishable) electrons over the M_s levels is

$$\frac{M_s!}{N_s!(M_s - N_s)!},$$

and the total number of arrangements of all the electrons is

$$W = \prod_s \frac{M_s!}{N_s!(M_s - N_s)!}, \qquad (\text{A } 1 \cdot 3)$$

where

$$N = \sum_s N_s \qquad (\text{A } 1 \cdot 4)$$

and

$$U = \sum_s N_s E_s. \qquad (\text{A } 1 \cdot 5)$$

To find the most probable distribution we make $\log W$ a maximum, subject to the conditions (A 1·4) and (A 1·5). Introducing two undetermined multipliers α, β and using Stirling's theorem in the approximate form $\log N_s! = N_s \log N_s - N_s$, we have, on varying N_s,

$$\delta(\log W - \alpha N - \beta U) = \delta \sum_s \{M_s \log M_s - N_s \log N_s$$
$$- (M_s - N_s) \log (M_s - N_s) - \alpha N_s - \beta N_s E_s\}$$
$$= \sum_s \delta N_s \left(\log \frac{M_s - N_s}{N_s} - \alpha - \beta E_s\right) = 0.$$

Hence
$$\frac{N_s}{M_s} = \frac{1}{e^{\alpha + \beta E_s} + 1}. \tag{A 1·6}$$

At this stage it is convenient to revert to the average number of electrons in each state, i.e. to use $\bar{n}_s = N_s/M_s$ instead of N_s. We then have

$$N = \sum_s \bar{n}_s = \sum_s \frac{1}{e^{\alpha + \beta E_s} + 1}, \quad U = \sum_s \bar{n}_s E_s = \sum_s \frac{E_s}{e^{\alpha + \beta E_s} + 1}, \tag{A 1·7}$$

and these equations determine the parameters α and β in terms of N and U. When the density of electrons is small, all the $\bar{n}_s$'s must be small, and in this case the first term $e^{\alpha + \beta E_s}$ in the denominators above must be large, so that in the first approximation the second term can be neglected. Then

$$\bar{n}_s = e^{-\alpha - \beta E_s},$$

which is the Boltzmann distribution function, and we see that $\beta = 1/kT$, where T is the absolute temperature and k is Boltzmann's constant, while α is determined by the number of electrons present and is a function of N and T.

If we wish to consider assemblies in which the electrons do not all have the same spin there are several different cases to consider.

(i) If each state can accommodate two electrons with opposite spins, each state can be considered to be doubly degenerate. We then have

$$\bar{n}_s = \frac{2}{e^{\alpha + \beta E_s} + 1}. \tag{A 1·8}$$

(ii) If the rth group of states is such that each state can accommodate one electron whose spin can have either direction, let N_r^+ and N_r^- be the numbers of electrons with positive and negative spins respectively. Then the N_r^+ electrons with positive spins can be distributed over the M_r states in

$$\frac{M_r!}{N_r^+! (M_r - N_r^+)!}$$

ways, while the N_r^- electrons with negative spins can be distributed over the remaining $M_r - N_r^+$ states in

$$\frac{(M_r - N_r^+)!}{N_r^-! (M_r - N_r^+ - N_r^-)!}$$

ways. The rth group of electrons therefore contribute the factor

$$W_r = \frac{M_r!}{N_r^+! \, N_r^-! \, (M_r - N_r^+ - N_r^-)!} \qquad (\text{A } 1\cdot 9)$$

to W, while their contribution to N and U is $N_r^+ + N_r^-$ and $(N_r^+ + N_r^-) E_r$ respectively. If we now vary N_r^+ and N_r^- independently, subject to N and U being constant, we find that $N_r^+ = N_r^-$ and

$$\log \frac{M_r - N_r^+ - N_r^-}{N_r^+} - \alpha - \beta E_r = 0,$$

i.e.
$$N_r = N_r^+ + N_r^- = \frac{M_r}{\frac{1}{2} e^{\alpha + \beta E_r} + 1}. \qquad (\text{A } 1\cdot 10)$$

(iii) If the rth group of states is such that each state can accommodate one or two electrons, but not less than one, as occurs, for example, if the second ionization potential is much larger than the first, let N_r^+ be the number of states which have a single electron with a positive spin and N_r^- the number which have a single electron with a negative spin. Then W_r, the contribution of the rth group to W, is given by (A 1·9), while the contribution to N is $2M_r - N_r^+ - N_r^-$ and to U is $(2M_r - N_r^+ - N_r^-) E_r$. Hence, varying N_r^+ and N_r^- independently subject to N and U remaining constant, we find that $N_r^+ = N_r^-$ and

$$\log \frac{M_r - N_r^+ - N_r^-}{N_r^+} + \alpha + \beta E_r = 0,$$

i.e.
$$N_r = N_r^+ + N_r^- = 2M_r \frac{e^{\alpha + \beta E_r}}{2 e^{\alpha + \beta E_r} + 1}, \qquad (\text{A } 1\cdot 11)$$

while the total number of electrons in the M_r states is

$$2M_r - N_r = 2M_r \frac{e^{\alpha + \beta E_r} + 1}{2 e^{\alpha + \beta E_r} + 1}. \qquad (\text{A } 1\cdot 12)$$

The corresponding number of electrons in excess of one per state is

$$(2M_n - N_n) - M_n = \frac{M_n}{2 e^{\alpha + \beta E_n} + 1}. \qquad (\text{A } 1\cdot 13)$$

The Thermodynamic Functions

A2. If X_r is the generalized force corresponding to a generalized coordinate x_r and if S is the entropy and μ the Gibbs thermodynamic potential, then in a general variation of the state of a thermodynamic system consisting of one type of particle only we have

$$dU = T \, dS - \sum_r X_r \, dx_r + \mu \, dN. \qquad (\text{A } 2\cdot 1)$$

Now X_r is the average value of $-\partial E_s/\partial x_r$ taken over the assembly, so that

$$dU + \sum_r X_r dx_r = dU - \sum_r \frac{\partial}{\partial x_r} \left(\sum_s \bar{n}_s E_s \right) dx_r. \qquad \text{(A 2·2)}$$

We now show that, if N is constant, β is an integrating factor of the differential form (A 2·2), and hence, by the second law of thermodynamics, $1/\beta$ is equal to a constant times the absolute temperature. (The constant can only be determined by considering a particular system, or, as in the preceding section, by passing to the classical limit.)

From (A 2·2) we have

$$\beta(dU + \sum_r X_r dx_r) = d(\beta U) - U\, d\beta + \beta \sum_r X_r dx_r$$

$$= d(\beta U) - \sum_s \frac{1}{e^{\alpha + \beta E_s} + 1} \left(d\alpha + E_s\, d\beta + \beta \sum_r \frac{\partial E_s}{\partial x_r} dx_r \right)$$

$$+ d \sum_s \frac{\alpha}{e^{\alpha + \beta E_s} + 1} - \alpha\, d \sum_s \frac{1}{e^{\alpha + \beta E_s} + 1}. \qquad \text{(A 2·3)}$$

Hence
$$k\beta(dU + \sum_r X_r dx_r) = dS - k\alpha\, dN, \qquad \text{(A 2·4)}$$

where
$$S = k\beta U + k \sum_s \log(1 + e^{-\alpha - \beta E_s}) + k\alpha N. \qquad \text{(A 2·5)}$$

Therefore, if N is constant, $k\beta$ is an integrating factor of $dU + \Sigma X_r dx_r$, so that $k\beta = 1/T$ and S is the entropy.

Also
$$dU + \sum_r X_r dx_r = \frac{dS}{k\beta} - \frac{\alpha}{\beta} dN,$$

which, by comparison with (A 2·1), shows that

$$\mu = -\alpha/\beta.$$

It is customary to use the symbol ζ instead of μ for the thermodynamic potential of the electrons, and we then have

$$\alpha = -\zeta/kT \qquad \text{(A 2·6)}$$

and
$$S = k \sum_s \log(1 + e^{(\zeta - E_s)/kT}) + (U - N\zeta)/T. \qquad \text{(A 2·7)}$$

A simpler and more useful thermodynamic function is the free energy F at constant volume. It is

$$F = U - ST = N\zeta - kT \sum_s \log(1 + e^{(\zeta - E_s)/kT}). \qquad \text{(A 2·8)}$$

The relation
$$\partial F/\partial \zeta = 0 \qquad \text{(A 2·9)}$$

is sometimes useful. It is equivalent to $N = \Sigma n_s$.

THE STATISTICS OF A PERFECT ELECTRON GAS

A 3. If we write
$$f_0(E) = \frac{1}{e^{(E-\zeta)/kT} + 1},\qquad\text{(A 3·1)}$$

it is shown in § 1·8 that for a perfect gas of electrons the number of electrons per unit volume with velocities lying in the range du, dv, dw is

$$2(m/h)^3 f_0(E)\,du\,dv\,dw,\qquad\text{(A 3·2)}$$

where the occurrence of the factor 2 is due to each translational state having weight 2 on account of the electronic spin. Hence if n is the number of electrons per unit volume, ζ is determined by the equation

$$n = 2\left(\frac{m}{h}\right)^3 \iiint f_0(E)\,du\,dv\,dw = \frac{8\sqrt{2}\,\pi m^{\frac{3}{2}}}{h^3} \int_0^\infty E^{\frac{1}{2}} f_0(E)\,dE. \qquad\text{(A 3·3)}$$

There are now two limiting cases to be considered, the 'high-temperature' and the 'low-temperature' cases.

(i) If n is sufficiently small and T sufficiently large, $f_0(E)$ must be effectively the Boltzmann function. Hence $e^{-\zeta/kT}$ must be large, and we can write as a first approximation

$$n = \frac{8\sqrt{2}\,\pi m^{\frac{3}{2}}}{h^3} \int_0^\infty E^{\frac{1}{2}} e^{(\zeta-E)/kT}\,dE = 2\frac{(2\pi mkT)^{\frac{3}{2}}}{h^3} e^{\zeta/kT}.$$

Hence
$$e^{-\zeta/kT} = 2(2\pi mkT)^{\frac{3}{2}}/(nh^3). \qquad\text{(A 3·4)}$$

The condition for this case to occur is that the right-hand side of (A 3·4) is large compared with unity.

(ii) If n is sufficiently large and T is sufficiently small, ζ/kT is large and positive. In this case $f_0(E)$ is effectively unity for $E < \zeta$ and practically zero for $E > \zeta$. As a first approximation we can therefore put

$$f_0(E) = 1 \quad (0 \leqslant E \leqslant \zeta), \qquad f_0(E) = 0 \quad (\zeta < E).$$

Equation (A 3·3) then becomes

$$n = \frac{8\sqrt{2}\,\pi m^{\frac{3}{2}}}{h^3} \int_0^\zeta E^{\frac{1}{2}}\,dE = \frac{16\sqrt{2}\,\pi m^{\frac{3}{2}}}{3h^3} \zeta^{\frac{3}{2}},$$

and so
$$\zeta = \frac{h^2}{8m}\left(\frac{3n}{\pi}\right)^{\frac{2}{3}}. \qquad\text{(A 3·5)}$$

If we define the 'degeneracy temperature' T_0 by the relation

$$kT_0 = \zeta,$$

the electron gas is said to be degenerate when $T_0 \gg T$.

INTEGRALS CONNECTED WITH THE FERMI FUNCTION

A 4. The evaluation of the integral in the preceding section for the two extreme limits of very high and very low temperatures is a particular case of a more general procedure. Consider the integral

$$I = -\int_0^\infty \phi(E) \frac{\partial f_0}{\partial E} dE = \int_0^\infty f_0(E) \frac{\partial \phi}{\partial E} dE, \tag{A 4·1}$$

where $\phi(0) = 0$. If $T \gg T_0$ we write

$$I = \int_0^\infty \frac{\partial \phi}{\partial E} \sum_{n=1}^\infty (-1)^{n+1} e^{-n(E-\zeta)/kT} dE, \tag{A 4·2}$$

and, if ϕ is expressible in terms of powers of E, I can be evaluated as a series of Γ functions.

If $T_0 \gg T$, we proceed as follows. Put $x = (E - \zeta)/kT$, and suppose that ϕ can be expanded in ascending powers of x:

$$\phi(E) = \Phi(x) = \sum_{n=0}^\infty \frac{x^n}{n!} \left(\frac{d^n \Phi}{dx^n}\right)_{x=0},$$

so that

$$I = \int_{-\zeta/kT}^\infty \sum_{n=0}^\infty \frac{1}{n!} \left(\frac{d^n \Phi}{dx^n}\right)_{x=0} \frac{x^n dx}{(e^x + 1)(1 + e^{-x})}. \tag{A 4·3}$$

If ϕ is such that the lower limit can be replaced by $-\infty$ with negligible error, the series can be evaluated by term-by-term integration. We then obtain

$$I = \sum_{n=0}^\infty \frac{I_{2n}}{(2n)!} \left(\frac{d^{2n} \Phi}{dx^{2n}}\right)_{x=0},$$

where, for $n > 0$,

$$I_{2n} = 2 \int_0^\infty \frac{x^{2n} e^{-x}}{(1 + e^{-x})^2} dx = 2 \sum_{l=1}^\infty (-1)^{l+1} l \int_0^\infty x^{2n} e^{-lx} dx$$

$$= 2(2n)! \sum_{l=1}^\infty \frac{(-1)^{l+1}}{l^{2n}} \tag{A 4·4}$$

and

$$I_0 = 2 \int_0^\infty \frac{e^{-x}}{(1 + e^{-x})^2} dx = 1. \tag{A 4·5}$$

Hence

$$-\int_0^\infty \phi(E) \frac{\partial f_0}{\partial E} dE = \phi(\zeta) + 2 \sum_{n=1}^\infty c_{2n} (kT)^{2n} \left(\frac{d^{2n}\phi}{dE^{2n}}\right)_{E=\zeta}, \tag{A 4·6}$$

where

$$c_{2n} = \sum_{l=1}^\infty \frac{(-1)^{l+1}}{l^{2n}}. \tag{A 4·7}$$

Alternative expressions for the coefficients are

$$c_{2n} = (2^{2n-1} - 1) \pi^{2n} B_n / (2n)! = (1 - 2^{1-2n}) \zeta(2n), \tag{A 4·8}$$

where B_n is the nth Bernoulli number and $\zeta(s)$ is the Riemann ζ-function. The following are the values of the first few coefficients:

$$c_2 = \frac{\pi^2}{12}, \quad c_4 = \frac{7\pi^4}{720}, \quad c_6 = \frac{31\pi^6}{30240}, \quad \dots \quad \text{(A 4·9)}$$

A 41. In general the most interesting temperature range is the medium and low temperature region where the asymptotic formula (A 4·6) can be used to evaluate the integrals. In some cases, however, the calculations have to be carried out in a temperature range where neither the low temperature nor the high temperature approximations are sufficiently accurate, and more exact methods of calculation are necessary. The most important cases are those in which $\phi(E)$ is a simple power, and the general type of integral to be evaluated is then

$$F_n(\xi) = \int_0^\infty \frac{x^n \, dx}{e^{x-\xi}+1} \quad (n > -1). \quad \text{(A 41·1)}$$

These integrals have been tabulated by McDougall and Stoner (1938) for half-integral values of n, and by Rhodes (1950) for integral values of n. For complete details of the computation the original papers must be consulted, but the following general remarks can be made.

(i) If $n > 0$, $\qquad\qquad F'_n(\xi) = nF_{n-1}(\xi). \qquad\qquad$ (A 41·2)

(ii) For $\xi \leqslant 0$ we can expand the denominator of F_n in powers of $e^{-x+\xi}$. We then obtain

$$F_n(\xi) = \Gamma(n+1) \sum_{s=1}^\infty (-1)^{s-1} \frac{e^{s\xi}}{s^{n+1}}. \quad \text{(A 41·3)}$$

(iii) For $\xi = 0$, the F_n's are Riemann ζ-functions, as is obvious from (A 41·3). To obtain $F_n(0)$ in a more usual form, we write

$$\frac{1}{e^x+1} = \frac{1}{e^x-1} - \frac{2}{e^{2x}-1}.$$

Then, if $n > 0$,

$$F_n(0) = \int_0^\infty \frac{x^n \, dx}{e^x-1} - 2\int_0^\infty \frac{x^n \, dx}{e^{2x}-1}$$

$$= (1-2^{-n}) \int_0^\infty \frac{x^n \, dx}{e^x-1} = (1-2^{-n}) \Gamma(n+1) \zeta(n+1). \quad \text{(A 41·4)}$$

Further, by successive applications of (A 41·2),

$$F_n^{(r)}(0) = (1-2^{r-n}) \Gamma(n+1) \zeta(n+1-r) \quad (n > r). \quad \text{(A 41·5)}$$

(iv) For $\xi \geqslant 1$, the asymptotic expansion (A 4·6) gives

$$F_n(\xi) = \frac{\xi^{n+1}}{n+1} \left[1 + \sum_{r=1}^s a_{2r} \xi^{-2r}\right] + R_{2s}, \quad \text{(A 41·6)}$$

where
$$a_{2r} = 2(n+1)\,n \ldots (n-2r+2)\,(1-2^{1-2r})\,\zeta(2r), \tag{A 41.7}$$

and R_{2s} can be shown to be less than $(2s+2)\,a_{2s+2}\xi^{n-2s-1}$.

(v) If n is an integer, the asymptotic series becomes a polynomial, and the remainder can be expressed in terms of $F_n(-\xi)$, which can be evaluated by means of (A 41.3).

To find the remainder, we write

$$(n+1)\,F_n(\xi) = \int_0^\infty \frac{x^{n+1}\,dx}{(e^{x-\xi}+1)\,(1+e^{-x+\xi})}$$

$$= \int_{-\infty}^\infty \frac{x^{n+1}\,dx}{(e^{x-\xi}+1)\,(1+e^{-x+\xi})} - \int_{-\infty}^0 \frac{x^{n+1}\,dx}{(e^{x-\xi}+1)\,(1+e^{-x+\xi})}$$

$$= (n+1)\,S_n(\xi) + (-1)^{n+2}\int_0^\infty \frac{x^{n+1}\,dx}{(e^{x+\xi}+1)\,(1+e^{-x-\xi})},$$

where
$$S_n(\xi) = \frac{1}{n+1}\int_{-\infty}^\infty \frac{x^{n+1}\,dx}{(e^{x-\xi}+1)\,(1+e^{-x+\xi})}. \tag{A 41.8}$$

Hence
$$F_n(\xi) + (-1)^{n+1}\,F_n(-\xi) = S_n(\xi). \tag{A 41.9}$$

The integral $S_n(\xi)$ can be evaluated exactly by the formula (A 4.6). Its value is

$$S_n(\xi) = \frac{\xi^{n+1}}{n+1}\left[1 + 2\sum_{r=1}^{[\frac{1}{2}n+\frac{1}{2}]}(n+1)\,n\ldots(n-2r+2)\,(1-2^{1-2r})\,\zeta(2r)\,\xi^{-2r}\right], \tag{A 41.10}$$

where $[p]$ denotes the integral part of p.

(vi) Tables are given by McDougall and Stoner for $F_{\frac{1}{2}}(\xi)$, $F_{\frac{3}{2}}(\xi)$ and by Rhodes for $F_1(\xi)$, $F_2(\xi)$, $F_3(\xi)$ and $F_4(\xi)$. In fig. A 1, the graphs of $\frac{2}{3}F_{\frac{3}{2}}(\xi)$ and of $F_{\frac{1}{2}}(\xi)$ are shown, together with the non-degenerate approximation for comparison, and it will be seen that there is a fairly wide range in which the functions cannot be represented either by the non-degenerate approximation or by the asymptotic series. The graphs of $F_1(\xi)$ and $\frac{1}{24}F_4(\xi)$ are given in fig. A 2. The curve for $\frac{1}{24}F_4(\xi)$ for $\xi < 0$ is almost indistinguishable

Table A 1. *Corresponding values of kT/ζ_0 and ξ*

kT/ζ_0	ξ	kT/ζ_0	ξ
0·05	19·959	1·0	−0·021
0·1	9·916	1·1	−0·199
0·2	4·823	1·2	−0·357
0·3	3·049	1·3	−0·500
0·4	2·101	1·4	−0·630
0·5	1·486	1·5	−0·749
0·6	1·041	1·6	−0·859
0·7	0·697	1·7	−0·961
0·8	0·416	1·8	−1·057
0·9	0·181	1·9	−1·146
1·0	−0·021	2·0	−1·231

from its 'high temperature approximation' e^{ξ}, while for $\xi > 2$ the polynomial approximation gives a good representation of both curves. In the range $0 < \xi < 2$ the polynomial approximation lies well above $F_1(\xi)$ and well below $\frac{1}{24}F_4(\xi)$.

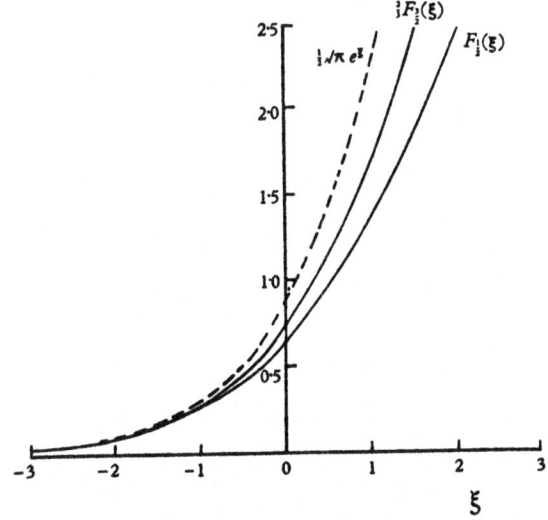

Fig. A1. The Fermi-Dirac functions $F_{\frac{1}{2}}(\xi)$ and $\frac{2}{3}F_{\frac{3}{2}}(\xi)$.

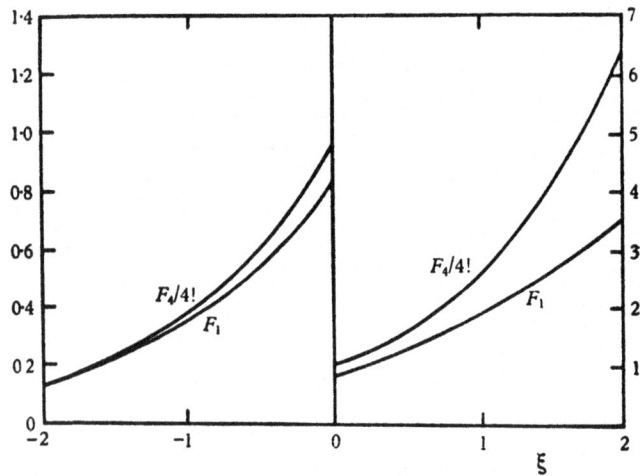

Fig. A2. The Fermi-Dirac functions of integral order.

(vii) The relation
$$(\zeta_0/kT)^{\frac{3}{2}} = \tfrac{3}{2}F_{\frac{1}{2}}(\xi)$$

derived on p. 147, gives the connexion between ξ and T for a band of standard form, ζ_0 being the value of $kT\xi$ for $T = 0$. Some values of ξ for values of kT/ζ_0 between 0·05 and 2 are given in Table A 1.

THE EVALUATION OF SOME ASSOCIATED INTEGRALS

A 5. (i) The integral
$$\alpha_m^{(n)} = -\int_{-\infty}^{\infty} E^n \eta^m \frac{\partial f_0}{\partial \eta} d\eta, \tag{A 5·1}$$

where m is an integer and $\eta = (E - \zeta)/kT$ can be evaluated as follows. Put $\gamma = kT/\zeta$ and expand E^n as the series

$$E^n = \zeta^n (1 + \gamma\eta)^n = \zeta^n \sum_{r=0}^{\infty} \binom{n}{r} (\gamma\eta)^r. \tag{A 5·2}$$

Hence, assuming that the series can be integrated term by term, we have

$$\alpha_m^{(n)} = -\zeta^n \sum_{r=0}^{\infty} \binom{n}{r} \gamma^r \int_{-\infty}^{\infty} \eta^{m+r} \frac{\partial f_0}{\partial \eta} d\eta. \tag{A 5·3}$$

Inserting the value of the integral from equation (A 4·4) we obtain

$$\alpha_0^{(n)} = \zeta^n + 2\zeta^n \sum_{r=2,4,\ldots}^{\infty} r! \binom{n}{r} c_r \left(\frac{kT}{\zeta}\right)^r, \tag{A 5·4}$$

$$\alpha_m^{(n)} = 2\zeta^n m! c_m + 2\zeta^n \sum_{r=2,4,\ldots}^{\infty} (m+r)! \binom{n}{r} c_{m+r} \left(\frac{kT}{\zeta}\right)^r \quad (m>0) \tag{A 5·5}$$

if m is even, and
$$\alpha_m^{(n)} = 2\zeta^n \sum_{r=1,3,\ldots}^{\infty} (m+r)! \binom{n}{r} c_{m+r} \left(\frac{kT}{\zeta}\right)^r \tag{A 5·6}$$
if m is odd.

(ii) To evaluate integrals of the form

$$\int_{-\infty}^{\infty} \frac{F(\eta)}{(e^{-\eta}+1)(e^{\eta+z}+1)} d\eta, \tag{A 5·7}$$

put
$$G(\eta) = \int^{\eta} F(\eta') d\eta' \tag{A 5·8}$$

and
$$\frac{1}{(e^{-\eta}+1)(e^{\eta+z}+1)} = \frac{1}{e^z-1}\left(\frac{1}{e^{\eta}+1} - \frac{1}{e^{\eta+z}+1}\right).$$

Then, on integrating by parts, we have

$$\int_{-\infty}^{\infty} \frac{F(\eta)}{(e^{-\eta}+1)(e^{\eta+z}+1)} d\eta = -\frac{1}{e^z-1}\int_{-\infty}^{\infty} G(\eta)\left(\frac{\partial f_0(\eta)}{\partial \eta} - \frac{\partial f_0(\eta+z)}{\partial \eta}\right) d\eta,$$

provided that $F(\eta)$ is such that $G(\eta)\{f_0(\eta) - f_0(\eta+z)\}$ vanishes at $\eta = \pm \infty$. Changing the variable in the second term on the right, we see that

$$\int_{-\infty}^{\infty} \frac{F(\eta)}{(e^{-\eta}+1)(e^{\eta+z}+1)} d\eta = -\frac{1}{e^z-1}\int_{-\infty}^{\infty} \{G(\eta) - G(\eta-z)\} \frac{\partial f_0}{\partial \eta} d\eta. \tag{A 5·9}$$

In the particular case in which $F(\eta) = \eta^s$, where s is an integer, denote the integral by I_s. Then

$$I_s = \int_{-\infty}^{\infty} \frac{\eta^s}{(e^{-\eta}+1)(e^{\eta+z}+1)}\,d\eta = \frac{1}{(s+1)(e^z-1)} \int_{-\infty}^{\infty} \frac{\eta^{s+1}-(\eta-z)^{s+1}}{(e^\eta+1)(e^{-\eta}+1)}\,d\eta.$$
(A 5·10)

Hence, on expanding $(\eta - z)^{s+1}$ and using (A 4·4), we obtain

$$I_s = \int_{-\infty}^{\infty} \frac{\eta^s}{(e^{-\eta}+1)(e^{\eta+z}+1)}\,d\eta$$

$$= \frac{2}{(s+1)(e^z-1)} \sum_{r=0,1,\ldots}^{[\frac12 s]} (-1)^{s-2r} \binom{s+1}{2r} (2r)!\,c_{2r}\,z^{s+1-2r}, \quad (A 5·11)$$

where $[\frac12 s]$ is the integral part of $\frac12 s$, and $c_0 = \frac12$.

(iii) To evaluate the integrals

$$\mathscr{I}_n(x) = \int_0^x \frac{z^n\,dz}{(e^z-1)(1-e^{-z})} = -\frac{x^n}{e^x-1} + n\int_0^x \frac{z^{n-1}}{e^z-1}\,dz, \quad (A 5·12)$$

where n is an integer, we consider separately the cases of large and small x.

If x is large we write

$$\int_0^x \frac{z^{n-1}}{e^z-1}\,dz = \int_0^\infty - \int_x^\infty \frac{z^{n-1}}{e^z-1}\,dz.$$

Then
$$\int_0^\infty \frac{z^{n-1}}{e^z-1}\,dz = \sum_{s=1}^\infty \int_0^\infty z^{n-1}e^{-sz}\,dz = (n-1)!\sum_{s=1}^\infty \frac{1}{s^n}$$

and
$$\int_x^\infty \frac{z^{n-1}}{e^z-1}\,dz = \sum_{s=1}^\infty \int_x^\infty z^{n-1}e^{-sz}\,dz$$

$$= (n-1)!\sum_{s=1}^\infty \frac{e^{-sx}}{s^n}\left(1+sx+\ldots+\frac{(sx)^{n-1}}{(n-1)!}\right).$$

Hence, for large x,

$$\mathscr{I}_n(x) = n!\sum_{s=1}^\infty \frac{1}{s^n} - \frac{x^n}{e^x-1} - n!\sum_{s=1}^\infty \frac{e^{-sx}}{s^n}\left(1+sx+\ldots+\frac{(sx)^{n-1}}{(n-1)!}\right).$$
(A 5·13)

For small x we use the expansion

$$\frac{z}{e^z-1} = 1 - \frac{z}{2} + \sum_{s=1}^\infty (-1)^{s+1}\frac{B_s}{(2s)!}\,z^{2s},$$

where the B's are the Bernoulli numbers. We then have

$$\int_0^x \frac{z^{n-1}}{e^z-1}\,dz = \frac{x^{n-1}}{n-1} - \frac{x^n}{2n} + \sum_{s=1}^\infty (-1)^{s-1}\frac{B_s}{(2s)!}\frac{x^{n+2s-1}}{n+2s-1}$$

and
$$\mathscr{I}_n(x) = x^{n-1}\left(\frac{1}{n-1} - \sum_{s=1}^\infty (-1)^{s-1}\frac{B_s}{(2s)!}\frac{2s-1}{n+2s-1}\,x^{2s}\right). \quad (A 5·14)$$

The following table of values of some of the $\mathscr{J}_n$'s has been given by Sondheimer (1950).

Table A2. *Values of* $\mathscr{J}_n(\Theta/T)$

T/Θ	5	7	9	11	13	15	17
0	124·43	5062·1	$3·6361 \times 10^5$	$3·9937 \times 10^7$	$6·2278 \times 10^9$	$1·3077 \times 10^{11}$	$3·5569 \times 10^{14}$
0·05	124·42	5078·2	$3·6180 \times 10^5$	$3·9083 \times 10^7$	$5·8160 \times 10^9$	$1·1030 \times 10^{11}$	$2·5004 \times 10^{14}$
0·076923	123·14	4809·8	$3·0344 \times 10^5$	$2·5647 \times 10^7$	$2·6594 \times 10^9$	$3·0916 \times 10^{11}$	$3·8962 \times 10^{13}$
0·1	116·38	3972·1	$1·9743 \times 10^5$	$1·2123 \times 10^7$	$8·4470 \times 10^8$	$6·3770 \times 10^{10}$	$5·0803 \times 10^{12}$
0·125	101·48	2796·8	$1·0353 \times 10^5$	$4·4849 \times 10^6$	$2·1340 \times 10^8$	$1·0785 \times 10^{10}$	$5·6799 \times 10^{11}$
0·16667	70·873	1328·9	31011	$8·1272 \times 10^5$	$2·2840 \times 10^7$	$6·7201 \times 10^8$	$2·0417 \times 10^{10}$
0·2	50·263	705·56	11953	$2·2377 \times 10^5$	$4·4520 \times 10^6$	$9·2226 \times 10^7$	$1·9661 \times 10^9$
0·25	29·488	281·75	3167·4	38845	$5·0260 \times 10^5$	$6·7413 \times 10^6$	$9·2781 \times 10^7$
0·33333	12·771	72·010	468·43	3291·4	24264	$1·8479 \times 10^5$	$1·4409 \times 10^6$
0·5	3·2293	8·3763	24·717	78·230	258·74	881·95	3072·9
0·66667	1·1199	1·6538	2·7649	4·9460	9·2329	17·748	34·851
0·83333	0·47907	0·45534	0·48883	0·56090	0·67116	0·82665	1·0399
1	0·23662	0·15665	0·11700	0·093343	0·077632	0·066444	0·058071
1·25	0·098845	0·041987	0·020100	0·010273	0·0054720	0·0029989	0·0016782

REFERENCES

McDougall, J. and Stoner, E. C. (1938). The computation of Fermi-Dirac functions. *Philos. Trans.* A, **237**, 67.

Rhodes, P. (1950). Fermi-Dirac functions of integral order. *Proc. Roy. Soc.* A, **204**, 396.

Sondheimer, E. H. (1950). The theory of the transport phenomena in metals. *Proc. Roy. Soc.* A, **203**, 75.

LIST OF IMPORTANT SYMBOLS

$\mathbf{a}_i$ $(i=1,2,3)$ Axes of the unit cell of a crystal.

a Lattice constant.

$\mathbf{A}$ Vector potential.

$\mathbf{b}_i$ $(i=1,2,3)$ Fundamental vectors of the reciprocal lattice defined by

$$\mathbf{b}_i = \frac{\mathbf{a}_j \times \mathbf{a}_k}{\mathbf{a}_i . (\mathbf{a}_j \times \mathbf{a}_k)}.$$

c Velocity of light.

c Length of hexagonal axis of a hexagonal crystal.

C_v Specific heat per unit volume.

Δ Volume of the unit cell.

$\mathscr{E}, \mathscr{E}$ Electric field in electrostatic units.

$E_\mathbf{k}$ Energy of the electronic state $\mathbf{k}$.

$-\epsilon$ Charge of the electron in electrostatic units.

$\eta = (E-\zeta)/kT$.

ζ The thermodynamic potential of the conduction electrons; the Fermi energy.

ζ_0 The value of ζ for $T=0$.

Θ The Debye characteristic temperature.

Θ Thermoelectric force.

f Distribution function.

f_0 Equilibrium distribution function.

F Free energy per unit volume.

$\mathbf{g}_a = g_1\mathbf{a}_1 + g_2\mathbf{a}_2 + g_3\mathbf{a}_3$ General lattice vector.

$\mathbf{g}_b = g_1\mathbf{b}_1 + g_2\mathbf{b}_2 + g_3\mathbf{b}_3$ Genera lvector in the reciprocal lattice.

h Planck's constant.

$\hbar = h/2\pi$.

$H, \mathbf{H}$ Magnetic field in magnetic units.

$\mathscr{H}$ Hamiltonian function.

$\mathbf{j}, \mathbf{J}$ Current density.

$\mathbf{k} = (k_1, k_2, k_3)$ Wave vector of an electronic wave function.

k Boltzmann's constant.

$k_0 = (2m\zeta_0)^{\frac{1}{2}}/\hbar$ Radius of the Fermi distribution for free electrons at $T=0$.

$K = |\mathbf{k}|$.

κ Thermal conductivity.

l Free path; mean free path.

$L, \mathscr{L}$ Collision operator.

L Lorenz number.

m	Mass of the electron.
m^*	Effective mass of a conduction electron.
m'	Zone quantum number.
M	Mass of an atom.
M_A	Atomic weight.
μ	Thomson coefficient.
$\mu_0 = \epsilon h/(4\pi mc)$	Bohr magneton.
n	Number of electrons per unit volume.
n_a	Number of atoms per unit volume.
$\mathfrak{n}(E)$	Density of the electronic states.
N	Total number of electrons.
ν	Frequency.
$\xi = \zeta/kT.$	
$\mathbf{q}$	Wave vector of the lattice vibrations.
$q = \lvert \mathbf{q} \rvert.$	
Q	Quantity of heat per unit volume.
$\mathbf{r}$	Position vector of an electron.
$\mathbf{R}$	Position vector of an atom.
R	Hall coefficient.
$r_0 = h^2/(4\pi^2 m\epsilon^2)$	The radius of the first Bohr orbit in hydrogen.
ρ	Electrical resistivity in electrostatic units.
S	Entropy per unit volume.
σ	Electrical conductivity in electrostatic units.
σ	Relative magnetization (in Chapter VII only).
T	Absolute temperature.
τ	Time of relaxation.
u_0	Velocity of sound in a crystal.
$u_{\mathbf{k}}$	The part of the electronic wave function which is periodic in the unit cell.
U	Internal energy per unit volume.
$\mathbf{v} = (u, v, w)$	Velocity of an electron.
V	Potential energy.
$\mathbf{w}$	Thermal current density.
χ	Magnetic susceptibility per unit volume.
$\psi_{\mathbf{k}} = e^{i\mathbf{k}\cdot\mathbf{r}} u_{\mathbf{k}}(\mathbf{r})$	Wave function of an electron in a crystal lattice.
ψ, Ψ, ϕ	Wave functions.
Z	Partition function.

UNITS

In theoretical work it is convenient to use Gaussian units and to measure extensive variables per unit volume. In experimental work it is customary to use practical units. The relations between the practical units and those used in this book are given below.

	Practical unit	Multiplying factor to change to Gaussian units
Electrical conductivity	Ohm^{-1} cm.$^{-1}$	9×10^{11}
Thermal conductivity	Watt cm.$^{-1}$ degree^{-1}	10^7
Lorenz number	(Volt/degree)^2	$1/(9 \times 10^4)$
Thermoelectric power	Volt/degree	$1/300$
Hall coefficient	Volt cm./ampere gauss	$1/(9 \times 10^{11})$

The Hall electric field is sometimes measured in electromagnetic units. The Hall coefficient is then expressed in cm.^{3}/ampere second, and the measured values must then be divided by 9×10^{19} to reduce them to Gaussian units. If the electric field and current are both expressed in electromagnetic units, the measured values must be divided by 9×10^{20} to reduce them to Gaussian units.

Specific heats and magnetic susceptibilities can be measured per unit volume, per unit mass or per gram atom. If ρ is the density, M_A the atomic weight, n_a the number of atoms per c.c. and $\mathscr{L}$ is the number of atoms per gram atom, the relations between the various quantities are as follows:

$$C_v = \rho(C_v)_m, \quad (C_v)_A = M_A(C_v)_m = (M_A/\rho)\, C_v = (\mathscr{L}/n_a)\, C_v,$$

$$\chi = \rho\chi_m, \quad \chi_A = M_A\chi_m = (M_A/\rho)\, \chi = (\mathscr{L}/n_a)\, \chi.$$

PHYSICAL CONSTANTS

ϵ	Electronic charge	$4\cdot802 \times 10^{-10}$ e.s.u.
m	Mass of electron	$9\cdot106 \times 10^{-28}$ g.
c	Velocity of light	$2\cdot998 \times 10^{10}$ cm./sec.
k	Boltzmann's constant	$1\cdot380 \times 10^{-16}$ erg/degree.
h	Planck's constant	$6\cdot623 \times 10^{-27}$ erg sec.
$\mathscr{L}$	Avogadro's number	$6\cdot024 \times 10^{23}$ atoms/gram atom.
μ_0	Bohr magneton	$9\cdot273 \times 10^{-21}$ erg/gauss.

INDEX OF SUBJECTS

INDEX OF NAMES

Printed by Printforce, United Kingdom